Gold Nanoparticles in Biomedical Applications

Gold Nanoparticles in Biomedical Applications

Lev Dykman
Nikolai Khlebtsov

CRC Press
Taylor & Francis Group
Boca Raton London New York

CRC Press is an imprint of the
Taylor & Francis Group, an **informa** business

CRC Press
Taylor & Francis Group
6000 Broken Sound Parkway NW, Suite 300
Boca Raton, FL 33487-2742

First issued in paperback 2019

ISBN-13: 978-1-138-56074-1 (hbk)
ISBN-13: 978-0-367-89221-0 (pbk)

Library of Congress Cataloging-in-Publication Data

Names: Dykman, L. A. (Lev Abramovich), author. | Khlebtsov, Nikolai G., author.
Title: Gold nanoparticles in biomedical applications / Lev A. Dykman and Nikolai Khlebtsov.
Description: Boca Raton : Taylor & Francis, 2017. | Includes bibliographical references and index.
Identifiers: LCCN 2017035815| ISBN 9781138560741 (hardback : alk. paper) | ISBN 9780203711507 (ebook)
Subjects: | MESH: Metal Nanoparticles | Gold--therapeutic use | Biomedical Engineering--methods
Classification: LCC R856 | NLM QT 36.5 | DDC 610.28--dc23
LC record available at https://lccn.loc.gov/2017035815

Visit the Taylor & Francis Web site at
http://www.taylorandfrancis.com

and the CRC Press Web site at
http://www.crcpress.com

Contents

Preface

The first mentions of colloidal gold go back to the fifth and fourth centuries BC in tracts by Chinese, Arabic, and Indian researchers, but the scientific history of gold particles is much younger and begins with the seminal works by Faraday (1857), Zsigmondy (1898), and Mie (1908). In spite of such a long story, the first usage of present-day "gold nanoparticles" wording appeared in two titles of scientific articles only in 1992 and two years after in a seminal article by M. Brust and coworkers (*Chem. Comm.* 1994). It comes as no surprise if one remembers that the U.S. National Nanotechnology Initiative started in 2000. Since the first usages in 1992–1994, the term "gold nanoparticles" can be found now in several million Internet documents (Google, 6,100,000)

In the past two decades, gold nanoparticles (GNPs) and gold-based nanocomposites have attracted strong interest from the scientific community owing to significant progress made in the following three directions. The first one is the development of robust fabrication technologies based on wet chemical and template synthesis and on top–down nanolithographic approaches. Top–down electron beam lithography technologies allow precise control over nanoparticle (NP) size/shape/assembly, but they are expensive and time-consuming and not scalable for practical needs. By current wet chemical methods, GNPs can be fabricated by fine tuning the particle size, shape, structure, heterogeneous composition, and plasmonic properties from green to near-infrared light. The fabrication progress has been accomplished by the development of new characterization methods based on time-resolved laser technologies in addition to traditional methods of optical spectroscopy, dynamic light scattering, and common electron microscopy. One of the most powerful evolved techniques is the electron three-dimensional tomography with atomic-scale resolution.

Second, classical electromagnetic theory, when combined with properly modified dielectric functions, provides a reliable optical description of plasmonic particles and assemblies. Twenty-five years ago, the classical theories by Mie, Rayleigh, and Gans were the most useful analytical solutions for small plasmonic spheres and ellipsoids. During the past two decades, significant progress in theoretical modeling has been achieved due to a combination of numerical electrodynamic methods with state-of-the-art computer capabilities based on parallel multiprocessor computations. Furthermore, new concepts have been developed to combine classical scattering and antenna theories with quantum mechanical models. These new theoretical tools do not suffer from geometrical and composition limitations and can be applied to huge arrays of optically interacting nonspherical and inhomogeneous nanostructures in a complex surrounding.

Third, there has been a tremendous growth of publications on biomedical applications of GNPs and nanocomposites. This trend is related to crucial advances in the functionalization of NPs with stabilizing, targeting, sensing, diagnostic, and therapeutic molecular agents, accompanied by rapid progress in the development of powerful biophotonic technologies such as optoacoustics imaging, fluorescence life-time imaging based on time-correlated single-photon counting, surface-enhanced Raman spectroscopy–based immunoassays for multiplexed biomarker sensing, laser printing of cells, etc.

As a result of current progress in nanotechnologies and nanobiotechnologies, new independent research fields have recently emerged: (1) plasmonics, which deals with the optical generation, manipulation, and controlled transmission of localized or propagating plasmonic excitations in metal nanostructures interacting with light; (2) theranostics, which deals with the fabrication and application of multifunctional NPs, combining therapeutic and diagnostic possibilities in a single nanostructure.

Because of safety and toxicity concerns, gold is the preferred metal in current nanobiotechnology even though silver demonstrates superior plasmonic properties in the visible spectrum. In addition, gold surface chemistry is well developed for functionalization with different ligands and biomolecules, whereas silver surface is easily oxidized. It should also be noted that the wet chemical synthesis of gold particles can be done with more efficient control over the geometrical and structure parameters of NPs. For example, high-quality gold nanorods can be easily fabricated thorough seed-mediated surfactant-assisted synthesis. Such nanorods can further be used as templates for silver coating, thus producing high-quality Au@Ag rod-like particles with remarkable plasmonic properties. All these reasons explain the subject of the present book.

The efficiency of biomedical applications of functionalized GNPs in therapy and diagnostics is determined by their intrinsic physicochemical parameters and by effective delivery of GNPs to target organs, cells, or subcellular components. Here, we discuss the optical properties of GNPs, their applications in biological studies, analysis of data on the *in vitro* and *in vivo* biodistribution, uptake into mammalian cells, immunological properties, and toxicity of most popular GNPs. A special chapter is devoted to applications of multifunctional gold-based nanocomposites.

To our knowledge, this is the first book that comprehensively discusses the published data on GNP biodistribution and toxicity providing the systematization of data over the particle types and parameters, their surface functionalization, animal and cell models, organs examined, doses applied, the type of particle administration and time for examination, assays for evaluation gold particle toxicity, and methods for determination of gold concentration in organs and distribution of particles over cells.

Although the potential of GNPs in nanobiotechnology has been recognized for the past years, new insights into the unique properties of multifunctional nanostructures have just recently started to emerge. Here, we discuss hybrid NP systems that combine different nanomaterials to create multifunctional structures with desired modalities. Such multifunctional structures enable simultaneous diagnostic and therapeutic functions, which can be physically and chemically tailored for a particular organ, disease, or patient.

As distinct from other published books, here, we discuss the immunological properties of GNPs to summarize what is known about their applications as an antigen carrier and adjuvant in immunization for the preparation of antibodies *in vivo*. In particular, the basic principles, recent advances, and current challenges are discussed to present a detailed analysis of data on the interaction of GNPs with immune cells. Emphasis is placed on the systematization of data over production of antibodies by using GNPs and adjuvant properties of GNPs.

We acknowledge financial support from the Russian Scientific Foundation through grant no. 15-14-00002 and technical help from Mr. D.N. Tychinin and Ms. A.A. Elbakyan. We are grateful to our colleagues and friends for their enthusiastic support of our collaborative research work and writing this book: Drs. V.A. Bogatyrev, B.N. Khlebtsov, V.A. Khanadeev, T.E. Pylaev, E.V. Panfilova, L. Yu. Matora, S.Yu. Shyogolev, O.A. Bibikova, S.A. Staroverov, A.M. Burov, O.I. Sokolov, and A.S. Fomin (all from IBPPM RAS), V.V. Tuchin, D.A. Gorin, D.N. Bratashov, E.S. Tuchina, G.S. Terentyk, A.N. Bashkatov, E.A. Genina (Saratov National Research State University), A.B. Bucharskaya, G.N. Maslyakova (Saratov Medical University), V.N. Bagratashvili, and M.Yu. Tsvetkov (Department of Advanced Laser Technologies ILIT RAS). Special thanks go to the CRC team, Michael Slaughter, Mario D'Agostino, Adel Rosario, and Joette Lynch, for their help and support in the book proposal and all publication stages. Finally, we would like to thank our wives Natalia and Margarita for their support and patience. One of us (N.K.) had a wonderful summer writing and relaxing with two grandchildren, Polina and Dasha.

Lev Dykman
Nikolai Khlebtsov
Saratov

Authors

Dr. Lev Dykman is a lead researcher at the Immunochemistry Laboratory of the Russian Academy of Sciences' Institute of Biochemistry and Physiology. He has published over 360 scientific works, including 2 monographs on colloidal gold nanoparticles and 7 book chapters. His current research interests are in immunochemistry, fabrication of gold nanoparticles, and their applications to biological and medical studies. In particular, his research is focused on the interaction of nanoparticles and conjugates with immunocompetent cells and on the delivery of engineered particles to target organs, tissues, and cells.

Professor Nikolai Khlebtsov is head of the Laboratory of Nanobiotechnology at the Institute of Biochemistry and Physiology, Russian Academy of Sciences, Saratov, Russia. He is also a full professor at the Department of Nano- and Biomedical Technologies, Saratov National Research State University. He has published 550 scientific works, including 200 peer-reviewed papers (90 in international scientific journals), 11 books and chapters in books, 96 papers in conference proceedings, and 10 patents within the Russian Federation. His current research interests are in fabrication, functionalization, optical properties, and biomedical applications of metal and hybrid plasmonic nanoparticles. He is currently an associate Editor of the *Journal of Quantitative Spectroscopy and Radiative Transfer* and a member of the Editorial Boards of the *Journal of Nanoscience* and *Journal of Biomedical Photonics and Engineering*.

Introduction

Gold is one of the first metals discovered by humans, and the history of its study and application is estimated to be a minimum of several thousand years. The first information on colloidal gold (CG) can be found in tracts by Chinese, Arabic, and Indian scientists who obtained CG as early as in the fifth and fourth centuries BC and used it, in particular, for medical purposes (the Chinese "gold solution" and the Indian "liquid gold"). In the Middle Ages, alchemists in Europe actively studied and used CG. Probably, wonderful color changes that accompany the condensation of gold atoms prepared by reduction of salt solutions led alchemists to believe in transformations of elements, CG being considered as a panacea. Specifically, Paracelsus wrote about the therapeutic properties of *quinta essentia auri*, which he obtained through reduction of auric chloride with alcohol or oil plant extracts. He used "potable gold" to treat some mental disorders and syphilis. Paracelsus once proclaimed that chemistry is for making medicines, not for making gold out of metals: "Many have said of Alchemy, that it is for the making of gold and silver. For me such is not the aim, but to consider only what virtue and power may lie in medicines." His contemporary Giovanni Andrea applied *aurum potabile* to the treatment of lepra, ulcer, epilepsy, and diarrhea. The first book on CG preserved to our days was published by philosopher and doctor of medicine Francisco Antonii in 1618 [1]. It contains information on the preparation of CG and on its medical applications, including practical suggestions. From the seventeenth century, CG was used for the production of red (ruby) glasses, decoration on porcelain (purple of Cassius), and silk coloration [2]. In 1633, alchemist David de Planis-Campy, surgeon to the King of France Louis XIII, recommended his "elixir of longevity"—an aqueous CG solution—as a means of life prolongation [3]. In 1712, Hans Heicher published a complete summary of gold's medicinal uses, which describes the solutions and the gold stabilization with boiled starch, which is an example of stabilization of CG with ligands [4].

The beginning of scientific research on CG dates back to the mid-nineteenth century, when Michael Faraday published an article [5] devoted to methods of synthesis and properties of CG. In this article, Faraday described, for the first time, aggregation of CG in the presence of electrolytes, the protective effect of gelatin and other high-molecular-mass compounds, and the properties of thin films of CG. CG solutions prepared by Faraday are still stored in the Royal Institution of Great Britain in London.

In 1898, Richard Zsigmondy published the fundamental paper on the properties of CG [6]. He was the first to describe methods of synthesis of CG with different particle sizes with the use of hydrogen peroxide, formaldehyde, and white phosphorus as reducing agents and report on important physicochemical (including optical) properties of gold sols. Zsigmondy used CG as the main experimental object when inventing (in collaboration with Siedentopf) an ultramicroscope. In 1925, Zsigmondy was awarded the Nobel Prize in Chemistry "for his demonstration of the heterogeneous nature of colloid solutions and for the methods he used, which have since become fundamental in modern colloid chemistry."

Studies by the Nobel Prize laureate Theodor Svedberg on the preparation, analysis of mechanisms of CG formation, and sedimentation properties of CG (with the use of the ultra-centrifuge he had invented) are also among classical studies [7]. Svedberg investigated the kinetics of reduction of gold halides and formulated the main concepts about the mechanism of formation (chemical condensation) of CG particles.

In 1880, a method was put forward for alcoholism treatment by intravenous injection of a CG solution ("gold cure") [8]. In 1927, the use of CG was proposed to ease the suffering of inoperable cancer patients [9]. CG in color reactions toward spinal-fluid and blood-serum proteins has been used since the first half of the twentieth century [10]. Colloidal solutions of the [198]Au gold isotope (half-life time, 65 h) were successfully used for therapeutic purposes in cancer care facilities [11].

Despite the centuries-old history, a "revolution in immunochemistry" [12], associated with the use of gold particles in biological research, took place in 1971, when British investigators W.P. Faulk and G.M. Taylor published an article titled "An immunocolloid method for the electron microscope" [13]. In that article, they described a technique for conjugating antibodies to CG for direct electron microscopic visualization of *Salmonella* surface antigens, representing the first time that a CG conjugate was used as an immunochemical marker. From this point on, the use of CG biospecific conjugates in various fields of biology and medicine became very active. There has been a wealth of reports dealing with the application of functionalized gold nanoparticles (GNPs) (conjugates with recognizing biomacromolecules, *e.g.*, antibodies, lectins, enzymes, or aptamers) [14–16] to the studies of biochemists, microbiologists, immunologists, cytologists, physiologists, morphologists, and many other specialist researchers.

The range of uses of GNPs in current medical and biological research is extremely broad. In particular, it includes genomics; biosensorics; immunoassay; clinical chemistry; detection and photothermolysis of microorganisms and cancer cells; targeted delivery of drugs, peptides, DNA, and antigens; and optical bioimaging and monitoring of cells and tissues with the use of state-of-the-art nanophotonic recording systems. It has been proposed that GNPs be used in practically all medical application, including diagnostics, therapy, prophylaxis, and hygiene. Extensive information on the most important aspects of preparation and use of CG in biology and medicine can be found elsewhere [17–71]. Such a broad range of applications is based on the use of the unique physical and chemical properties of GNPs. Specifically, the optical properties of GNPs are determined by their plasmon resonance, which is associated with the collective excitation of conduction electrons and is localized in a wide region (from visible to infrared, depending on particle size, shape, and structure) [72].

The present book summarizes data on optical properties of GNPs, their applications in biological studies, analysis of data on the *in vitro* and *in vivo* biodistribution, uptake into mammalian cells, immunological properties, and the toxicity of most popular GNPs. A special section is devoted to applications of multifunctional gold-based nanocomposites.

References

1. Antonii F. *Panacea Aurea-Auro Potabile*. Hamburg: Bibliopolio Frobeniano. 1618.
2. Antonio Neri R.P. *L'arte Vetraria*. Firenze: Nella Stamperia de'Giunti. 1612.
3. de Planis-Campy D. *Traicté de la vraye, unique, grande, et universelle médecine des anciens dite des recens, or potabl*e. Paris: François Targa. 1633.
4. Heicher H.H. *Aurum Potabile, oder Gold-Tinctur*. Breslau, Lepzig: J. Herbord Klossen. 1712.
5. Faraday M. Experimental relations of gold (and others metals) to light // *Phil. Trans. Royal. Soc. (Lond)*. 1857. V. 147. P. 145–181.
6. Zsigmondy R. Ueber wassrige Lösungen metallischen Goldes // *Ann. Chem.* 1898. Bd. 301. S. 29–54.
7. Svedberg T. *Die Methoden zur Herstellung kolloider Lösungen anorganischer Stoffe*. Dresden: Theodor Steinkopff. 1909.
8. The Keeley "gold cure" for inebriety // *Br. Med. J.* 1892. V. 2. P. 85–86.
9. Ochsner E.H. The use of colloidal gold in inoperable cancer // *Int. J. Med. Surg.* 1927. V. 40. P. 100–104.
10. Lange C. Die Ausflockung kolloidalen Goldes durch Zerebrospinalflüssigkeit bei luetischen Affektionen des Zentralnervensystems // *Ztschr. f. Chemotherap.* 1912. Bd. 1. S. 44–78.
11. Rogoff E.E., Romano R., Hahn E.W. The prevention of Ehrlich ascites tumor using intraperitoneal colloidal [198]Au // *Radiology.* 1975. V. 114. P. 225–226.
12. Beesley J.E. Colloidal gold: A new revolution in marking cytochemistry // *Proc. R. Microsc. Soc.* 1985. V. 20. P. 187–197.
13. Faulk W., Taylor G. An immunocolloid method for the electron microscope // *Immunochemistry.* 1971. V. 8. P. 1081–1083.
14. Hermanson G.T. *Bioconjugate Techniques*. San Diego: Academic Press. 1996.
15. Glomm W.R. Functionalized gold nanoparticles for applications in bionanotechnology // *J. Dispers. Sci. Technol.* 2005. V. 26. P. 389–414.
16. Thanh N.T.K., Green L.A.W. Functionalisation of nanoparticles for biomedical applications // *Nano Today.* 2010. V. 5. P. 213–230.
17. *Techniques in Immunocytochemistry* / Eds. Bullock G.R., Petrusz P. London: Acad. Press, 1982.
18. *Colloidal Gold: Principles, Methods, and Applications* / Ed. Hayat M.A. San Diego: Acad. Press, 1989.
19. Dykman L.A., Bogatyrev V.A., Shchegolev S.Y., Khlebtsov N.G. *Gold Nanoparticles: Synthesis, Properties, and Biomedical Applications*. Moscow: Nauka. 2008 (in Russian).
20. *Gold Nanoparticles: Properties, Characterization and Fabrication* / Ed. Chow P.E. New York: Nova Science Publisher. 2010.

21. De Mey J., Moeremans M. The preparation of colloidal gold probes and their use as marker in electron microscopy // In: *Advanced Techniques in Biological Electron Microscopy* / Ed. Koehler J.K. Berlin: Springer-Verlag. 1986. V. 3. P. 229–271.

22. Horisberger M. Colloidal gold and its application in cell biology // *Int. Rev. Cytol.* 1992. V. 136. P. 227–287.

23. Roth J. The silver anniversary of gold: 25 years of the colloidal gold marker system for immunocytochemistry and histochemistry // *Histochem. Cell Biol.* 1996. V. 106. P. 1–8.

24. Dykman L.A., Bogatyrev V.A. Colloidal gold in solid-phase assays. A review // *Biochemistry (Moscow).* 1997. V. 62. P. 350–356.

25. Daniel M.C., Astruc D. Gold nanoparticles: Assembly, supramolecular chemistry, quantum-size-related properties, and applications toward biology, catalysis, and nanotechnology // *Chem. Rev.* 2004. V. 104. P. 293–346.

26. Dykman L.A., Bogatyrev V.A. Gold nanoparticles: Preparation, functionalisation and applications in biochemistry and immunochemistry // *Russ. Chem. Rev.* 2007. V. 76. P. 181–194.

27. Khlebtsov N.G., Bogatyrev V.A., Dykman L.A., Khlebtsov B.N. Gold plasmon resonant nanostructures for biomedical applications // *Nanotechnol. Russia.* 2007. V. 2. P. 69–86.

28. Chen P.C., Mwakwari S.C., Oyelere A.K. Gold nanoparticles: From nanomedicine to nanosensing // *Nanotechnol. Sci. Appl.* 2008. V. 1. P. 45–66.

29. Giljohann D.A., Seferos D.S., Daniel W.L., Massich M.D., Patel P.C., Mirkin C.A. Gold nanoparticles for biology and medicine // *Angew. Chem. Int. Ed.* 2010. V. 49. P. 3280–3294.

30. Edgar J.A., Cortie M.B. Nanotechnological application of gold // In: *Gold: Science and Applications* / Eds. Corti C., Holliday R. Boca Raton: CRC Press. 2010. P. 369–397.

31. Dykman L.A., Staroverov S.A., Bogatyrev V.A., Shchyogolev S.Y. Adjuvant properties of gold nanoparticles // *Nanotechnol. Russia.* 2010. V. 5. P. 748–761.

32. Khlebtsov N.G., Dykman L.A. Plasmonic nanoparticles: Fabrication, optical properties, and biomedical applications // In: *Handbook of Photonics for Biomedical Science* / Ed. Tuchin V.V. Boca Raton: CRC Press. 2010. P. 37–86.

33. Khlebtsov N.G., Dykman L.A. Biodistribution and toxicity of engineered gold nanoparticles: A review of *in vitro* and *in vivo* studies // *Chem. Soc. Rev.* 2011. V. 40. P. 1647–1671.

34. Jelveh S., Chithrani D.B. Gold nanostructures as a platform for combinational therapy in future cancer therapeutics // *Cancers.* 2011. V. 3. P. 1081–1110.

35. Dykman L.A., Khlebtsov N.G. Gold nanoparticles in biology and medicine: Recent advances and prospects // *Acta Naturae.* 2011. V. 3. P. 36–58.

36. Yeh Y.-C., Creran B., Rotello V.M. Gold nanoparticles: Preparation, properties, and applications in bionanotechnology // *Nanoscale.* 2012. V. 4. P. 1871–1880.

37. Dykman L.A., Khlebtsov N.G. Gold nanoparticles in biomedical applications: Recent advances and perspectives // *Chem. Soc. Rev.* 2012. V. 41. P. 2256–2282.

38. Dreaden E.C., Alkilany A.M., Huang X., Murphy C.J., El-Sayed M.A. The golden age: Gold nanoparticles for biomedicine // *Chem. Soc. Rev.* 2012. V. 41. P. 2740–2779.

39. Sau T.P., Goia D. Biomedical application of gold nanoparticles // In: *Fine Particles in Medicine and Pharmacy* / Ed. Matijević E. New York: Springer. 2012. P. 101–145.

40. Louis C., Pluchery O. *Gold Nanoparticles for Physics, Chemistry and Biology.* Singapore: Imperial College Press. 2012.

41. Jiang X.-M., Wang L.-M., Wang J., Chen C.-Y. Gold nanomaterials: Preparation, chemical modification, biomedical applications and potential risk assessment // *Appl. Biochem. Biotechnol.* 2012. V. 166. P. 1533–1551.

42. Wang H.-H., Su C.-H, Wu Y.-J., Lin C.-A.J., Lee C.-H., Shen J.-L., Chan W.-H., Chang W.H., Yeh H.-I. Application of gold in biomedicine: Past, present and future // *Int. J. Gerontol.* 2012. V. 6. P. 1–4.

43. Kim D., Jon S. Gold nanoparticles in image-guided cancer therapy // *Inorg. Chim. Acta.* 2012. V. 393. P. 154–164.

44. Ahmad M.Z., Akhter S., Rahman Z., Akhter S., Anwar M., Mallik N., Ahmad F.J. Nanometric gold in cancer nanotechnology: Current status and future prospect // *J. Pharm. Pharmacol.* 2013. V. 65. P. 634–651.

45. Khan M.S., Vishakante G.D., Siddaramaiah H. Gold nanoparticles: A paradigm shift in biomedical applications // *Adv. Colloid Interface Sci.* 2013. V. 199–200. P. 44–58.

46. Liu A., Ye B. Application of gold nanoparticles in biomedical researches and diagnosis // *Clin. Lab.* 2013. V. 59. P. 23–36.

47. Kumar D., Saini N., Jain N., Sareen R., Pandit V. Gold nanoparticles: An era in bionanotechnology // *Expert Opin. Drug Deliv.* 2013. V. 10. P. 397–409.

48. Loomba L., Scarabelli T. Metallic nanoparticles and their medicinal potential. Part I: gold and silver colloids // *Ther. Deliv.* 2013. V. 4. P. 859–873.

49. Lin M., Pei H., Yang F., Fan C., Zuo X. Applications of gold nanoparticles in the detection and identification of infectious diseases and biothreats // *Adv. Mater.* 2013. V. 25. P. 3490–3496.

50. Khlebtsov N.G., Bogatyrev V.A., Dykman L.A., Khlebtsov B.N., Staroverov S.A., Shirokov A.A., Matora L.Y., Khanadeev V.A., Pylaev T.E., Tsyganova N.A., Terentyuk G.S. Analytical and theranostic applications of gold nanoparticles and multifunctional nanocomposites // *Theranostics.* 2013. V. 3. P. 167–180.

51. Dykman L.A., Khlebtsov N.G. Uptake of engineered gold nanoparticles into mammalian cells // *Chem. Rev.* 2014. V. 114. P. 1258–1288.

52. Cao-Milán R., Liz-Marzán L.M. Gold nanoparticle conjugates: Recent advances toward clinical applications // *Expert Opin. Drug Deliv.* 2014. V. 11. P. 741–752.

53. *Gold Clusters, Colloids and Nanoparticles* / Ed. Mingos M.P. New York: Springer. 2014.

54. Shah M., Badwaik V.D., Dakshinamurthy R. Biological applications of gold nanoparticles // *J. Nanosci. Nanotechnol.* 2014. V. 14. P. 344–362.

55. Kobayashi K., Wei J., Iida R., Ijiro K., Niikura K. Surface engineering of nanoparticles for therapeutic applications // *Polymer J.* 2014. V. 46. P. 460–468.

56. Khlebtsov N.G., Dykman L.A., Khlebtsov B.N., Khanadeev V.A., Panfilova E.V. Fabrication, optical properties and biomedical applications of conjugates of metallic and composite multifunctional nanoparticles with controllable parameters of plasmon resonance and surface functionalization by probing molecules // *RFBR J.* 2014. No. 4. P. 18–33 (in Russian).

57. Pedrosa P., Vinhas R., Fernandes A., Baptista P.V Gold nanotheranostics: Proof-of-concept or clinical tool? // *Nanomaterials.* 2015. V. 5. P. 1853–1879.

58. Kim E.Y., Kumar D., Khang G., Lim D.-K. Recent advances in gold nanoparticle-based bioengineering applications // *J. Mater. Chem. B.* 2015. V. 3. P. 8433–8444.

59. Sasidharan A., Monteiro-Riviere N.A. Biomedical applications of gold nano-materials: Opportunities and challenges // *Wiley Interdiscip. Rev. Nanomed. Nanobiotechnol.* 2015. V. 7. P. 779–796.

60. Zhang X. Gold nanoparticles: Recent advances in the biomedical applications // *Cell Biochem. Biophys.* 2015. V. 72. P. 771–775.

61. Heath J.R. Nanotechnologies for biomedical science and translational medicine // *Proc. Natl. Acad. Sci. USA.* 2015. V. 112. P. 14436–14443.

62. Yang X., Yang M., Pang B., Vara M., Xia Y. Gold nanomaterials at work in bio-medicine // *Chem. Rev.* 2015. V. 115. P. 10410–10488.

63. Khlebtsov N.G., Dykman L.A., Khlebtsov B.N., Bogatyrev V.A. Gold and compos-ite nanoparticles for analytical and theransotic applications // In: *Basic Science for Medicine: Biophysical Medical Technology* / Eds. Grigoriev A.I., Vladimirov Y.A. Moscow: MAKS Press. 2015. V. 2. P. 6–64 (in Russian).

64. Ashraf S., Pelaz B., del Pino P., Carril M., Escudero A., Parak W.J., Soliman M.G., Zhang Q., Carrillo-Carrion C. Gold-based nanomaterials for applications in nanomedicine // *Top. Curr. Chem.* 2016. V. 370. P. 169–202.

65. Daraee H., Eatemadi A., Abbasi E., Aval S.F., Kouhi M., Akbarzadeh A. Application of gold nanoparticles in biomedical and drug delivery // *Artif. Cells Nanomed. Biotechnol.* 2016. V. 44. P. 410–422.

66. Versiani A.F., Andrade L.M., Martins E.M.N., Scalzo S., Geraldo J.M., Chaves C.R., Ferreira D.C., Ladeira M., Guatimosim S., Ladeira L.O., da Fonseca F.G. Gold nanoparticles and their applications in biomedicine // *Fut. Virol.* 2016. V. 11. P. 293–309.

67. Dykman L.A., Khlebtsov N.G. Multifunctional gold-based nanocomposites for theranostics // *Biomaterials.* 2016. V. 108. P. 13–34.

68. Dykman L.A., Khlebtsov N.G. Biomedical applications of multifunctional gold-based nanocomposites // *Biochemistry (Moscow).* 2016. V. 81. P. 1771–1789.

69. Carneiro M.F.H., Barbosa F., Jr. Gold nanoparticles: A critical review of therapeu-tic applications and toxicological aspects // *J. Tox. Environ. Health B.* 2016. V. 19. P. 129–148.

70. Chen W., Zhang S., Yu Y., Zhang H., He Q. Structural-engineering rationales of gold nanoparticles for cancer theranostics // *Adv. Mater.* 2016. V. 28. P. 8567–8585.

71. Dykman L.A., Khlebtsov N.G. Immunological properties of gold nanoparticles // *Chem. Sci.* 2017. V. 8. P. 1719–1735.

72. Khlebtsov N.G., Dykman L.A. Optical properties and biomedical applications of plasmonic nanoparticles // *J. Quant. Spectrosc. Radiat. Transfer.* 2010. V. 111. P. 1–35.

1

Optical Properties of Gold Nanoparticles

1.1 Chemical Wet Synthesis of Gold Nanoparticles

Methods of synthesis of colloidal gold (CG) (and other metal colloids) can be arbitrarily divided into the following two large groups: dispersion methods (metal dispersion) and condensation methods (reduction of the corresponding metal salts).

Dispersion methods for the preparation of CG are based on destruction of the crystal lattice of metallic gold in high-voltage electric field [1] or laser ablation [2]. If an electric arc is created in a liquid between two gold electrodes under electric field, its blazing leads to the mass transfer between electrodes accompanied by CG formation. The yield and shape of gold particles formed under electric current depend not only on the voltage between electrodes and the current strength but also on the presence of electrolytes in solution. The use of direct current leads to the formation of nonuniform gold particles. The addition of even very small amounts of alkalis or chlorides and the use of high-frequency alternating current for dispersion substantially improve the quality of gold hydrosols.

Condensation methods are more commonly employed than dispersion methods. CG is most often prepared by reduction of gold halides (for example, of $HAuCl_4$) with the use of chemical reducing agents and/or irradiation (ultrasonic and ultraviolet [UV] irradiation, pulse or laser radiolysis). Various organic and inorganic compounds (more than 100) serve as chemical reducing agents. At present, the most popular colloidal synthesis protocols involve the citrate reduction of $HAuCl_4$, as suggested by Borowskaja [3], Turkevich *et al.* [4], and Frens [5], to produce relatively monodisperse particles with controlled average equivolume diameters of 10 to 100 nm. For smaller particles, other reducing agents can be used, *e.g.*, a mixture of borohydride and ethylenediaminetetra acetic acid (EDTA) [6] (5 nm) or sodium or potassium thiocyanate [7] (1 to 2 nm). The shape of particles larger than 100 nm can be improved by the method described by Goia and Matijević [8]. A sol with a narrow Au particle size distribution can be prepared by adding presynthesized Au seeds, on which condensation will occur, to an $HAuCl_4$ solution. In this method, sodium citrate or hydroxylamine can serve as reducing agents [9]. Isodisperse and isomorphous sols can be prepared only if the formation of new nuclei is prevented. As a rule, it is achieved by performing the process in two steps. Initially, a new phase nucleates, and then weak supersaturation is created in the sol, due to which new nuclei are not produced any more and only the already formed nuclei grow [10].

Despite more than half a century of study and application recovery citrate $HAuCl_4$, the chemistry of the reaction and the process of formation and development of gold particles continues to interest researchers [11–18].

Gold nanoparticles (GNPs) can also be prepared by the two-phase microemulsion method. In the first step, metal-containing reagents are transferred from an aqueous to an organic phase. After the addition of a surfactant solution to this system, a microemulsion, *i.e.*, a dispersion of two immiscible liquids, is formed. The reduction reaction proceeds in a dispersed phase, in which the drop size is at most 100 nm. As a result, virtually monodispersed sols are formed [19]. Later, a single-phase method without the use of aqueous solutions was developed [20]. In microemulsion methods of synthesis of CG, alkanethiols are often added to the reaction solution, and these additives form dense self-assembled monolayers on the gold surface. This method was employed for the preparation of self-assembled two-dimensional (2D) and three-dimensional (3D) ensembles of GNPs [21,22].

The application of biopolymers, microorganisms, fungi, plant, and animal cells for the CG synthesis is a new line of investigations in nanobiotechnology [23–26].

Perhaps, Yu *et al.* [27] were the first to suggest an electrochemical oxidation/reduction procedure in combination with surfactant additives and to prepare CG nanorods (GNRs) with an average thickness of about 10 nm and with aspect ratios ranging from 1.5 to 11. Murphy's [28] and El-Sayed's [29] groups introduced a two-step seed-mediated CTAB-assisted protocol (CTAB designates cetyltrimethylammonium bromide) in which fine seed gold particles are formed (2 to 4 nm) in step 1 and then GNRs are grown by adding the gold seeds to a growth solution containing CTAB, silver nitrate, $HAuCl_4$, and ascorbic acid.

For synthesis of silica/gold nanoshells (GNSs), Halas and coworkers [30] developed a two-step protocol involving the fabrication of silica nanospheres followed by their amination, the attachment of fine (2 to 4 nm) gold seeds, and the formation of a complete gold shell by reducing $HAuCl_4$ on seeds. A summary of ten-year development of this approach is given by Brinson *et al.* [31].

Xia *et al.* introduced a new protocol based on galvanic replacement of silver ions by reducing gold atoms [32,33]. This approach can produce a variety of particle shapes and structures, beginning from silver cubes and ending with gold nanocages [34]. To date, one can find a lot of published protocols for fabrication of nanoparticles with various shapes and structures. In particular, the reducing of $HAuCl_4$ on 10–15-nm gold seed leads to formation of gold nanostars [35].

A summary of recent advantages in spherical GNPs synthesis can be found in reviews in Refs. [36–38] and in anisotropic GNPs and nanopowders in Refs. [39–44]. In conclusion, we show a gallery of particles mentioned in this section in Figure 1.1.

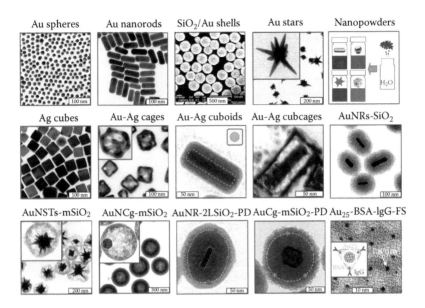

FIGURE 1.1 Examples of plasmonic GNPs and nanocomposites: 16-nm Au nanospheres; GNRs; SiO_2/Au nanoshells; gold nanostars; plasmonic nanopowders of gold nanospheres, nanorods, nanostars, and gold/silver nanocages; silver nanocubes; gold–silver nanocages; Au(core)/Ag(shell) nanocuboids and cubcages; nanocomposites containing a GNR core and a mesoporous silica shell, nanorattles consisting of a hollow mesoporous silica shell with embedded gold nanostar or Au-Ag cage; nanocomposites containing a GNR or nanocage core and a mesoporous silica shell dopped with photodynamic (PD) dye (here, hematoporphyrin); fluorescent Au-BSA-IgG-PS complexes consisting of an Au_{25} cluster stabilized by BSA molecules that are dopped with Photosens PD dye and human antistaphylococcal immunoglobulin, IgG.

1.2 Basic Physical Principles

Collective excitations of conductive electrons in metals are called "plasmons" [45,46]. Depending on the boundary conditions, it is commonly accepted to distinguish bulk plasmons (3D plasma), surface propagating plasmons or surface plasmon polaritons (2D films), and surface localized plasmons (nanoparticles) (Figure 1.2).

A quantum of bulk plasmons $\hbar\omega_p$ is about 10 eV for noble metals. Because of their longitudinal nature, the bulk plasmons cannot be excited by visible light. The surface plasmon polaritons propagate along metal surfaces in a waveguide-like fashion. Here, we consider only the nanoparticle localized plasmons. In this case, the electric component of an external optical field exerts a force on the conductive electrons and displaces them from their equilibrium positions to create uncompensated charges at the nanoparticle surface (Figure 1.2c). As the main effect producing the restoring force is the polarization of the particle surface, these oscillations are called "surface" plasmons, which have a well-defined resonance frequency.

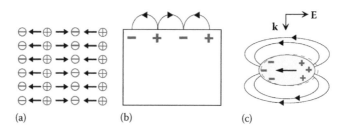

FIGURE 1.2 Schematic representation of the bulk (a), surface propagating (b), and surface localized (c) plasmons. The dashed line shows the electron cloud displacement. (Adapted from Khlebtsov, N.G., Dykman, L.A., *J. Quant. Spectrosc. Radiat. Transfer*, 111, 1–35, 2010. With permission.)

The most common physical approach to linear nanoparticle plasmonics is based on the linear local approximation:

$$\mathbf{D}(\mathbf{r},\omega) = \varepsilon(\mathbf{r},\omega)\mathbf{E}(\mathbf{r},\omega), \tag{1.1}$$

where ω is the angular frequency and the electric displacement $\mathbf{D}(\mathbf{r}, \omega)$ depends on the electric field $\mathbf{E}(\mathbf{r}, \omega)$ and on the dielectric function $\varepsilon(\mathbf{r}, \omega)$ only at the same position $\mathbf{r}$. Then, the liner optical response can be found by solving the Maxwell equations with appropriate boundary conditions. Actually, in the local approximation, the indicating of position vector in dielectric functions $\varepsilon(\mathbf{r}, \omega)$ may be omitted.

As the particle size a is decreased to the value comparable with the electron mean free path ($a \sim L_{eff}$), deviations of the phenomenological dielectric function $\varepsilon(\omega, a)$ of the particle from the bulk values $\varepsilon(\omega) = \varepsilon(\omega, a \gg L_{eff})$ can be expected. A general recipe for the inclusion of macroscopic tabulated data and size effects to the size-dependent dielectric function consists in the following [45,47]. Let $\varepsilon_b(\omega)$ be the macroscopic dielectric function, which can be found in the literature from measurements with massive samples [48]. Then, the size-dependent dielectric function of a particle is

$$\varepsilon(\omega,a) = \varepsilon_b(\omega) + \Delta\varepsilon(\omega,a), \tag{1.2}$$

where the correction $\Delta\varepsilon(\omega, a)$ takes into account the contribution of size-dependent scattering of electrons to the Drude part of the dielectric function described by the expression

$$\Delta\varepsilon(\omega,a) = \varepsilon_b^{Drude}(\omega) - \varepsilon_p^{Drude}(\omega,a) = \frac{\omega_p^2}{\omega(\omega+i\gamma_b)} - \frac{\omega_{p,a}^2}{\omega(\omega+i\gamma_p)}. \tag{1.3}$$

Here, $\gamma_b = \tau_b^{-1}$ is the volume decay constant; τ_b is the electron free path time in a massive metal; $\omega_{p,a}$ is the plasma frequency for a particle of diameter a (we assume here that $\omega_{p,a} \simeq \omega_p$);

$$\gamma_p = \tau_p^{-1} = \gamma_b + \gamma_s = \gamma_b + A_s \frac{v_F}{L_{eff}} \tag{1.4}$$

is the size-dependent decay constant equal to the inverse electron mean transit time $\gamma_p = \tau_p^{-1}$ in a particle; L_{eff} is the effective electron mean free path; γ_s is the size-dependent contribution to the decay constant; and A_s is a dimensionless parameter determined by the details of scattering of electrons by the particle surface [49,50] (which is often simply set equal to 1).

To solve the Maxwell equation, various analytical or numerical methods can be used. The most popular and simple analytical method is the dipole (electrostatic) approximation [51], developed in classic works by Rayleigh [52], Mie [53], and Gans [54] (actually, the Gans solution for ellipsoid had earlier been done by Rayleigh [55]). Among the many numerical methods available, the discrete dipole approximation (DDA) [56], the boundary element method (BEM) [60], the finite element and finite difference time domain methods (FEM and FDTDM) [57], the multiple multipole method (MMP) [58], and the T-matrix method [59] are most popular. For details, relevant references, and discussion of the advantages and drawbacks of various methods, the readers are referred to reviews [47,60–67] and books [59,68].

Understanding of the interaction of intense laser pulses with plasmon resonant nanoparticles requires going beyond linear response description. For instance, the second and third harmonic generation is widely known example [60]. Over the past decade, many transient transmission experiments [69] have been carried out to elucidate the time-dependent physics of nonlinear responses. These experiments involve an intense femtosecond pump laser pulse followed by measuring the transmittance of a weaker probe laser beam. Figure 1.3 shows a flowchart of the processes [60] accompanied the nonlinear particle response after irradiation with pump pulse. These experiments lead to better understanding the size-dependent dynamical effects (such as electron relaxation) and the temperature-dependent models of the particle dielectric function [60].

In some cases, an analytical approximation of the Drude type should be used instead of tabulated bulk $\varepsilon_b(\lambda)$ data. Such a situation appears, for example, in FDTDM method, where one needs to calculate an integral convolution [57]. This integration could be performed analytically at each step of FDTDM only by using a few simple models for $\varepsilon(\omega)$, including the Drude formula. Specifically, there exists several sets of interpolation parameters for Johnson and Christy's data [70] or other tabulated data [47], including the interpolation sets given by Oubre and Nordlander [71]: $\varepsilon_{ib} = 9.5$, $\omega_p = 8.95$ eV, $\gamma_b = 0.0691$ (gold) and $\varepsilon_{ib} = 5.0$, $\omega_p = 9.5$ eV и $\gamma_b = 0.0987$ eV (silver).

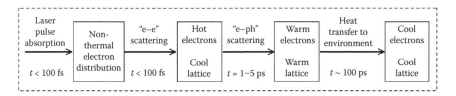

FIGURE 1.3 Schematic of a transient nanoparticle response to absorption of a laser pulse followed by the electron–electron (e–e) and electron–phonon (e–ph) scattering and the heat exchange with environment. (Adapted from Khlebtsov, N.G., Dykman, L.A., *J. Quant. Spectrosc. Radiat. Transfer*, 111, 1–35, 2010. With permission.)

1.3 Surface-Chemical and Quantum-Size Effects

Although the basic optics of plasmon resonant metal nanoparticles can be reasonably well explained with simple models discussed in Section 1.2, there are large discrepancies between the optical properties of metal sols prepared in water, particularly those of silver, and nanoparticle ensembles prepared in other matrices. These differences can be attributed to the unique double layer presented at the metal-water interface. In his excellent review, Mulvaney [72] gave a thorough consideration of surface physicochemical effects such as cathodic or anodic polarization, chemisorption, metal adatom deposition, and alloying. These processes can alter the surface plasmon absorption band of aqueous metal colloids, which is sensitive to electrochemical processes occurring at metal particle surfaces. Phenomenological description implies deviation of the surface scattering constant A_s from 1. A theoretical study to understand the origin of the increase in A_s was given by Persson [73], who considered the influence of a surface layer of adsorbed molecules on the damping rate of the localized surface plasmon. In general, the widths of the plasmon resonant spectra of particles, being deposited on a substrate or embedded in a matrix, have to be interpreted with care since the net size-limiting effect may be less effective than other mechanisms and can be obscured by the chemical interface damping.

The unique properties of metal nanoparticles could help in answering the old question: "How many atoms does it take to make a solid?" Indeed, a typical metal nanoparticle can be thought as an aggregate composed of a relatively small number of atoms, starting with clusters consisting of a few atoms to large nanoparticles with more than 10^5 atoms. Thus, these objects are intermediate in size between the domain of atoms and small molecules, which require a full quantum mechanical (QM) treatment, and the bulk materials describing by classical electrodynamics and solid-state physics. Metal nanoparticles are the objects at mesoscopic scale between the microscopic and macroscopic worlds. So, they may exhibit a wide number of new phenomena [74], including the discrete spectrum of the electronic states and coherent motion of electrons as far as electron can propagate through the whole system without experiencing inelastic or phase-breaking scattering. In terms of quantum mechanic language, a metallic nanoparticle subject to an external driving field reveals a collective electronic excitation, the so-called surface plasmon.

The question of primary importance concerns the applicability of phenomenological modification of the Drude's. As pointed out by Kawabata and Kubo in their pioneering work [75], the interpretation of the surface plasmon linewidth in terms of the free path effect is "too naive, if not entirely incorrect, because in such a small particle the electron states are quantized into discrete levels which are determined by the boundary conditions at the surface." Using the linear response theory and fluctuation-dissipation theorem [75], Kawabata and Kubo arrived at the following final expression for the size-dependent contribution to the damping constant (see Equation 1.5):

$$\gamma_s = \frac{3}{4}\frac{v_F}{a}g_{kk}\left(\frac{\hbar\omega_0}{\varepsilon_F}\right), \tag{1.5}$$

where $g_{kk}(x)$ is a decreasing function of the surface plasmon energy $\hbar\omega_0$ normalized to the Fermy energy ε_F, and $g(0) = 1$. Surprisingly enough, the first-principle treatment has led to the same $\gamma_s \sim 1/L_{eff} \sim 1/a$ law that was established from simple physical [45] or even pure geometrical [49] approaches.

Since the 1966 Kawabata and Kubo paper [75], which included significant approximations, various refinements have been published (see, *e.g.*, references in [45,47,76]), but all these studies confirmed the same $\gamma_s \sim 1/L_{eff} \sim 1/a$ law with some minor modifications. As a matter of fact, all theoretical calculations reproduced this damping law (Equation 1.4), but with different parameter A_s.

In addition to the surface scattering problem, there exist other quantum effects that have been omitted in our short consideration. For example, the so-called "spill-out" effect [76] means that the electron wave function extends outside of the geometrical particle boundary, resulting in a decrease in the effective electron density and in a corresponding decreasing the apparent plasma frequency. Nevertheless, the existing experimental data for noble metal nanoparticles and the theoretical studies based on the time-dependent local density approximation for nanoshells [77] and rods [78] prove the validity of using the classical electrodynamics combined with properly size-corrected dielectric function. At this point, we agree with Bohren and Huffman's [51] notion that "surface modes in small particles are adequately and economically described in their essentials by simple classical theories" provided that the possible quantum effects have been included in the dielectric function.

It is also interesting to note that the plasmons in metal nanostructures exhibit some analogy to electron wave functions of simple atomic and molecular orbitals [79]. It has been recently realized that this analogy and the "plasmon hybridization" concept [80] can be exploited in the design of various nanostructures and understanding their optical properties.

Although the local approximation and frequency-dependent dielectric function give reasonable agreement between theory and experiment, the nonlocal effects can play an important role at small distances in the nanometer region [81]. As the first-principle quantum-mechanical calculations of the optical response [76] are available only for relatively simple systems (hundreds of noble metal atoms) and are insufficient for real nanoparticles, several compromise approximations have been developed to describe quantum-sized and nonlocal effects [77,81]. In particular, a specular reflection model [81] has been applied to describe the optical properties of small gold and silver spheres, bispheres, and nanoshells. It was shown that the nonlocal effects produce significant blue shift of plasmon resonance and its broadening in comparison with the commonly accepted local approach.

By contrast to local relationship (Equation 1.1), the nonlocal response is described by integral convolution

$$\mathbf{D}(\mathbf{r},\omega) = \int d\mathbf{r}' \varepsilon(\mathbf{r}',\mathbf{r},\omega)\mathbf{E}(\mathbf{r}',\omega). \qquad (1.6)$$

For homogeneous media, $\varepsilon(\mathbf{r'}, \mathbf{r}, \omega) = \varepsilon(|\mathbf{r'} - \mathbf{r}|, \omega)$, and after Fourier transformation, we have the following momentum-space ($\mathbf{q}$) representation:

$$\mathbf{D}(q,\omega) = \varepsilon(q,\omega)\mathbf{E}(q,\omega). \tag{1.7}$$

The local approximation means a transition to the $q \to 0$ limit $\varepsilon(q, \omega) \to \varepsilon^{loc}(\omega)$. An approximate recipe to account for the nonlocal effects can be written as a modification of Equation 1.2 [81]:

$$\varepsilon(q,\omega) = \varepsilon(\omega) + \Delta\varepsilon(q,\omega) \equiv \varepsilon(\omega) + \varepsilon^{nl}(q,\omega) - \varepsilon^{Drude}(\omega), \tag{1.8}$$

where $\varepsilon(\omega)$ is the experimental (local) tabulated dielectric function, $\varepsilon^{Drude}(\omega)$ is the Drude function, and $\varepsilon^{nl}(q, \omega)$ is the nonlocal valence-electrons response as introduced by Lindhard and Mermin [82]. With the nonlocal dielectric function in hand, further calculations can be made in the same fashion as with the local approximation. Several illustrative examples are presented in the previously cited paper [81].

The previous consideration is valid only for relatively large atom clusters. At sizes close to the Fermi wavelength of an electron, discrete nanocluster energy levels become accessible, thus causing significant change in optical properties. An instructive example is Au_8 quantum dots (QDs) described in Refs. [83,84]. Such small metal nanoclusters exhibit molecule-like transitions as the density of states is small to merge the conduction and valence bands. The luminescence from Au_8 QDs is thought to arise from electron transitions between filled "d" and "sp" conduction bands. As QDs size decreases, the level-energy spacing increases, leading to a blue shift in fluorescence relative to that from larger QDs. Therefore, in contrast to small metal nanoparticles, Au and Ag QDs [85] demonstrate QM behavior, which cannot be treated in terms of classical electrodynamics.

For real metal nanostructures, the first principle QM simulation of realistic surface-enhanced Raman spectroscopy systems remains a formidable challenge because of complex interaction between electronically localized molecules and delocalized metal electrons [86,87]. For example, in the case of plasmonic dimers with a nanometer gap between particles or for palsmonic nanomatryoshkas with a subnanometer gap between core and shell [88], the physical background of classical EM treatment becomes questionable. Specifically, on a subnanometer scale, the classical local approach fails to account for (1) a smooth spatial profile of the electron density at the metal-dielectric boundary and the nonlocal displacement of the screening charge centroid with respect to the geometrical interface [89]; (2) QM tunneling of electrons through a subnanometer gaps [90,91], thus increasing the effective gap conductivity before direct physical contact. To address both points and to avoid the shortcomings of the first-principle QM simulations, two semiclassical approaches have been suggested.

First, a nonlocal hydrodynamic model [92,93] was developed to incorporate the Drude-like nonlocal longitudinal component of dielectric tensor [94]. The second important QM correction to classical EM treatment is expected for nanogaps below 0.3–0.5 nm [95], *i.e.*, below the crossover region between the classical local and semiclassical nonlocal treatments [91]. The semiclassical quantum corrected model (QCM) [90] uses Maxwell equations with an effective dielectric function that accounts for electron tunneling across the gap with an "effective" local separation. For a more detailed discussion and literature information, the readers are referred to Ref. [88].

1.4 Plasmon Resonances

In the general case, the eigenfrequency of a "collective" plasmon oscillator does not coincide with the wave frequency and is determined by many factors, including the concentration and effective mass of conductive electrons; the shape, structure, and size of particles; the interaction between particles; and the influence of the environment. However, for an elementary description of the nanoparticle optics, it is sufficient to use a combination of the usual dipole (Rayleigh) approximation and the Drude theory [51]. In this case, the absorption and scattering of light by a small particle are determined by its electrostatic polarizability α, which can be calculated by using the optical dielectric function $\varepsilon(\omega)$ or $\varepsilon(\lambda)$, where λ is the wavelength of light in vacuum.

Let us consider three types of small particles shown in Figure 1.4: a nanosphere of a radius a, a nanorod with semiaxes a and b, and a core-shell particle with the corresponding radii a_c and a. All the particles are assumed to have dipole-like scattering and absorption properties. The core-shell particle is replaced with a homogeneous sphere of a dielectric permittivity ε_{av} according to the dipole equivalence principle [96,97].

For a small particle with an equivolume radius $a_{ev} = (3V/4\pi)^{1/3}$ embedded in a homogeneous dielectric medium with permittivity ε_m (Figure 1.4), we have the following expressions for the extinction, absorption and scattering cross-sections [47]

$$C_{ext} = C_{abs} + C_{sca} = \frac{12\pi k\,\varepsilon_m\,\mathrm{Im}(\varepsilon)}{a_{ev}^3\,|\varepsilon - \varepsilon_m|^2}|\alpha|^2 + \frac{8\pi}{3}k^4\,|\alpha|^2 \simeq 4\pi k\,\mathrm{Im}(\alpha), \tag{1.9}$$

where $k = 2\pi\sqrt{\varepsilon_m}/\lambda$ is the wave number in the surrounding medium and α is the renormalized particle polarizability [47]:

$$\alpha = \frac{\alpha_0}{1 + f_{rd}(ka_v)a_{ev}^{-3}\alpha_0}, \tag{1.10}$$

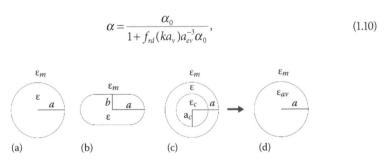

(a)   (b)   (c)   (d)

FIGURE 1.4 Three examples of small dipole-scattering metal particles: a nanosphere of a radius a, a nanorod with semiaxes a and b (the s-cylinder model [97]), and a core-shell particle with the corresponding radii a_c and a. The dielectric functions of metal, surrounding medium, and the dielectric core are ε, ε_m, and ε_c, respectively. The core-shell particle is replaced with a homogeneous sphere of a dielectric permittivity ε_{av} according to the "dipole equivalence principle." (Adapted from Khlebtsov, B.N., Khlebtsov, N.G., *J. Quant. Spectr. Radiat. Transfer*, 106, 154–169, 2007. With permission.)

α_0 is the electrostatic polarizability, and the correction function $f_{rd}(ka_{ev})$ accounts for the radiative damping effects [98]. For a sphere of a radius $a_{ev} = a$, we have

$$f_{rd}(ka) = 2 + 2(ika - 1)\exp(ika). \tag{1.11}$$

For simplicity, we will not distinguish the electrostatic polarizability α_0 from the renormalized polarizability α. In this approximation, $\alpha = \alpha_0$ and the extinction of a small particle is determined by its absorption $C_{abs} = C_{ext} = 4\pi k \, \mathrm{Im}(\alpha)$, whereas and scattering contribution can be neglected. For all three types of particles, the polarizability α can be written in following general form:

$$\alpha = \frac{3V}{4\pi} \frac{\varepsilon - \varepsilon_m}{\varepsilon + \varphi\varepsilon_m} = \frac{3V}{4\pi} \frac{\varepsilon_{av} - \varepsilon_m}{\varepsilon_{av} + \varphi\varepsilon_m}, \tag{1.12}$$

where the second expression is written for the metal nanoshell and the parameter φ depends on the particle shape and structure. Specifically, $\varphi = 2$ for spheres, $\varphi = (1/L_a - 1)$ for spheroids (L_a and $L_b = (1 - L_a)/2$ are the geometrical depolarization factors [51]), and

$$\varphi = \frac{1}{2}\left[p_0 + \left(p_0^2 - (\varepsilon_c/\varepsilon_m)\right)^{1/2}\right], p_0 = \frac{\varepsilon_c}{\varepsilon_m}\left(\frac{3}{4f_s} - \frac{1}{2}\right) + \frac{3}{2f_s} - \frac{1}{2} \tag{1.13}$$

for metal nanoshells on a dielectric core [96]. Here, $f_s = (a_s/a)^3 = 1 - (a_c/a)^3$ is the volume fraction of the metal shell. For thin shells, we have [96]

$$\varphi = \frac{3}{f_s}(1 + \varepsilon_c/2\varepsilon_m). \tag{1.14}$$

One can see from the expressions presented earlier that the resonance condition for polarizability and optical cross-sections reads

$$\varepsilon(\omega_{max} \equiv \omega_0) = \varepsilon(\lambda_{max}) = -\varphi\varepsilon_m. \tag{1.15}$$

The plasmon resonance frequency can be estimated from the Drude approximation for dielectric function

$$\varepsilon(\omega) = \varepsilon_{ib} - \frac{\omega_p^2}{\omega(\omega + i\gamma_b)}, \tag{1.16}$$

where ε_{ib} is the interband contribution, ω_p is the frequency of volume plasma oscillations of free electrons, γ_b is the volume decay constant related to the electron mean free path

l_b and the Fermi velocity v_F by the expression $\gamma_b = l_b/v_F$. By combining previous equations, one can obtain the following expressions for the resonance plasmon frequency and wavelength [47]:

$$\omega_{max} \equiv \omega_0 = \frac{\omega_p}{\sqrt{\varepsilon_{ib} + \varphi\varepsilon_m}}, \quad \lambda_{max} \equiv \lambda_0 = \lambda_p\sqrt{\varepsilon_{ib} + \varphi\varepsilon_m}. \tag{1.17}$$

Here, $\lambda_p = 2\pi c/\omega_p$ is the wavelength of volume oscillations of the metal electron plasma. For gold, $\lambda_p \approx 131$ nm.

Two additional important notes are in order here. First, it follows from Equation 1.17 that the dipole resonances of small gold or silver spheres ($5 \le a \le 20$ nm) in water is localized near 520 nm and 380 nm and do not depend on their size. By contrast, the dipole resonance of rods and nanoshells can be easily tuned through variation in their aspect ratio ($L_a = 1/3$(sphere) $\rightarrow$ 0(needle)) or the ratio shell thickness/core radius ($f_s = 1$ for a homogeneous sphere and $f_s \rightarrow 0$ for a thin shell). Second, Equation 1.17 determines the very first ($n = 1$) dipole resonance of a spherical particle. In addition to the dipole resonance, higher multipoles and corresponding multipole (quadrupole, *etc.*) resonances can be also excited in larger particles. For each multipole mode, the resonance condition exists, which is similar to Equation 1.15 and corresponds to the resonance of the quadrupole, octupole, and so on, contributions. For spherical particles, these conditions correspond to the resonance relations for the partial Mie coefficients [51] $\omega_n = \omega_p(\varepsilon_{ib} + \varepsilon_m(n + 1)/n)^{-1/2}$, where n is the mode (resonance) number. With an increase in the sphere size, the multipole frequency decreases [60,99].

It is important to distinguish two possible scenarios for excitation of multipole resonances. The first case corresponds to a small nonspherical particle of irregular or uneven shape, when the distribution of induced surface charges is strongly inhomogeneous and does not correspond to the dipole distribution. This inhomogeneous distribution generates high multipoles even in the case when the system size is certainly much smaller than the light wavelength. Typical examples are cubic particles [100] or two contacting spheres [101], where the field distribution near the contact point is so inhomogeneous that multipole expansions converge very slowly or diverge at all. The second scenario of high multipole excitation is realized with increasing the particle size, when the transition from the quasi-stationary to radiative regime is realized, and the contribution of higher spherical harmonics should be taken into account in the Mie series (or another multipole expansion). For example, while the extinction spectrum of a silver 30-nm nanoparticle is completely determined by the dipole contribution and has one resonance, the spectrum of a 60-nm sphere exhibits a distinct high-frequency quadrupole peak in addition to the low-frequency dipole peak.

1.5 Metal Spheres

The optical properties of a small sphere are completely determined by their scattering amplitude matrix [51]. Here, we are interested mainly in its extinction, absorption, and scattering cross-sections $C_{ext,abs,sca}$ or efficiencies $Q_{ext,abs,sca} = C_{ext,abs,sca}/S_{geom}$. From an experimental point of view, we shall consider the absorbance (extinction) and the scattering intensity at a given metal concentration c:

$$A_{ext} = 0.326 \frac{cl}{\rho} \frac{Q_{ext}}{a_{ev}}, \quad I_{90}(\lambda, \theta_{sca}) = 0.326 \frac{cl}{\rho a_{ev}} \left[\frac{16 S_{11}(ka_{ev}, \theta_{sca})}{3(ka_{ev})^2} \right], \quad (1.18)$$

where $S_{11}(ka_{ev}, \theta)$ is the normalized intensity of scattering at an θ_{sca} angle (the first element of the Mueller scattering matrix [51]) and ρ and $d = 2a$ are the metal density and the particle equivolume diameter. The expression in square brackets is normalized so that it is equal to the scattering efficiency of small particles. For metal nanoshells, Equation 1.18 should be slightly modified [102].

Figure 1.5 shows well-known theoretical extinction and scattering spectra calculated by Mie theory for gold and silver colloids at a constant concentration of gold (57 µg/ml, corresponds to complete reduction of 0.01% HAuCl$_4$) and silver (5 µg/ml). Note that for larger particles, the extinction spectra reveal quadrupole peaks and significant contribution of scattering. With an increase in the particle diameter, the spectra become red-shifted and broadened. This effect can be used for fast and convenient particle sizing [44,103,104]. The solid line in Figure 1.5c shows the corresponding calibration curve based on long-term set of the literature experimental data (shown by different symbols) [103]:

$$d = \begin{cases} 3 + 7.5 \times 10^{-5} X^4, & X < 23 \\ [\sqrt{X - 17} - 1]/0.06, & X \geq 23 \end{cases}, X = \lambda_{max} - 500. \quad (1.19)$$

In practice, this calibration can be used for particles larger than 5–10 nm. For smaller particles, the resonance shifting is negligible, whereas the resonance broadening becomes quite evident because of size-limiting effects and surface electron scattering [47]. The extinction spectra width also can be used for sizing of fine particles with diameters less than 5 nm [105].

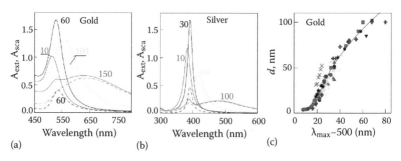

FIGURE 1.5 Extinction (solid lines) and scattering (dashed lines) spectra of gold (a) and silver (b) spheres in water. Numbers near curves correspond to the particle diameter. Panel (c) shows a calibration curve for spectrophotometric determination of the average diameter of CG particles. The symbols present experimental data taken from 14 sources (see references in Ref. [103]). (Adapted from Khlebtsov, N.G., Dykman, L.A., *J. Quant. Spectrosc. Radiat. Transfer*, 111, 1–35, 2010. With permission.)

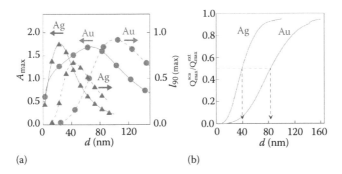

(a)   (b)

FIGURE 1.6 Dependences of the resonance extinction of suspensions and the intensity of scattering at 90 degrees on the particle diameter at constant weight concentrations of gold (57 µg/ml) and silver (5 µg/ml) (a). Panel (b) shows the ratio of the resonance scattering and extinction efficiencies as a function of particle diameter. (Adapted from Khlebtsov, N.G., Dykman, L.A., *J. Quant. Spectrosc. Radiat. Transfer*, 111, 1–35, 2010. With permission.)

Figure 1.6a presents the dependences of the maximal extinction and scattering intensity on the diameter of silver and gold particles. For a constant metal concentration, the maximal extinction is achieved for silver and gold particles of diameters about 25 and 70 nm, respectively.

The maximal scattering per unit metal mass is observed for 40-nm silver and 100-nm gold particles, respectively. Figure 1.6b presents the size dependence of the integral albedo at resonance conditions. Small particles mainly absorb light, whereas large particles mainly scatter it. The contributions of scattering and absorption to the total extinction become equal for 40-nm silver and 80-nm gold particles, respectively.

1.6 Metal Nanorods

1.6.1 Extinction and Scattering Spectra

Figure 1.7 shows the extinction and scattering spectra (in terms of the single-particle efficiencies) of randomly oriented gold (a, c) and silver (b, d) nanorods with the equivolume diameter of 20 nm and the aspect ratio from 1 to 6.

Calculations were carried out by the T-matrix method for cylinders with semispherical ends. We see that the optical properties of rods depend very strongly on the metal nature. First, as the aspect ratio of GNR is increased, the resonance extinction increases approximately by a factor of five and the Q factor also increases. For silver, vice versa, the highest Q factor is observed for spheres, whereas the resonance extinction for rods is lower. Second, for the same volume and axial ratio, the extinction and scattering of light by silver rods are considerably more efficient. The resonance scattering efficiencies of silver particles are approximately five times larger than those for gold particles. Third, the relative intensity of the transverse plasmon resonance of silver particles with the aspect ratio above 2 is noticeably larger than that for gold particles, where this resonance can be simply neglected. Finally, principal differences are revealed for moderate nonspherical particles (Figure 1.7c and d). The resonance for gold particles shifts to the red and

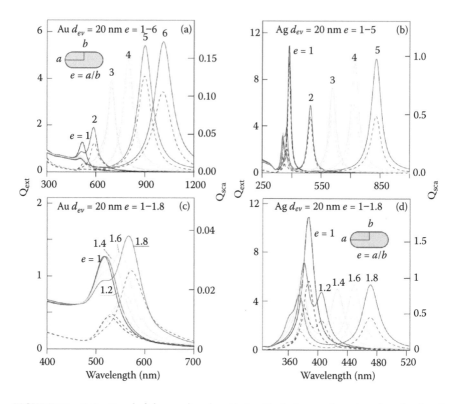

FIGURE 1.7 Extinction (solid curves) and scattering (dashed curves) spectra of randomly oriented gold (a, c) and silver (b, d) s-cylinders with the equivolume diameter 20 nm and the aspect ratio from 1 to 6. Panels (c) and (d) show the transformation of spectra at small deviations of the particle shape from spherical. (Adapted from Khlebtsov, N.G., Dykman, L.A., *J. Quant. Spectrosc. Radiat. Transfer*, 111, 1–35, 2010. With permission.)

gradually splits into two bands with dominating absorption in the red region. The scattering band shifts to the red and its intensity increases. For silver rods, the situation is quite different. The short-wavelength resonance shifts to the blue, its intensity decreases, and it splits into two distinct bands. In this case, the intensity of the long-wavelength extinction band remains approximately constant, it is comparable with the short wavelength band intensity and shifts to the red with increasing nonsphericity. The integrated scattering and absorption spectra approximately reproduce these features.

The position of the longitudinal long-wavelength resonance can be predicted from the axial ratio of particles and, vice versa, the average aspect ratio can be quite accurately estimated from the resonance position. Figure 1.8 presents calibration dependences for measuring the axial ratio of particles from the longitudinal plasmon resonance (PR) position. Along with the T-matrix calculations for s-cylinders of different diameters, we present our and literature experimental data (for details, see Ref. [47]). In the dipolar limit, the plasmon resonance wavelength is completely determined by the particle shape. However, when the particle diameter is increased, the results of a rigorous solution

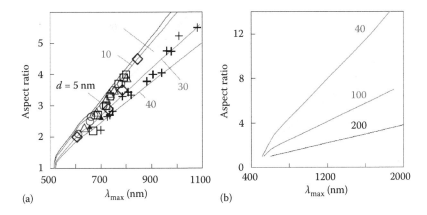

FIGURE 1.8 (a) Calibration plots for determination of the particle aspect ratio through the longitudinal resonance wavelength. Calculations for randomly oriented 5–40-nm gold s-cylinders in water by T-matrix method [106]. The symbols show experimental points (for details, see [97]). (b) Calibration plots for thick 40–200-nm Au s-cylinders calculated by the BEM at perpendicular rod orientation and TM excitation. (Adapted from Bryant, G.W., García de Abajo, F.J., Aizpurua, J., *Nano Lett.*, 8, 631–636, 2008. With permission.)

noticeably differ from the limiting electrostatic curve obtained in the dipole approximation (Figure 1.8). This fact was shown for the first time for s-cylinders [106] by the T-matrix method and for right cylinders [107] by the DDA method. Analogous calculations have been extended for larger gold s-cylinders with using the T-matrix method [108] and the BEM in a full electromagnetic calculation [109] (see also a close study [110]).

1.6.2 Depolarized Light Scattering

If colloidal nonspherical particles are preferentially oriented by an external field, the suspension exhibits anisotropic properties such as dichroism, birefringence, and orientation-dependent variations in turbidity and in light scattering. Moreover, even for *randomly* oriented particles, there remain some principal differences between light scattering from nanospheres and that from nanorods. Indeed, when randomly oriented nonspherical particles are illuminated by linearly polarized light, the cross-polarized scattering intensity occurs [59,106], whereas for spheres, this quantity equals zero.

The cross-polarized scattered intensity I_{vh} can be characterized by the depolarization ratio $\Delta_{vh} = I_{vh}/I_{vv}$, where the subscripts "v" and "h" stand for vertical and horizontal polarization with respect to the scattering plane. According to the theory of light scattering by small particles, the maximal value of the depolarization ratio Δ_{vh} cannot exceed 1/3 and 1/8 for dielectric rods and disks with positive values of the real and imaginary parts of dielectric permeability. However, the dielectric limit 1/3 does not hold for plasmon-resonant nanorods whose theoretical depolarization limit equals 3/4 [106].

The first measurements of the depolarized light scattering spectra from suspensions of GNRs have been reported for 400–900 nm spectral interval [111]. For separated, highly monodisperse and monomorphic samples, we observed depolarized light

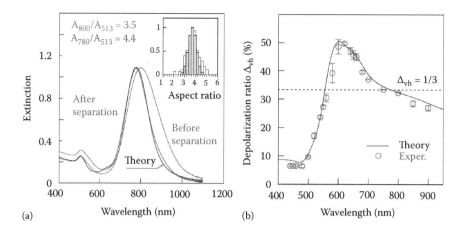

FIGURE 1.9 (a): Extinction spectra of the GNR-780 sample before and after separation, together with T-matrix calculations based on best-fitting data (e_{av} = 3.7, σ = 0.1, W_c = 0.06). The inset shows a comparison of TEM aspect distribution (light columns) with normal best-fitting distribution (dark columns) for the same sample taken after separation. (b) Experimental and simulated depolarization spectra. The error bars correspond to three independent runs, with analog and photon-counting data included. The dashed line shows the dielectric-needle limit 1/3. (Adapted from Khlebtsov, N.G., Dykman, L.A., *J. Quant. Spectrosc. Radiat. Transfer*, 111, 1–35, 2010. With permission.)

scattering spectra with unprecedented depolarization ratio of about 50% at wavelengths of 600 to 650 nm, below the long-wavelength 780-nm extinction peak (Figure 1.9). These unusual depolarization ratios are between 1/3 and 3/4 theoretical limits established for small dielectric and plasmon-resonant needles, respectively.

To simulate the experimental extinction and depolarization spectra, we introduce a model that included two particle populations: (1) the major nanorod population and (2) a by-product particle population with the weight fraction $0 \leq W_b \leq 0.2$. The optical parameters were calculated for both populations and then were summed with the corresponding weights W_b and $W_{rods} = 1 - W_b$. The nanorod population was modeled by $N = 10 - 20$ fractions of rods possessing a constant thickness $d = 2b$, whereas their aspect ratios were supposed to have normal distribution $\sim\exp[-(e/e_{av} - 1)^2/2\sigma^2]$. The average value e_{av} and the normalized dispersion σ were obtained from transmission electron microscope (TEM) data and were also considered to be fitting parameters for best agreement between measured and calculated extinction and depolarization spectra. T-matrix calculations with these parameters resulted in excellent agreement between measured and simulated spectra (Figure 1.9). Thus, we have developed a fitting procedure based on simultaneous consideration of the extinction and depolarization spectra. This method gives an aspect ratio distribution for the rods in solution from which the average value and the standard deviation can be accurately determined in a more convenient and less expensive way than by the traditional TEM analysis.

For larger particles, our T-matrix simulations predict multiple-peak depolarization spectra and unique depolarization ratios exceeding the upper dipolar limit (3/4) because of multipole depolarization contributions (Figure 1.10).

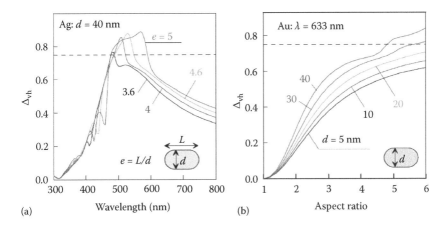

(a) Wavelength (nm)   (b) Aspect ratio

FIGURE 1.10 Spectral dependences of the depolarization ratio for randomly oriented silver s-cylinders of diameter d = 40 nm and aspect ratios 3.6–5 (a) and the dependence of the depolarization ratio on the aspect ratio calculated for gold s-cylinders in water at different particle diameters from 5 to 40 nm (b). (Adapted from Khlebtsov, N.G., Dykman, L.A., *J. Quant. Spectrosc. Radiat. Transfer*, 111, 1–35, 2010. With permission.)

To elucidate the physical origin of strong depolarization [112], let us consider the scattering geometry depicted in Figure 1.11, where the incident x-polarized light travels along the positive z direction and the scattered light is observed in the plane (x, z) in the x-direction. If the particle symmetry axis is directed along the x, y, or z axis, then no depolarization occurs because of evident symmetry constraints. Moreover, no depolarization occurs either for any particle located in the (x, y) or (x, z) plane. Thus, the maximal depolarization contribution is expected from particles located in the (y, z) plane.

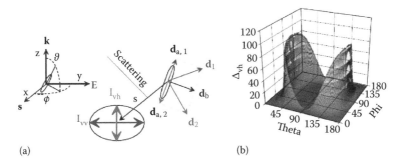

(a)   (b)

FIGURE 1.11 (a) Scheme to explain the physical origin of enhanced depolarization [112]. The incident light propagates along the z axis and is scattered along the x-axis, the particle orientation is specified by spherical angles θ and φ. (b) 3D plot of the depolarization ratio as a function of angles θ and φ. Calculations by the extended precision T-matrix code for gold s-cylinder in water. The rod's diameter and length are 20 and 80 nm, respectively, and the wavelength of 694 nm corresponds to the spectral depolarization ratio maximum. (Adapted from Khlebtsov, N.G., Dykman, L.A., *J. Quant. Spectrosc. Radiat. Transfer*, 111, 1–35, 2010. With permission.)

For usual dielectric rods, the induced dipoles $\mathbf{d}_b$ and $\mathbf{d}_{a,1}$ oscillate in phase. Accordingly, the deviation of the resultant dipole $\mathbf{d}_1$ from the exciting electric field direction is small. Thus, the depolarized scattering should be weak. However, for metal nanorods, the perpendicular ($\mathbf{d}_b$) and longitudinal ($\mathbf{d}_{a,2}$) dipoles can be excited in opposite phases. In this case, the direction of the resultant dipole $\mathbf{d}_2$ can be close to the z-axis direction, thus causing the appearance of significant depolarization.

Figure 1.11b shows the orientation dependence of the depolarized light scattering calculated for 20 × 80-nm gold s-cylinder in water. The scattering geometry is depicted in panel (a). The wavelength of 694 nm corresponds to the maximum of the depolarization spectrum calculated at optimal orientation angles $\varphi = 20°$ and $\varphi = 96°$. It follows from Figure 1.11b that there exists two orientation planes ($\varphi = 20°$ and $\varphi = 150°$) where GNRs can produce unusual depolarization ratio up to 100% provided that the particle azimuth is close to 90°. Thus, the experimental 50% depolarization maximum is caused by particles located near these optimal orientations.

1.6.3 Multipole Plasmon Resonances

Compared to numerous data on the dipole properties of nanorods published in the last years, studies on multipole resonances are quite limited. The first theoretical studies concerned the quadrupole modes excited in silver and gold spheroids, right circular and s-cylinders, as well as in nanolithographic 2D structures such as silver and gold semi-spheres and nanoprisms [47]. Recently, the first observations of multipole PRs in silver nanowires on a dielectric substrate [113] and CG [114] and gold–silver bimetal nanorods [115] were reported.

To elucidate the size and shape dependence of the multipole plasmons, two studies have been performed by using T-matrix formalism [108] and the BEM [109]. Specifically, an extended-precision T-matrix code was developed to simulate the electrodynamic response of gold and silver nanorods, whose shape can be modeled by prolate spheroids and cylinders with flat or semispherical ends (s-cylinders). Here, we present a brief summary of the most important results [108].

The multipole contributions to the extinction, scattering, and absorption spectra can be expressed in terms of the corresponding normalized cross-sections (or efficiencies):

$$Q_{ext,sca,abs} = \sum_{l=1}^{N} q_{ext,sca,abs}^{l}. \tag{1.20}$$

For randomly oriented particles, the multipole partial contributions are given by the equation

$$q_{ext}^{l} = \left(2/k^2 R_{ev}^2\right) \text{Spur}_l(T_{\sigma\sigma}), \tag{1.21}$$

where $\sigma \equiv (l, m, p)$ stands for the T-matrix multi-index and the trace (Spur) is taken over all T-matrix indices except for the multipole order l. For particles at a particular

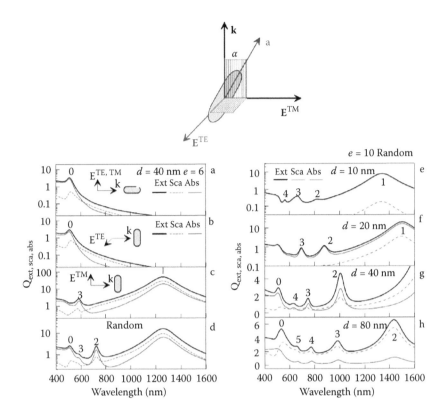

FIGURE 1.12 Two basic TE and TM polarizations of incident light. The orientation of a particle is specified by an angle α between the wave vector K and the particle symmetry axis A. The panels (a–d) show the extinction, scattering, and absorption spectra of gold s-cylinders for longitudinal, perpendicular TE, perpendicular TM, and random orientations of GNR with respect to the incident polarized light. Panels (e–h) show variation of spectra with an increase in the rod diameter at a constant aspect ratio e = 10. The numbers near the curves designate multipole orders according to the T-matrix assignment. (Adapted from Khlebtsov, N.G., Dykman, L.A., *J. Quant. Spectrosc. Radiat. Transfer*, 111, 1–35, 2010. With permission.)

fixed orientation (specified by the angle α between the vector **k** and the nanorod axis **a**), we consider two fundamental cross-sections corresponding to the transverse magnetic (TM) and transverse electric (TE) plane wave configurations, where $\mathbf{E} \in (\mathbf{k}, \mathbf{a})$ plane in the TM case and $\mathbf{E} \perp (\mathbf{k}, \mathbf{a})$ plane in the TE case (Figure 1.12).

Figure 1.12 shows the scattering geometry and the extinction, scattering, and absorption spectra of a gold s-cylinder (diameter 40 nm, aspect ratio 6) for the longitudinal, perpendicular (TM polarization), and random orientations with respect to the incident polarized light (a–d). The spectral resonances are numbered according to the designation $Q_{ext}^n \equiv Q_{ext}(\lambda_n)$, $n = 1, 2,...$ from far-infrared (IR) to visible, whereas the symbol "0" designates the shortest wavelength resonance located near the TE mode.

Three important remarks can be made on the basis of the plots in Figure 1.12a–d. First, the maximal red-shifted resonance for random orientations is due to TM dipole

($n = 1$) excitation of particles, whereas the short-wavelength "zero" resonance is excited by TE polarization of the incident light. Second, both scattering and absorption resonances have the same multipole order, are located at the same wavelengths, and give comparable contributions to the extinction for the size and aspect ratio under consideration. Finally, for basic longitudinal or perpendicular orientations, some multipoles are forbidden because of the symmetry constrains. Consider now the case of strongly elongated particles with an aspect ratio $e = 10$ (panels e–g). In this case, the random orientation spectra exhibit a rich multipole structure that includes multipole resonances up to $n = 5$. As in the previous case, at a minimal thickness $d = 20$ nm, the scattering contribution is negligible, and all resonances are caused by the absorption multipoles. Note also that the short-wavelength resonance has negligible amplitude. With an increase in the particle thickness from 20 to 40 and further to 80 nm, all resonances move to the red region, and new scattering and absorption high-order multipoles appear. Again, the scattering contributions dominate for large particles, whereas at moderate thicknesses, the scattering contribution is comparable to or less than the absorption contribution, except for the long-wavelength dipole resonance of number 1.

Figure 1.13 shows the extinction spectra Q^{ext} (λ) of randomly oriented gold spheroids with a minor axis $d = 80$ nm and an aspect ratio $e = 10$ (data for s-cylinders see in Ref. [108]). Besides the usual long-wavelength dipole resonance (not shown), five additional multipole resonances can be identified (the sixth resonance looks like a weak shoulder, but it is clearly seen on the multipole contribution curve, designated q_6). A close inspection of spectra leads to the following conclusions: (1) The parity of multipole contributions coincides with the parity of the total resonance. (2) For a given spectral resonance

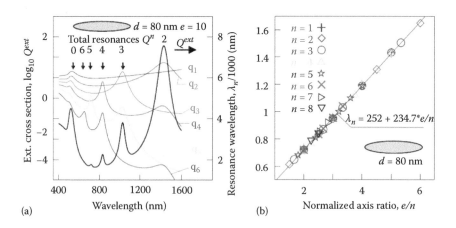

FIGURE 1.13 (a) Extinction spectra Q^{ext} (λ) of randomly oriented gold spheroids with a thickness $d = 80$ nm and an aspect ratio $e = 10$. The curves $q_1 - q_6$ show the spectra of multipole contributions. The upper row of numbers $n = 2 - 6$ designates the total resonances $Q^n = Q_{ext}$ (λ_n). The resonance Q^1 is located in the far IR region and is not shown. (b): Linear scaling of the multipole resonance wavelengths λ_n vs. the normalized aspect ratio e/n. The regression scaling equation is shown in the plot. (Adapted from Khlebtsov, N.G., Dykman, L.A., *J. Quant. Spectrosc. Radiat. Transfer*, 111, 1–35, 2010. With permission.)

number n, the number of partial multipole contributions l is equal to or greater than n. This means that the multipole q_1 does not contribute to the resonances $Q^{2n+1}(n \geq 1)$, the multipole q_2 does not contribute to the resonances Q^{2n} ($n \geq 1$), and so on. Note that the second statement of the multipole contribution rule is, in general, shape dependent [108]. To elucidate the scaling properties of the multipole resonances, we carried out extensive calculations for various nanorod diameters and aspect ratios. The general course of λ_n vs. the e plots show an almost linear shift of the multipole resonances with an increase in aspect ratio:

$$\lambda_n = f(e/n) \simeq A_0 + A\frac{e}{n} \qquad (1.22)$$

Figure 1.13b illustrates the scaling law (Equation 1.22) for randomly oriented gold spheroids ($d = 80$ nm, $e = 2 - 20$) in water. It is evident that for resonance numbers $n = 1 - 8$, all data collapse into single linear functions. Physically, this scaling can be explained in terms of the standing plasmon wave concept and the plasmon wave dispersion law, as introduced for metal nanoantennas [116]. A very simple estimation of the resonance position is given by the equation

$$n\frac{\pi}{L} = \frac{2\pi}{\lambda_n^{eff}} = q^{eff}(\omega_n), \qquad (1.23)$$

where $q^{eff}(\omega_n)$ is an effective wave number corresponding to the resonance frequency ω_n. At a constant particle thickness, d, Equation 1.23 predicts a linear scaling $\lambda_n^{eff} \sim L/n \sim de/n \sim e/n$, in full accord with our finding given by Equation 1.22.

In our classification of plasmonic peaks of GNRs, we implied that the multipole peaks have predominant polarization and 2^l–pole character. For randomly oriented particles, the parity of a given spectral resonance number n coincided with the parity of their multipole contributions l, where l is equal to or greater than n and the total resonance magnitude is determined by the lowest multipole contribution. Therefore, the multipole peaks in the extinction spectrum of a random ensemble can be interpreted in terms of dipole, quadrupole, *etc.*, contributions. Although it is roughly true for randomly oriented GNRs, this assumption is not accurate in general, especially for silver rods. Actually, the classification of multipole plasmonic peaks is determined by the particle symmetry and polarization-scattering configuration. This important question has been studied by Gantzounis [117] by using T-matrix theory and group theory analysis. Such a methodology allows an unambiguous assessment of the eigenmodes for nonspherical particles, according to the irreducible representations of the appropriate point symmetry group, and it provides a consistent explanation of relevant extinction spectra.

1.7 Coupled Plasmons

Along with the optics of individual plasmonic particles, the collective behavior of the interacting plasmon resonant particles is of great interest for nanobiotechnology [102,118]. The analysis of its features includes the study of various structures, beginning from one-dimensional chains with unusual optical properties [119]. Another example is the optics of two-dimensional arrays [102,118,120,121], in particular, clusters of spherical particles on a substrate and two-dimensional planar ensembles formed by usual gold or polymer-coated spheres [118]. The unusual properties of monolayers of silver nanoparticles in a polymer film [122,123] and gold or silver nanoparticles and nanoshells in water or on a glass substrate [124] have been recently discovered. A review of 3D sphere-cluster optics can be found in Refs. [45,101,118].

Apart from direct numerical simulations (*e.g.*, MMP simulation of disk pairs on a substrate [125] and FEM simulations of silver sphere-clusters [126]) or approximate electrostatic considerations [127], a new concept called "the plasmon hybridization model" [79,80] has been developed to elucidate basic physics behind plasmon coupling between optically interacting particles.

A more complex assembly of high-aspect-ratio GNRs on a substrate has been studied in Ref. [128]. The long axes of GNRs were oriented perpendicularly to the substrate. The optical response of such an assembly was governed by a collective plasmonic mode resulting from the strong electromagnetic coupling between the dipolar longitudinal plasmons supported by individual GNRs. The spectral position of this coupled plasmon resonance and the associated electromagnetic field distribution in the nanorod array were shown to be strongly dependent on the interrod coupling strength.

The unique properties of strongly coupled metal particles are expected to find interesting applications in such fields as the manipulation of light in nanoscale waveguide devices, sensing and nonlinearity enhancement applications, and the subwavelength imaging. As an example, we can refer to recent work by Reinhard and coworkers [129]. These authors extended the conventional gold particle tracking with the capability to probe distances below the diffraction limit using the distance dependent near-field interactions between individual particles. This technology was used for precise monitoring the interaction between individual GNPs on the plasma membrane of living HeLa cells.

As coupled plasmonics is an extremely fast growing field, the number of related works exceeds the scope of this chapter. Therefore, we have to restrict our consideration to only several instructive examples, including gold and silver interacting bispheres, gold linear chains, and a monolayer of interacting nanoparticles on a substrate.

1.7.1 Metal Bispheres

Figure 1.14a and b shows the absorption spectra calculated by multipole method [130] for two gold particle diameters $d = 15, 30$ (data for 60 nm, see Ref. [101]) separated by a variable distance s. When the interparticle separation satisfies the condition $s/d \geq 0.5$, the absorption efficiencies approach the single-particle quantities and the dipole and multipole calculations give identical results [101]. However, the situation changes dramatically when the relative separation s/d is about several percent. The coupled spectra demonstrate quite evident splitting into two components with pronounced red-shifting of the long-wavelength resonance when the spheres approach each other.

In the case of silver bispheres (Figure 1.14c and d), the resonance light scattering of 60-nm clusters exceeds the resonance absorption so that any comparison between the dipole and the multipole approaches becomes incorrect unless both scattering and absorption are taken into account for the total extinction. That is why we show

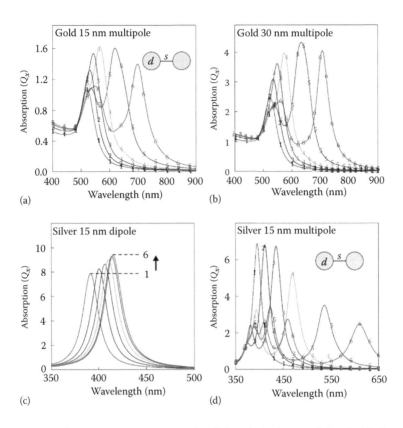

FIGURE 1.14 Absorption spectra at an incident light polarization parallel to the bisphere axis ($x \equiv \parallel$). Calculations by the exact GMM multipole code (a, b, d) and dipole approximation (c) for gold (a, b) and silver (c, d) particles with diameters d = 15 (a, c, d) and 30 nm (b) and the relative interparticle separations s/d = 0.5 (1), 0.2 (2), 0.1(3), 0.05 (4), 0.02 (5), 0.01 (6). (Adapted from Khlebtsov, N.G., Dykman, L.A., *J. Quant. Spectrosc. Radiat. Transfer*, 111, 1–35, 2010. With permission.)

only the calculated data for silver nanospheres with d = 15 nm. At moderate separations (s/d > 0.05), the independent-particle spectrum splits into two modes; therefore, the data of Figure 1.14c and d are in great part analogous to those for independent particles. However, at smaller separations s/d < 0.05, we observe the appearance of four plasmon resonances related to the quadrupole and the next high-order multipole excitations. The exact multipole approach predicts the well-known enormous theoretical [118] and experimental [131] red-shifting of spectra and their splitting [132] into two modes, whereas the dipole spectra show only a minor red shift (Figure 1.14c).

The dependence of the extinction spectra on the total multipole order N_M was studied in Refs. [101,118]. Note that N_M means the maximal order of vector spherical harmonics retained in the coupled equations rather than the number of multipoles involved in the final calculations of optical characteristics. According to computer experiments (Figure 1.15), one has to include extra-high single-particle multipole orders (up to 30–40) into coupled

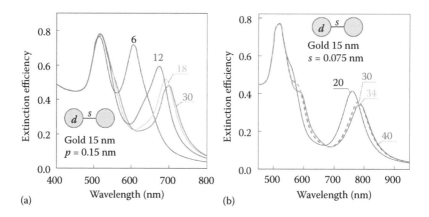

FIGURE 1.15 Extinction spectra of 15-nm randomly oriented gold bispheres in water calculated by the exact T-matrix method for separation distances between spheres 0.15 nm (a) and 0.075 nm (b). The numbers near the curves designate the multipole orders included in the single-particle field expansions of coupled equations. (Adapted from Khlebtsov, N.G., Dykman, L.A., *J. Quant. Spectrosc. Radiat. Transfer*, 111, 1–35, 2010. With permission.)

equations to calculate correctly the extinction spectra of 15-nm gold spheres separated by a 0.5%–1% relative distance s/d. The need to retain high multipoles for small spheres, which are themselves well within the dipole approximation, seems to be somewhat counterintuitive. It should again be emphasized that the final calculations involve a rather small number of multipoles (as a rule, fewer than 6). However, to find these small-order contributions correctly, one needs to include many more multipoles into coupled equations.

The physical origin of this unusual electrodynamic coupling was first established by Mackowski [133] for small soot bispheres. He showed that the electric-field intensity can be highly inhomogeneous in the vicinity of contact point between the spheres even if the external filed is homogeneous on the scale of bisphere size. Evidently, the same physics holds in our case, as the imaginary part of the dielectric permittivity is the main parameter that determines the spatial electric-field distribution near the contact bisphere point.

1.7.2 Linear Chains and Square Lattices

To gain insight into the mechanisms determining the effect of particle aggregation on the absorption efficiency, it is useful to consider simple 1D linear chains and 2D lattice square arrays from metal (here, gold) spheres of a diameter d and separated by the interparticle distance s. In both cases, the incident TE and TM electromagnetic waves are directed perpendicularly to the cluster axis or the cluster plane. For a linear chain, TM polarization means that the electric vector of an EM wave is directed along the chain axis, whereas for 2D square arrays, both polarizations are equivalent. The absorption spectra of 1D and 2D clusters can be characterized in terms of the *normalized cross-section* defined as

$$F(\lambda) = C_{abs}(\lambda)/NC_{abs1}(\lambda_{max}),\qquad(1.24)$$

where λ is the wavelength in vacuum, N is the number of cluster particles, and $C_{abs1}(\lambda_{max})$ is the absorption cross-section of a single sphere at the plasmon resonance wavelength ($\lambda_{max} = 520$ nm for small gold spheres). In the case of a specific irradiation wavelength λ_0 (e.g., $\lambda_0 = 800$ nm), we also used the *amplification factor* at a particular wavelength, defined as

$$F_a(\lambda_0) = C_{abs}(\lambda_0)/NC_{abs1}(\lambda_0). \tag{1.25}$$

Figure 1.16 shows normalized spectra of linear gold sphere chains in terms of the normalized cross-section, as defined by Equation 1.24. These spectra are for two particle numbers 4 and 32 (other examples can be found in Ref. [102]) and for different relative interparticle distances s/d, beginning with the noninteracting particles to closely packed chains with $s/d = 0.025$.

In the general case, there are two characteristic absorption maxima of linear chains. The short wavelength maximum corresponds to the plasmon resonance of monomers excited by the TE mode of incident light (the electric vector oscillates perpendicularly to the chain). The other, long-wavelength resonance appears at the electric excitation of the chain along its axis and is due to the strong electrodynamic coupling of chain spheres. In the case of unpolarized incident light, both modes are seen in the spectra. Analogous properties have been found for silver nanoshell chains [134].

A comparison with Figure 1.7 reveals an evident analogy between the longitudinal plasmon resonance of metal nanorods and the longitudinal resonance of linear chains. One can say that a linear chain of coupling spheres absorbs and scatters light like an "equivalent" solid metal rod does. However, unlike solid rods, the linear chains exhibit the extinction spectra show of richer structure. For instance, the longitudinal peak position depends not only on the aspect ratio but also on the separation between neighboring

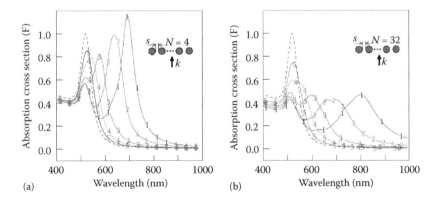

FIGURE 1.16 Normalized absorption spectra of linear chains in water at perpendicular excitation by an unpolarized incident wave. The cluster particle numbers are 4 (a) and 32 (d); the particle diameter is 40 nm; and the normalized interparticle distances s/d are 0.025 (1), 0.05 (2), 0.125 (3), 0.25 (4), and 0.5 (5). The dashed curve corresponds to the noninteracting spheres. (Adapted from Khlebtsov, N.G., Dykman, L.A., *J. Quant. Spectrosc. Radiat. Transfer*, 111, 1–35, 2010. With permission.)

particles. Besides, the single-sphere resonance is clearly seen in Figure 1.16, whereas the corresponding rod resonance shows as a spectral shoulder. It is evident from computer simulations that the maximum possible increase in absorption per particle is achieved at longitudinal excitation, and the desired tissue-optic tuning to 800 nm may be obtained for chains consisting of as many as 10 particles. In this case, the absorption amplification is about 50%–120%.

We also observed some saturation in the absorption amplification for long chains. This effect is more evident from Figure 1.17, where the extinction and scattering spectra

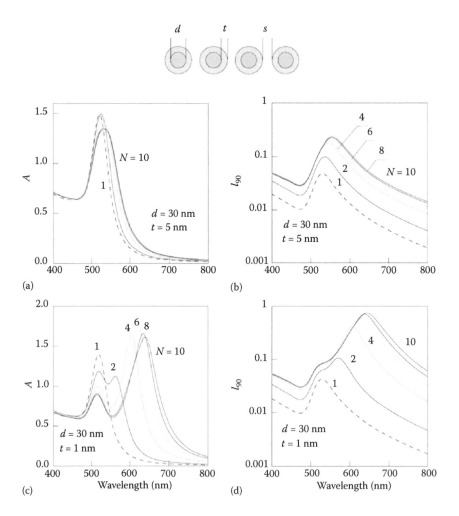

FIGURE 1.17 Extinction and scattering spectra of randomly oriented linear chains of two-layered conjugates with different conjugate number $N = 1 - 10$. Calculations for the core diameter $d = 30$ nm, shell thickness $t = 5$ (a, b) and 1 nm (c, d), and separation distance $s = 0$. (Adapted from Khlebtsov, N.G., Dykman, L.A., *J. Quant. Spectrosc. Radiat. Transfer*, 111, 1–35, 2010. With permission.)

are shown for linear chains of two-layered nanospheres with a gold core and a dielectric shell [118]. Such particles serve as a model of nanoparticle conjugates [135]. The number of conjugates in Figure 1.17 is the variable parameter of curves. It is worth noting a principal difference between the spectra of densely packed metal spheres (shell thickness $t = 1\,nm$, strong binary coupling, Figure 1.17c and d) and rare metallic chains ($t = 5\,nm$, weak binary coupling, Figure 1.17a and b). The extinction spectra do not change essentially for the case $t = 5\,nm$ (Figure 1.17a), while the plasmon resonance peak in the scattering spectra increases significantly due to constructive far-field interference. However, the peak position does not shift significantly. For a thin dielectric shell of about 1 nm, the transformation of extinction and scattering spectra are due to both the electrodynamic coupling and the far-field interference. Again, we observe a rapid saturation of spectra with an increase in the chain particle number. In other words, beginning from a particular number of interacting spheres (say, about of 10), the extinction or scattering spectra of a long linear chain become insensitive to the total number of constitutive spheres. This can be explained by an effective electrodynamic interaction between monomers, which belong to a finite conjugate group [118].

The dimensionless amplification factor (Equation 1.25) is defined as the ratio of cluster absorption to the sum of single-particle absorptions. Figure 1.18a shows that this parameter is determined mainly by the relative interparticle distance in a universal manner and that the calculated curves F_a (s/d) collapse into a universal dependence if the number of particles is greater than 4. Similar results have been obtained for lattice 2D arrays (Figure 1.18b). It follows from these graphs that a significant increase in the amplification factor can be achieved by decreasing the relative interparticle distance down to $s/d < 0.1$. It should be stressed that the drop in amplification of well-separated chains is due primarily to the resonance moving away from 800 nm.

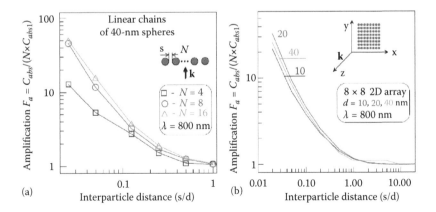

FIGURE 1.18 Dependence of absorption amplification at 800 nm on the normalized interparticle distance of linear 4-, 8-, and 16-particle chains of 40-nm spheres (a) and 8 × 8 2D arrays of 10-, 20-, and 40-nm spheres (b). (Adapted from Khlebtsov, B.N., Zharov, V.P., Melnikov, A.G., Tuchin, V.V, Khlebtsov, N.G. *Nanotechnology*, 17, 5167–5179, 2006. With permission.)

1.7.3 Universal Plasmon Ruler

With use of GNP-DNA conjugates, Reinhard *et al.* [136] reported a detailed calibration of the plasmon peak position versus bisphere separation for 42- and 87-nm-diameter gold particles. The experimental resonance wavelengths as a function of interparticle separation were found to be in agreement with the T-matrix calculations by Wei *et al.* [137] and with the so-called universal plasmon ruler (UPR) equation (see, *e.g.*, [138] and references therein):

$$\Delta\lambda = A\exp[-B(s/L)]+C, \tag{1.26}$$

where $\Delta\lambda$ is the plasmon resonance shift caused by the particle optical coupling, L is a characteristic size (*e.g.*, L is a sphere diameter or nanorod length), and A, B, and C are the fitting constants. In essence, the UPR means a universal dependence of the couple resonance shift determined only by the separation distance normalized to the characteristic particle size. In particular, the UPR has been confirmed in simulations [125,138].

Figure 1.19a shows experimental points and UPR curve (adapted from Ref. [136,139]) together with our T-matrix calculations (adapted from Ref. [101]). In accord with Ref. [136], we found good agreement between the experimental data and the T-matrix curve within the separation range $0.2 \le s/d \le 2$. However, the single-exponential fit (Equation 1.26) does not match the general character of T-matrix curves at small separations $s/d \le 0.2$.

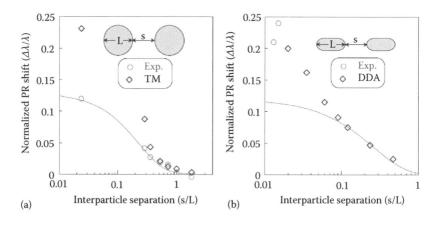

(a)   (b)

FIGURE 1.19 Normalized plasmon resonance shift of two 48-nm gold spheres (a) and gold s-cylinders (b) as a function of interparticle distance scaled for sphere diameter or nanorod length. Experimental points are adapted from Refs. [136] (a) and [139] (b). Simulations were carried out by T-matrix [101] (a) and DDA [139] (b) methods. Solid curves show the UPR fits. Medium refractive index is 1.6 (a) and 1.56 (b). (Adapted from Khlebtsov, N.G., Dykman, L.A., *J. Quant. Spectrosc. Radiat. Transfer*, 111, 1–35, 2010. With permission.)

Mulvaney and coworkers [139] reported experimental scattering spectra of GNR dimers arranged in different orientations and separations. The spectra exhibited both red- and blue-shifted surface plasmon resonances, consistent with both the DDA simulations and the plasmon hybridization model. Figure 1.19b shows experimental points and DDA simulations for the normalized plasmon resonance shift caused by the interaction of two GNRs (adapted from Ref. [139]). The solid curve shows UPR fit over the first four DDA points at maximal separations. Clearly, both pictures (a) and (b) demonstrate very close behavior. These plots show that the approximation of the coupling distance dependence according to the dipole–dipole approximation or as an exponential may be not valid for normalized separations less than 0.1. Perhaps, the previous deviations from UPR are related to the strong multipole interactions excited at extra small separation distances. For a more detailed theoretical treatment, see Ref. [140].

1.7.4 Monolayers of Metal Nanoparticles and Nanoshells

Recently, Chumanov *et al.* [122,123] found an unusual behavior of the extinction spectra for a monolayer of interacting silver nanoparticles embedded in a polymer film. Specifically, they showed an intense sharpening of the quadrupole extinction peak resulting from selective suppression of the coupled dipole mode as the interparticle distance becomes smaller. This phenomenon was explained qualitatively by using simple symmetry considerations.

A comprehensive theoretical analysis of a monolayer consisting of metal or metallodielectric nanoparticles with the dipole and quadrupole single-particle resonances has been done in Ref. [124]. The theoretical models included spherical gold and silver particles and also gold and silver nanoshells on silica and polystyrene (PS) cores, forming 2D random clusters or square-lattice arrays on a dielectric substrate (glass in water). The parameters of individual particles were chosen so that a quadrupole plasmon resonance could be observed along with the dipole scattering band. By using an exact multipole cluster-on-a-substrate solution, it has been shown that particle-substrate coupling can be neglected in calculation of the monolayer extinction spectra, at least for the glass-in-water configuration. When the surface particle density in the monolayer was increased, the dipole resonance became suppressed and the spectrum for the cooperative system was determined only by the quadrupole plasmon. The dependence of this effect on the single-particle parameters and on the cluster structure was examined in detail. In particular, the selective suppression of the long-wavelength extinction band was shown to arise from the cooperative suppression of the dipole scattering mode, whereas the short-wavelength absorption spectrum for the monolayer was shown to be little different from the single-particle spectrum.

Here, we provide only several illustrative examples. As a simple monolayer model, we used a square lattice with a period $p = d_e (1 + s)$, where d_e is the external diameter of particles and s is the relative interparticle distance. Another monolayer model was obtained by randomly filling a square of side L/d_e with a given number of particles N. Then, the relative coordinates of the particles X_i were transformed as $x_i = X_i d_e (1 + s)$, where the parameter s controls the minimal interparticle distance. The structure of the resultant monolayer is characterized by the particle number N and the average surface particle density $\rho = N S_{geom}/[L^2(1 + s)^2]$.

The interparticle distance is a crucial parameter determining the electrodynamic particle coupling and the cooperative spectral properties of an ensemble. Therefore, we first investigated the influence of the interparticle-distance parameter s on the suppression of the dipole mode. Figure 1.20 shows a comparison of the extinction spectra of lattice and random clusters. All particle and cluster parameters are indicated in the figure caption. It can be seen that in both cases, the spectra for the lattice clusters with an s parameter of 0.2 are much the same as the spectra for the random clusters. Closer agreement between the spectra for the 36-particle clusters can be obtained if the s parameter is 0.35 in the lattice case. Beginning with parameter s values about 0.2, there was effective suppression of the dipole mode, so that the resonance was determined by only the quadrupole mode. A twofold decrease in the s parameter (to as low as 0.1) brought about little change in the system's spectrum. These conclusions are general and depend little on the properties of particles themselves.

To gain an insight into the physical mechanisms responsible for suppression of the coupled dipole mode, one also has to investigate the influence of particle interactions on the cooperative absorption and scattering of light. This question has been studied for several models, including silver and gold spheres as well as silver nanoshells and GNSs on dielectric cores (PS and silica). In the case of silver spheres, it has been found that the suppression in a monolayer of the dipole extinction band is determined entirely by the decrease in *dipole resonance scattering* that occurs when strongly scattering particles with a dipole and a quadrupole resonance move closer together. The electrodynamic interparticle interaction almost does not change the absorption spectrum, including its fine structure in the short-wavelength region.

Figure 1.21 shows the dependence of the extinction, scattering, and absorption spectra for random clusters of thirty-six 20-nm silver nanoshells on 110-nm PS spheres. The calculated results were averaged over five independent cluster generations. It can be seen that

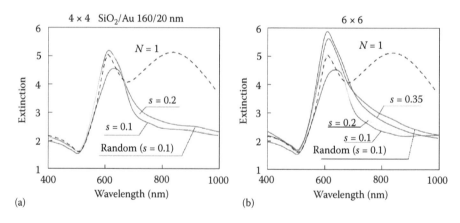

FIGURE 1.20 Comparison of the extinction spectra for lattice (4×4, 6×6) and random clusters made up of SiO_2/Au (160/20 nm) nanoshells with particle numbers of 16 (a) and 36 (b). Also shown are the spectra for isolated particles ($N = 1$). The interparticle-distance parameter is 0.1, 0.2, and 0.35 for the lattice clusters and equals 0.1 for the random clusters. The average particle density for the random clusters is 0.415 (a) and 0.36 (b) for 16 and 36 particles, respectively. (Adapted from Khlebtsov, N.G., Dykman, L.A., *J. Quant. Spectrosc. Radiat. Transfer*, 111, 1–35, 2010. With permission.)

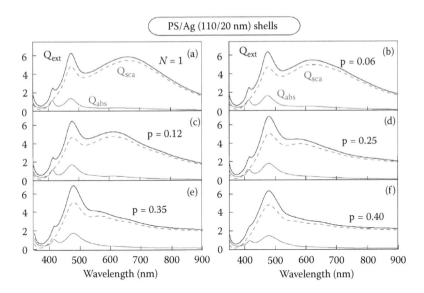

FIGURE 1.21 Extinction, scattering, and absorption spectra for random clusters made up of 36 PS/Ag-type silver nanoshells (core diameter, 110 nm; shell thickness, 20 nm). The minimal-distance parameter s is 0.05, and the average particle density is 0 (a, a single particle), 0.06 (b), 0.12 (c), 0.25 (d), 0.35 (e), and 0.4 (f). The calculated results were averaged over five statistical realizations. (Adapted from Khlebtsov, N.G., Dykman, L.A., *J. Quant. Spectrosc. Radiat. Transfer*, 111, 1–35, 2010. With permission.)

the transformation of the extinction, scattering, and absorption spectra occurring with increasing particle density in the monolayer is similar to that for silver sphere monolayer [124]. Namely, the clear-cut quadrupole and octupole absorption peaks in the short-wavelength portion of the spectrum almost do not depend on the average particle density in the monolayer. As in the case of solid silver spheres, the disappearance of the dipole extinction band is associated with the suppression of the dominant dipole scattering band.

For experimental studies [124], the silica/GNS monolayers were fabricated by the deposition of nanoshells on a glass substrate functionalized by silane-thiol cross-linkers. The measured single-particle and monolayer extinction spectra were in reasonable agreement with simulations based on the nanoshell geometrical parameters (scanning electron microscopy data). Finally, the sensitivity of the coupled quadrupole resonance to the dielectric environment was evaluated to find a universal linear relation between the relative shift in the coupled-quadrupole-resonance wavelength and the relative increment in the environment refractive index.

We conclude this section with Figure 1.22, which shows a scanning electron microscope (SEM) image of fabricated monolayer, the single-particle extinction spectra, and the coupled spectra of the monolayer.

Calculations with fitting size parameters reproduce the spectral positions of both quadrupole and dipole bands. The quadrupole extinction peak for a particle suspension is located near 760 nm, and the dipole scattering band lies in the near-IR region (about

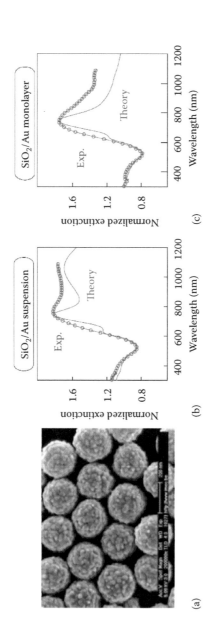

FIGURE 1.22 SEM image of a monolayer portion showing self-assembled silica/GNSs on a silane-functionalized quartz substrate. The average nanoparticle diameter is 245 ± 12 nm, and the average silica core diameter is 200 ± 10 nm (a). Normalized experimental and theoretical extinction spectra for a suspension of nanoshells (b) and for a monolayer on a silane-functionalized quartz substrate in water (c). Details of theoretical fitting can be found elsewhere [124]. (Adapted from Khlebtsov, N.G., Dykman, L.A., *J. Quant. Spectrosc. Radiat. Transfer*, 111, 1–35, 2010. With permission.)

1100 nm). The theoretical single-particle spectrum reveals an octupole resonance near 650 nm, which is not seen, however, in the experimental plots because of the polydispersity and surface roughness effects. A separate calculation of extinction, absorption, and scattering spectra allows us to attribute the octupole peak to the dominant absorption resonance, the quadrupole peak to both the scattering and the absorption contribution, and the dipole peak to the dominant scattering resonance. The particle interaction in the experimental monolayer brought about a noticeable decrease in the extinction shoulder (800–1100 nm) because of suppression of the scattering resonance. To simulate the monolayer spectrum, we used the fitting average core diameter and gold shell thickness (230 and 15 nm) and five independent calculations for a random array ($N = 36$) with the average surface particle density $\rho = 0.25$ (the s parameter equals 0.05). In general, the model calculations agree well with the measurement. In any case, the suppression of dipole peak is clearly seen in both the experimental and simulated spectra.

1.8 Concluding Remarks

The geometrical and structural parameters of nanoparticles determine the individual linear optical properties, which can now be characterized at the single particle level. At present, the first principle QM calculations (*e.g.*, by the time-domain density-functional theory [91]) can capture both nonlocal and quantum tunneling effects, but such simulations are limited to a small size of systems of about few nanometers. For realistic systems with sizes of several tens nanometers, more simpler and less computationally demanding approaches are based on hydrodynamic nonlocal and QCM Drude-like models, as discussed in Section 1.3. Thus, the classical or QM corrected EM approach can be used in the majority of practical nanostructures. Current development of the electromagnetic simulation tools allows for modeling various complex assemblies of bare or functionalized metal nanoparticles and assemblies of coupled nanoparticles. For instance, we have shown that T-matrix simulations corrected for the mean electron free path afford a quantitative description of a long-term set of data on the particle-size dependence of the plasmon resonance position of GNPs. It has been also shown [104] that both the size and the concentration of GNPs can be accurately determined from UV–Vis extinction spectra, provided that shape effects are taken into consideration. The size polydispersity of particles can be neglected in both T-matrix and Mie-theory simulations, unless the standard normal-distribution parameter σ is greater than 0.1. For larger polydispersities and for mean diameters greater than 20 nm, there appear significant deviations from simple monodisperse models. From a practical point of view, we believe that our discussion can help readers to understand the possible limitations related with application of a simple monodisperse Mie model to determination of the size and concentration of GNPs in laboratory samples.

References

1. Bredig G. Darstellung colloidaler Metallösungen durch elektrische Zerstäubung // *Z. Angew. Chem.* 1898. Bd. 11. S. 951–954.
2. Zhang J., Claverie J., Chaker M., Ma D. Colloidal metal nanoparticles prepared by laser ablation and their applications // *ChemPhysChem*. 2017. V. 18. P. 986–1006.

3. Borowskaja D.P. Zur Methodik der Goldsolbereitung // *Ztschr. Immunitatsforsch. Exp. Ther.* 1934. V. 82. P. 178–182.
4. Turkevich J., Stevenson P.C., Hillier J. A study of the nucleation and growth processes in the synthesis of colloidal gold // *Discuss. Faraday Soc.* 1951. V. 11. P. 55–75.
5. Frens G. Controlled nucleation for the regulation of the particle size in monodisperse gold suspensions // *Nat. Phys. Sci.* 1973. V. 241. P. 20–22.
6. Khlebtsov N.G., Bogatyrev V.A., Dykman L.A., Melnikov A.G. Spectral extinction of colloidal gold and its biospecific conjugates // *J. Colloid Interface Sci.* 1996. V. 180. P. 436–445.
7. Baschong W., Lucocq J.M., Roth J. "Thiocyanate gold:" Small (2–3 nm) colloidal gold for affinity cytochemical labeling in electron microscopy // *Histochemistry.* 1985. V. 83. P. 409–411.
8. Goia D.V., Matijević E. Tailoring the particle size of monodispersed colloidal gold // *Colloids Surf. A.* 1999. V. 146. P. 139–152.
9. Brown K.R., Walter D.G., Natan M.J. Seeding of colloidal Au nanoparticles solutions. 2. Improved control of particle size and shape // *Chem. Mater.* 2000. V. 12. P. 306–313.
10. Thanh N.T., Maclean N., Mahiddine S. Mechanisms of nucleation and growth of nanoparticles in solution // *Chem. Rev.* 2014. V. 114. P. 7610–7630.
11. Ziegler C., Eychmüller A. Seeded growth synthesis of uniform gold nanoparticles with diameters of 15–300 nm // *J. Phys. Chem. C.* 2011. V. 115. P. 4502–4506.
12. Bastús N.G., Comenge J., Puntes V. Kinetically controlled seeded growth synthesis of citrate-stabilized gold nanoparticles of up to 200 nm: Size focusing versus Ostwald ripening // *Langmuir.* 2011. V. 27. P. 11098–11105.
13. Schulz F., Homolka T., Bastús N.G., Puntes V.F., Weller H., Vossmeyer T. Little adjustments significantly improve the Turkevich synthesis of gold nanoparticles // *Langmuir.* 2014. V. 30. P. 10779–10784.
14. Leng W., Pati P., Vikesland P.J. Room temperature seed mediated growth of gold nanoparticles: Mechanistic investigations and life cycle assessment // *Environ. Sci. Nano.* 2015. V. 2. P. 440–453.
15. J. Piella, N. G Bastús, V. Puntes Size-controlled synthesis of sub-10 nm citrate-stabilized gold nanoparticles and related optical properties // *Chem. Mater.* 2016. V. 28. P. 1066–1075.
16. Xia H., Xiahou Y., Zhang P., Ding W., Wang D. Revitalizing the Frens method to synthesize uniform, quasi-spherical gold nanoparticles with deliberately regulated sizes from 2 to 330 nm // *Langmuir.* 2016. V. 32. P. 5870–5880.
17. Tyagi H., Kushwaha A., Kumar A., Aslam M. A facile pH controlled citrate-based reduction method for gold nanoparticle synthesis at room temperature // *Nanoscale Res. Lett.* 2016. V. 11. 362.
18. Shi L., Buhler E., Boué F., Carn F. How does the size of gold nanoparticles depend on citrate to gold ratio in Turkevich synthesis? Final answer to a debated question // *J. Colloid Interface Sci.* 2017. V. 492. P. 191–198.
19. Hirai H., Aizawa H. Preparation of stable dispersions of colloidal gold in hexanes by phase transfer // *J. Colloid Interface Sci.* 1993. V. 161. P. 471–474.

20. Green M., O'Brien P. A simple one phase preparation of organically capped gold nanocrystals // *Chem. Commun.* 2000. No. 3. P. 183–184.
21. Giersig M., Mulvaney P. Preparation of ordered colloid monolayers by electrophoretic deposition // *Langmuir.* 1993. V. 9. P. 3408–3413.
22. Brust M., Walker D., Bethell D., Schiffrin D.J., Whyman R. Synthesis of thiol-derivatised gold nanoparticles in a two-phase liquid-liquid system // *J. Chem. Soc. Chem. Commun.* 1994. No. 7. P. 801–802.
23. Dykman L.A., Lyakhov A.A, Bogatyrev V.A., Shchyogolev S.Y. Synthesis of colloidal gold using high-molecular-weight reducing agents // *Colloid J.* 1998. V. 60. P. 700–704.
24. Vetchinkina E.P., Loshchinina E.A., Burov A.M., Dykman L.A., Nikitina V.E. Enzymatic formation of gold nanoparticles by submerged culture of the basidio-mycete *Lentinus edodes* // *J. Biotechnol.* 2014. V. 182–183. P. 37–45.
25. Adil S.F., Assal M.E., Khan M., Al-Warthan A., Siddiquia M.R.H., Liz-Marzán L.M. Biogenic synthesis of metallic nanoparticles and prospects toward green chemistry // *Dalton Trans.* 2015. V. 44. P. 9709–9717.
26. Duan H., Wang D., Li Y. Green chemistry for nanoparticle synthesis // *Chem. Soc. Rev.* 2015. V. 44. P. 5778–5792.
27. Yu Y.-Y., Chang S.-S., Lee C.-L., Wang C.R.C. Gold nanorods: Electrochemical synthesis and optical properties // *J. Phys. Chem. B.* 1997. V. 101. P. 6661–6664.
28. Jana N.R., Gearheart L., Murphy C.J. Wet chemical synthesis of high aspect ratio cylindrical gold nanorods // *J. Phys. Chem.* 2001. V. 105. P. 4065–4067.
29. Nikoobakht B., El-Sayed M.A. Preparation and growth mechanism of gold nanorods (NRs) using seed-mediated growth method // *Chem. Mater.* 2003. V. 15. P. 1957–1962.
30. Oldenburg S., Averitt R.D., Westcott S., Halas N.J. Nanoengineering of optical resonances // *Chem. Phys. Lett.* 1998. V. 288. P. 243–247.
31. Brinson B.E., Lassiter J.B., Levin C.S., Bardhan R., Mirin N., Halas N.J. Nanoshells made easy: Improving Au layer growth on nanoparticle surfaces // *Langmuir.* 2008. V. 24. P. 14166–14171.
32. Sun Y., Mayers B.T., Xia Y. Template-engaged replacement reaction: A one-step approach to the large-scale synthesis of metal nanostructures with hollow interiors // *Nano Lett.* 2002. V. 2. P. 481–485.
33. Sun Y., Xia Y. Alloying and dealloying processes involved in the preparation of metal nanoshells through a galvanic replacement reaction // *Nano Lett.* 2003. V. 3. P. 1569–1572.
34. Chen J., McLellan J.M., Siekkinen A., Xiong Y., Li Z.-Y., Xia Y. Facile synthesis of gold-silver nanocages with controllable pores on the surface // *J. Am. Chem. Soc.* 2006. P. 128. P. 14776–14777.
35. Nehl C.L., Liao H., Hafner J.H. Optical properties of star-shaped gold nanoparticles // *Nano Lett.* 2006. V. 6. P. 683–688.
36. Herizchi R., Abbasi E., Milani M., Akbarzadeh A. Current methods for synthesis of gold nanoparticles // *Artif. Cells Nanomed. Biotechnol.* 2016. V. 44. P. 596–602.
37. Alex S., Tiwari A. Functionalized gold nanoparticles: Synthesis, properties and applications—A review // *J. Nanosci. Nanotechnol.* 2015. V. 15. P. 1869–1894.

38. Slepička P., Elashnikov R., Slepičková Kasálková N., Siegel J., Řezníčková A., Kolská Z., Švorčík V. Preparation and characterization of gold nanoparticles in liquid solutions // *Nanotechnol. Res. J.* 2015. V. 8. P. 163–207.

39. Sharma V., Park K., Srinivasarao M. Colloidal dispersion of gold nanorods: Historical background, optical properties, seed-mediated synthesis, shape separation and self-assembly // *Mater. Sci. Eng. R.* 2009. V. 65. P. 1–38.

40. Li N., Zhao P., Astruc D. Anisotropic gold nanoparticles: Synthesis, properties, applications, and toxicity // *Angew. Chem., Int. Ed.* 2014. V. 53. P. 1756–1789.

41. Niidome Y., Haine A.T., Niidome T. Preparation, properties and applications of anisotropic gold-based nanoparticles // *Chem. Lett.* 2016. V. 45. P. 488–498.

42. Liu A., Wang G., Wang F., Zhang Y. Gold nanostructures with near-infrared plasmonic resonance: Synthesis and surface functionalization // *Coord. Chem. Rev.* 2017. V. 336. P. 28–42.

43. Khlebtsov B.N., Khanadeev V.A., Panfilova E.V., Pylaev T.E., Bibikova O.A., Staroverov S.A., Bogatyrev V.A., Dykman L.A., Khlebtsov N.G. New types of nanomaterials: Powders of gold nanospheres, nanorods, nanostars, and gold-silver nanocages // *Nanotechnol. Russia.* 2013. V. 8. P. 209–219.

44. Khlebtsov N.G., Bogatyrev V.A., Dykman L.A., Khlebtsov B.N., Staroverov S.A., Shirokov A.A., Matora L.Y., Khanadeev V.A., Pylaev T.E., Tsyganova N.A., Terentyuk G.S. Analytical and theranostic applications of gold nanoparticles and multifunctional nanocomposites // *Theranostics.* 2013. V. 3. P. 167–180.

45. Kreibig U., Vollmer M. *Optical Properties of Metal Clusters.* Berlin: Springer-Verlag. 1995.

46. Khlebtsov N.G., Dykman L.A. Optical properties and biomedical applications of plasmonic nanoparticles // *J. Quant. Spectrosc. Radiat. Transfer.* 2010. V. 111. P. 1–35.

47. Khlebtsov N.G. Optics and biophotonics of nanoparticles with a plasmon resonance // *Quant. Electron.* 2008. V. 38. P. 504–529.

48. *Handbook of Optical Constants of Solids* / Ed. Palik E.D. New York: Academic Press. 1998.

49. Coronado E.A., Schatz G.C. Surface plasmon broadening for arbitrary shape nanoparticles: A geometrical probability approach // *J. Chem. Phys.* 2003. V. 119. P. 3926–3934.

50. Moroz A. Electron mean free path in a spherical shell geometry // *J. Phys. Chem. C.* 2008. V. 112. P. 10641–10652.

51. Bohren C.F., Huffman D.R. *Absorption and Scattering of Light by Small Particles.* New York: John Wiley & Sons. 1983.

52. Rayleigh D.W. On the scattering of light by small particles // *Phil. Mag.* 1871. V. 41. P. 447–454.

53. Mie G. Beitrage zur Optik truber Medien, speziell kolloidaler Metallösungen // *Ann. Phys.* 1908. Bd. 25. S. 377–445.

54. Gans R. Über die Form ultramikroskopischer Goldteilchen // *Ann. Phys.* 1912. Bd. 37. S. 881–900.

55. Rayleigh D.W. On the incidence of aerial and electric waves upon small obstacles in the form of ellipsoids or elliptic cylinders, and on the passage of electric waves through a circular aperture in a conducting screen // *Phil. Mag.* 1897. V. 44. P. 28–52.

56. Draine B.T. The discrete dipole approximation for light scattering by irregular targets // In: *Light Scattering by Nonspherical Particles: Theory, Measurements, and Applications* / Eds. Mishchenko M.I., Hovenier J.W., Travis L.D. San Diego: Academic Press. 2000. P. 131–145.

57. *Advances in Computational Electrodynamics: The Finite-Difference–Time-Domain Method* / Ed. Talfove A. Boston: Artech House. 1998.

58. Hafner C. *Post-Modern Electromagnetics: Using Intelligent Maxwell Solvers.* New York: John Wiley & Sons. 1999.

59. Mishchenko M.I., Travis L.D., Lacis A.A. *Scattering, Absorption, and Emission of Light by Small Particles.* Cambridge: University Press. 2002.

60. Pelton M., Aizpurua J., Bryant G. Metal-nanoparticle plasmonics // *Laser Photon. Rev.* 2008. V. 2. P. 136–159.

61. Myroshnychenko V., Rodríguez-Fernandez J., Pastoriza-Santos I., Funston A.M., Novo C., Mulvaney P., Liz-Marzán L.M., García de Abajo F.J. Modelling the optical response of gold nanoparticles // *Chem. Soc. Rev.* 2008. V. 37. P. 1792–1805.

62. Schultz S., Smith D.R., Mock J.J., Schultz D.A. Single-target molecule detection with nonbleaching multicolor optical immunolabels // *Proc. Natl. Acad. Sci. USA.* 2000. V. 97. P. 996–1001.

63. Chen H., McMahon J.M., Ratner M.A., Schatz G.C. Classical electrodynamics coupled to quantum mechanics for calculation of molecular optical properties: A RT-TDDFT/FDTD approach // *J. Phys. Chem. C.* 2010. V. 114. P. 14384–14392.

64. Harris N., Li S., Schatz G.C. Nanoparticles and theory // *AIP Conf. Proc.* 2012. V. 1504. P. 31–42.

65. Battie Y., Resano-Garcia A., Chaoui N., En Naciri A. Optical properties of plasmonic nanoparticles distributed in size determined from a modified Maxwell-Garnett-Mie theory // *Phys. Status Solidi C.* 2015. V. 12. P. 142–146.

66. Olson J., Dominguez-Medina S., Hoggard A., Wang L.-Y., Chang W.-S., Link S. Optical characterization of single plasmonic nanoparticles // *Chem. Soc. Rev.* 2015. V. 44. P. 40–57.

67. Ammari H., Deng Y., Millien P. Surface plasmon resonance of nanoparticles and applications in imaging // *Arch. Rational Mech. Anal.* 2016. V. 220. P. 109–153.

68. *Handbook of Photonics for Biomedical Science* / Ed. Tuchin V.V. Boca Raton: CRC Press. 2010.

69. Link S., El-Sayed M.A. Optical properties and ultrafast dynamics of metallic nanocrystals // *Annu. Rev. Phys. Chem.* 2003. V. 54. P. 331–366.

70. Johnson P.B., Christy R.W. Optical constants of noble metals // *Phys. Rev. B.* 1972. V. 6. P. 4370–4379.

71. Oubre C., Nordlander P. Optical properties of metallodielectric nanostructures calculated using the finite difference time domain method // *J. Phys. Chem. B.* 2004. V. 108. P. 17740–17747.

72. Mulvaney P. Surface plasmon spectroscopy of nanosized metal particles // *Langmuir.* 1996. V. 12. P. 788–800.

73. Persson B.N.J. Polarizability of small spherical metal particles: Influence of the matrix environment // *Surface Sci.* 1993. V. 281. P. 153–162.

74. Halperin W.P. Quantum size effects in metal particles // *Rev. Mod. Phys.* 1986. V. 58. P. 533–605.

75. Kawabata A., Kubo R. Electronic properties of fine metallic particles. II. Plasma resonance absorption // *J. Phys. Soc. Japan.* 1966. V. 21. P. 1765–1772.

76. Weick G. Quantum dissipation and decoherence of collective excitations in metallic nanoparticles. PhD Thesis, Universite Louis Pasteur, Strasbourg and Universitat Augsburg, 2006.

77. Prodan E., Nordlander P. Electronic structure and polarizability of metallic nanoshells // *Chem. Phys. Lett.* 2002. V. 352. P. 140–146.

78. Bruzzone S., Arrighini G.P., Guidotti C. Some spectroscopic properties of gold nanorods according to a schematic quantum model founded on the dielectric behavior of the electron-gas confined in a box // *Chem. Phys.* 2003. V. 291. P. 125–140.

79. Prodan E., Radloff C., Halas N.J., Nordlander P. A hybridization model for the plasmon response of complex nanostructures // *Science.* 2003. V. 302. P. 419–422.

80. Wang H., Brandl D.W., Nordlander P., Halas N.J. Plasmonic nanostructures: Artificial molecules // *Acc. Chem. Res.* 2007. V. 40. P. 53–62.

81. Garcıa de Abajo F.J. Nonlocal effects in the plasmons of strongly interacting nanoparticles, dimers, and waveguides // *J. Phys. Chem. C.* 2008. V. 112. P. 17983–17987.

82. Mermin N.D. Lindhard dielectric function in the relaxation-time approximation // *Phys. Rev. B.* 1970. V. 1. P. 2362–2363.

83. Zheng J., Petty J.T., Dickson R.M. High quantum yield blue emission from water-soluble Au_8 nanodots // *J. Am. Chem. Soc.* 2003. V. 125. P. 7780–7781.

84. Zheng J., Zhang C., Dickson R.M. Highly fluorescent, water-soluble, size-tunable gold quantum dots // *Phys. Rev. Lett.* 2004. V. 93. 077402.

85. Patel S.A., Richards C.I., Hsiang J.-C., Dickson R.M. Water-soluble Ag nanoclusters exhibit strong two-photon-induced fluorescence // *J. Am. Chem. Soc.* 2008. V. 130. P. 11602–11603.

86. Zuloaga J., Prodan E., Nordlander P. Quantum description of the plasmon resonances of a nanoparticle dimmer // *Nano Lett.* 2009. V. 9. P. 887–891.

87. Payton J.L., Morton S.M., Moore J.E., Jensen L. A hybrid atomistic electrodynamics–quantum mechanical approach for simulating surface-enhanced Raman scattering // *Acc. Chem. Res.* 2014. V. 47. P. 88–99.

88. Khlebtsov N.G., Khlebtsov B.N. Optimal design of gold nanomatryoshkas with embedded Raman reporters // *J. Quant. Spectrosc. Radiat. Transfer.* 2017. V. 190. P. 89–102.

89. Ciracì C., Hill R.T., Mock J.J., Urzhumov Y., Fernández-Domínguez A.I., Maier S.A., Pendry J.B., Chilkoti A., Smith D.R. Probing the ultimate limits of plasmonic enhancement. // *Science.* 2012. V. 337. P. 1072–1074.

90. Esteban R., Borisov A.G., Nordlander P., Aizpurua J. Bridging quantum and classical plasmonics with a quantum-corrected model // *Nat. Commun.* 2012. V. 3. 825.

91. Zhu W., Esteban R., Borisov A.G., Baumberg J.J., Nordlander P., Lezec H.J., Aizpurua J., Crozier Kenneth B. Quantum mechanical effects in plasmonic structures with subnanometre gaps // *Nat. Commun.* 2016. V. 7. 11495.

92. Eguiluz A., Quinn J.J. Hydrodynamic model for surface plasmons in metals and degenerate semiconductors // *Phys. Rev. B*. 1976. V. 14. P. 1347–1361.

93. Fuchs R., Francisco C. Multipolar response of small metallic spheres: Nonlocal theory // *Phys. Rev. B*. 1987. V. 35. P. 3722–3727.

94. Wubs M. Classification of scalar and dyadic nonlocal optical response models // *Opt. Express*. 2015. V. 23. P. 31296–31312.

95. Esteban R., Zugarramurdi A., Zhang P., Nordlander P., García-Vidal F.J., Borisov A.G., Aizpurua J. A classical treatment of optical tunneling in plasmonic gaps: Extending the quantum corrected model to practical situations // *Faraday Discuss*. 2015. V. 178. P. 151–183.

96. Khlebtsov B.N., Khlebtsov N.G. Biosensing potential of silica/gold nanoshells: Sensitivity of plasmon resonance to the local dielectric environment // *J. Quant. Spectr. Radiat. Transfer*. 2007. V. 106. P. 154–169.

97. Alekseeva A.V., Bogatyrev V.A., Khlebtsov B.N., Melnikov A.G., Dykman L.A., Khlebtsov N.G. Gold nanorods: Synthesis and optical properties // *Colloid J*. 2006. V. 68. P. 661–678.

98. Lakhtakia A. Strong and weak forms of the method of moments and the coupled dipole method for scattering of time-harmonic electromagnetic fields // *Int. J. Mod. Phys*. 1992. V. 3. P. 583–603.

99. Kolwas K., Demianiuk S., Kolwas M. Optical excitation of radius-dependent plasmon resonances in large metal clusters // *J. Phys. B*. 1996. V. 29. P. 4761–4770.

100. Noguez C.J. Surface plasmons on metal nanoparticles: The influence of shape and physical environment // *J. Phys. Chem. C*. 2007. V. 111. P. 3806–3819.

101. Khlebtsov B.N., Melnikov A.G., Zharov V.P., Khlebtsov N.G. Absorption and scattering of light by a dimer of metal nanospheres: Comparison of dipole and multipole approaches // *Nanotechnology*. 2006. V. 17. P. 1437–1445.

102. Khlebtsov B.N., Zharov V.P., Melnikov A.G., Tuchin V.V, Khlebtsov N.G. Optical amplification of photothermal therapy with gold nanoparticles and nanoclusters // *Nanotechnology*. 2006. V. 17. P. 5167–5179.

103. Haiss W., Thanh N.T.K., Aveard J., Fernig D.G. Determination of size and concentration of gold nanoparticles from UV–VIS spectra // *Anal. Chem*. 2007. V. 79. P. 4215–4221.

104. Khlebtsov N.G. Determination of size and concentration of gold nanoparticles from extinction spectra // *Anal. Chem*. 2008. V. 80. P. 6620–6625.

105. Sancho-Parramon J. Surface plasmon resonance broadening of metallic particles in the quasi-static approximation: A numerical study of size confinement and interparticle interaction effects // *Nanotechnology*. 2009. V. 20. 235706.

106. Khlebtsov N.G., Melnikov A.G., Bogatyrev V.A., Dykman L.A., Alekseeva A.V., Trachuk L.A., Khlebtsov B.N. Can the light scattering depolarization ratio of small particles be greater than 1/3? // *J. Phys. Chem. B*. 2005. V. 109. P. 13578–13584.

107. Brioude A., Jiang X.C., Pileni M.P. Optical properties of gold nanorods: DDA simulations supported by experiments // *J. Phys. Chem. B*. 2005. V. 109. P. 13138–13142.

108. Khlebtsov B.N., Khlebtsov N.G. Multipole plasmons in metal nanorods: Scaling properties and dependence on the particle size, shape, orientation, and dielectric environment // *J. Phys. Chem. C*. 2007. V. 111. P. 11516–11527.

109. Bryant G.W., García de Abajo F.J., Aizpurua J. Mapping the plasmon resonances of metallic nanoantennas // *Nano Lett.* 2008. V. 8. P. 631–636.

110. Prescott S.W., Mulvaney P. Gold nanorod extinction spectra // *J. Appl. Phys.* 2006. V. 99. 123504.

111. Khlebtsov B.N., Khanadeev V.A., Khlebtsov N.G. Observation of extra-high depolarized light scattering spectra from gold nanorods // *J. Phys. Chem. C.* 2008. V. 112. P. 12760–12768.

112. Calander N., Gryczynski I., Gryczynski Z. Interference of surface plasmon resonances causes enhanced depolarized light scattering from metal nanoparticles // *Chem. Phys. Lett.* 2007. V. 434. P. 326–330.

113. Laurent G., Felidj N., Aubard J., Levi G., Krenn J.R., Hohenau A., Schider G., Leitner A., Aussenegg F.R. Evidence of multipolar excitations in surface enhanced Raman scattering // *J. Chem. Phys.* 2005. V. 122. 011102.

114. Payne E.K., Shuford K.L., Park S., Schatz G.C., Mirkin C.A. Multipole plasmon resonances in gold nanorods // *J. Phys. Chem. B.* 2006. V. 110, P. 2150–2154.

115. Kim S., Kim S.K., Park S. Bimetallic gold-silver nanorods produce multiple surface plasmon bands // *J. Am. Chem. Soc.* 2009. V. 131. P. 8380–8381.

116. Schider G., Krenn J.R., Hohenau A., Ditlbacher H., Leitner A., Aussenegg F.R., Schaich W.L., Puscasu I., Monacelli B., Boreman G. Plasmon dispersion relation of Au and Ag nanowires // *Phys. Rev. B.* 2003. P. 68. 155427.

117. Gantzounis G. Plasmon modes of axisymmetric metallic nanoparticles: A group theory analysis // *J. Phys. Chem. C.* 2009. V. 113. P. 21560–21565.

118. Khlebtsov N.G., Melnikov A.G., Dykman L.A., Bogatyrev V.A. Optical properties and biomedical applications of nanostructures based on gold and silver bioconjugates // In: *Photopolarimetry in Remote Sensing* / Eds. Videen G., Yatskiv Y.S., Mishchenko M.I. Dordrecht: Kluwer Academic Publishers. 2004. P. 265–308.

119. Markel V.A. Divergence of dipole sums and the nature of non-Lorentzian exponentially narrow resonances in one-dimensional periodic arrays of nanospheres // *J. Phys. B.* 2005. V. 8. P. L115–L121.

120. Lamprecht B., Schider G., Lechner R.T., Ditlbacher H., Krenn J.R., Leitner A., Aussenegg F.R. Metal nanoparticle gratings: Influence of dipolar particle interaction on the plasmon resonance // *Phys. Rev. Lett.* 2000. V. 84. P. 4721–4724.

121. Zhao L.L., Kelly K.L., Schatz G.C. The extinction spectra of silver nanoparticle arrays: Influence of array structure on plasmon resonance wavelength and widths // *J. Phys. Chem. B.* 2003. V. 107. P. 7343–7350.

122. Chumanov G., Sokolov K., Cotton T.M. Unusual extinction spectra of nanometer-sized silver particles arranged in two-dimensional array // *J. Phys. Chem.* 1999. V. 100. P. 5166–5168.

123. Malynych S., Chumanov G. Light-induced coherent interactions between silver nanoparticles in two-dimensional arrays // *J. Am. Chem. Soc.* 2003. V. 125. P. 2896–2898.

124. Khlebtsov B.N., Khanadeyev V.A., Ye J., Mackowski D.W., Borghs G., Khlebtsov N.G. Coupled plasmon resonances in monolayers of metal nanoparticles and nanoshells // *Phys. Rev. B.* 2008. V. 77. 035440.

125. Härtling T., Alaverdyan Y., Hille A., Wenzel M.T., Käll M., Eng L.M. Optically controlled interparticle distance tuning and welding of single gold nanoparticle pairs by photochemical metal deposition // *Opt. Express.* 2008. V. 16. P. 12362–12371.

126. Chen M.W., Chau Y.-F., Tsai D.P. Three-dimensional analysis of scattering field interactions and surface plasmon resonance in coupled silver nanospheres // *Plasmonics.* 2008. V. 3. P. 157–164.

127. Davis T.J., Vernon K.C., Gómez D.E. Designing plasmonic systems using optical coupling between nanoparticles // *Phys. Rev. B.* 2009. V. 79. 155423.

128. Wurtz G.A., Dickson W., O'Connor D., Atkinson R., Hendren W., Evans P., Pollard R., Zayats A.V. Guided plasmonic modes in nanorod assemblies: Strong electromagnetic coupling regime // *Opt. Express.* 2008. V. 16. P. 7460–7470.

129. Rong G., Wang H., Skewis L.R., Reinhard B.M. Resolving sub-diffraction limit encounters in nanoparticle tracking using live cell plasmon coupling microscopy // *Nano Lett.* 2008. V. 8. P. 3386–3393.

130. Xu Y.-l., Khlebtsov N.G. Orientation-averaged cross sections of an aggregate of particles // *J. Quant. Spectr. Radiat. Transfer.* 2003. V. 78–80. P. 1121–1137.

131. Su K.-H., Wei Q.-H., Zhang X., Mock J.J., Smith D.R., Schultz S. Interparticle coupling effects on plasmon resonances of nanogold particles // *Nano Lett.* 2003. V. 3. P. 1087–1090.

132. Lamprecht B., Leitner A., Aussenegg F.R. SHG studies of plasmon dephasing in nanoparticles // *Appl. Phys. B.* 1999. V. 68. P. 419–423.

133. Mackowski D. Electrostatics analysis of radiative absorption by sphere clusters in the Rayleigh limit: Application to soot particles // *Appl. Opt. 1995.* V. 34. P. 3535–3545.

134. Pinchuk A.O., Schatz G.C. Collective surface plasmon resonance coupling in silver nanoshell arrays // *Appl. Phys. B.* 2008. V. 93. P. 31–38.

135. Khlebtsov N.G. Optical models for conjugates of gold and silver nanoparticles with biomacromolecules // *J. Quant. Spectr. Radiat. Transfer.* 2004, V. 89. P. 143–152.

136. Reinhard M.B., Siu M., Agarwal H., Alivisatos A.P., Liphardt J. Calibration of dynamic molecular rulers based on plasmon coupling between gold nanoparticles // *Nano Lett.* 2005. V. 5. P. 2246–2252.

137. Wei Q.H., Su K.H., Durant S., Zhang X. Plasmon resonance of finite one-dimensional Au nanoparticle chains // *Nano Lett.* 2004. V. 4. P. 1067–1071.

138. Jain P.K., El-Sayed M.A. Surface plasmon coupling and its universal size scaling in metal nanostructures of complex geometry: Elongated particle pairs and nanosphere trimers // *J. Phys. Chem. C.* 2008. V. 112. P. 4954–4960.

139. Funston A.M., Novo C., Davis T.J., Mulvaney P. Plasmon coupling of gold nanorods at short distances and in different geometries // *Nano Lett.* 2009. V. 9, P. 1651–1658.

140. Aizpurua J., Bryant G.W., Richter L.J., García de Abajo F.J. Optical properties of coupled metallic nanorods for field-enhanced spectroscopy // *Phys. Rev. B.* 2005. V. 71. 235420.

2

Gold Nanoparticles in Biology and Medicine

2.1 Functionalization of Gold Nanoparticles

In nanobiotechnology, metal nanoparticles are used in combination with recognizing biomacromolecules attached to their surface by means of physical adsorption or coupling through the Au–S bond. Such nanostructures are called bioconjugates [1], while the attachment of biomacromolecules to the nanoparticle surface is often called "functionalization." Thus, a probe conjugate molecule is used for unique coupling with a target, while a metal core serves as an optical label.

Metal hydrosols are typical lyophobic colloids [2–4], which are thermodynamically unstable and require special stabilization of particles. For instance, citrate colloidal gold (CG) particles are stabilized due to negative charge, while the stability of the seed-mediated gold nanorods (GNRs) is ensured by cetyltrimethylammonium bromide (CTAB) protection. Over last decade, a lot of functionalization protocols have been developed to include various nanoparticles and biomolecules [5,6].

The functionalization of gold nanoparticles (GNPs) with surface molecules is aimed at fabricating multifunctional nanoparticle bioconjugates possessing various modalities, such as active biosensing, enhanced imaging contrast, drug delivery, and tumor targeting. At present, bioconjugation chemistry, recently reviewed comprehensively by Medintz and coworkers [6], can be considered a separate specific branch of nanobiotechnology. The review [6] is based on 2081 literature sources and covers a broad range of particle types (including GNPs) and a great variety of functionalization technologies. For more detailed information, we point readers to this section; what follows is a description of only some general principles.

There are two basic approaches to functionalizing GNPs—the adsorptional and the chemosorptional. The adsorptional approach is based on the passive adsorption of a polymer on the surface of a particle through hydrophobic and electrostatic interactions or through sulfur bonding (Figure 2.1). Specifically, proteins bind to gold colloids through sulfur bonding (cysteine and methionine), charge–charge attraction (lysine), and hydrophobic attraction (tryptophan) [7]. For example, electrostatic interactions was reported to occur between the H_2N groups of lysine and the citrate ions at the surface of GNPs produced by the method of Frens [1]. Another example is the SH groups

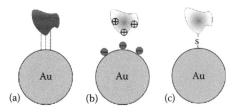

FIGURE 2.1 Schematic representation of adsorptional approaches to functionalizing GNPs with proteins. Proteins bind to the GNP surface by hydrophobic attractions (red: hydrophilic regions; blue: hydrophobic regions), which depend on the amino acids exposed to the particle surface (a), electrostatic interactions (b), and sulfur bonding (c).

of cysteine molecules [8], which are important for protein binding to the gold-particle surface. The strong point of the adsorptional approach is that the alterations in macromolecular structure are minimal, and consequently, the functional properties of the attached macromolecule are preserved. The weak points are possible desorption and competition with the binding sites on the target molecules. Typically, high-molecular-weight substances are adsorbed in amounts much larger than those needed for the formation of a monomolecular layer on the particle surface. In the case of dilute polymer solutions, the adsorption isotherms do not have inflections, the presence of which would attest to the formation of a discrete monolayer, as is observed for low-molecular-weight surfactants. In general, the adsorptional conjugation is irreversible, although long-term storage at high pH or in a buffer-containing surfactant may cause some proteins to dissociate [7].

The chemical attachment of biomolecules to GNPs is based on the classical chemistry of thiols and thiolated linkers or macromolecules. It is known that gold and sulfur atoms can form a dative bond [9]. This property is widely employed in bioconjugation chemistry to form alkanethiol linkers for covalent attachment of various biomolecules to GNPs (the chemosorption method). The chemical formula of alkanethiols can be represented as $HS(CH_2)nR$, where R stands for $-COOH$, $-OH$, or $-SO_3H$ and the number of groups n usually varies from 11 to 22 [10]. Interaction of alkanethiols with gold gives rise to Au(I) thiolates, $nAu0 \times Au + S - (CH_2)nR$ [11], which are organized as a monolayer on the surface of a particle. In 1996, Mirkin *et al.* proposed a technique for attaching 3′-thiolated oligonucleotides to 13-nm gold particles [12]. As a result, stable conjugates were prepared that could aggregate and disaggregate at low and high temperature, respectively; this principle was used by the authors for colorimetric detection of DNA in solution. The same group also suggested the use of cyclic disulfides [13] and trihexylthiol linkers [14] to prepare DNA conjugates with 30- and 100-nm GNPs, respectively. Walton *et al.* [15] and Ghosh *et al.* [16] used thiolated proteins (immunoglobulins and avidin) to obtain conjugates with GNPs. In addition to alkanethiols, the conjugation linkers can include phosphine-, amine-, and carboxyl-containing ligands. The advantages of chemical attachment are evident for linear molecules (like DNA) and are determined by C-terminal immobilization, which ensures a strict spatial orientation of probes. The type of bond between GNPs and functional molecules is, in particular, responsible for the release of the target substance inside a cell [17]. Thus, the presence on the GNP surface of polymers and biological macromolecules that differ in chemical makeup and properties leads to differences in the cellular uptake of functionalized particles.

A summary of recent advantages in functionalization of GNPs can be found in reviews [6,18–24].

2.2 GNPs in Diagnostics

2.2.1 Methods for Visualization and Bioimaging

GNPs have been actively used in various visualization and bioimaging methods for the identification of chemical and biological agents [25–28]. Historically, electron microscopy (mainly its transmission version, transmission electron microscope [TEM]) has, for a long time, (starting in 1971 [29]) been the principal method employed for the detection of biospecific interaction with the help of CG particles (owing to their high electron density). Although GNPs can intensely scatter and emit secondary electrons, they have not received equally wide acceptance in scanning electron microscopy [30]. It is no mere chance that the first three-volume book on the use of CG [31] is devoted mostly to the application of GNPs in TEM. A peculiarity of the current use of the electron microscopic technique is the application of high-resolution TEM and systems for the digital recording and processing of images [32,33]. The major application of immuno-electron microscopy in present-day medical and biological research is the identification of infectious agents and their surface antigens [34–36] (Figure 2.2a). The techniques often employed for the same purposes include scanning atomic-force [37] (Figure 2.2b), scanning electron [38], and fluorescence [39] microscopies.

Thanks to their porous structure, gold nanocages (GNCs) can be easily discriminated from particles of a similar size (and moreover from smaller or larger particles), while analyzing TEM images with nanocage-based biomarkers. This property suggests

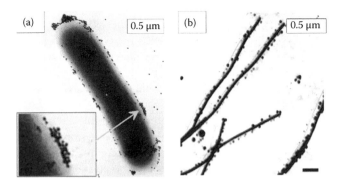

FIGURE 2.2 TEM image of a *Listeria monocytogenes* cell labeled with an antibody–CG conjugate (a) and scanning atomic force microscopy image of the tobacco mosaic virus labeled with an antibody–CG conjugate (b). (Adapted from Dykman, L.A., Khlebtsov, N.G., *Chem. Soc. Rev.*, 41, 2256–2282, 2012. With permission.)

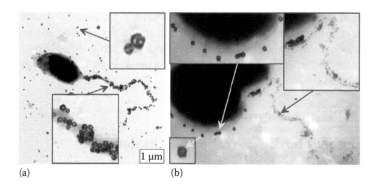

(a)   (b)

FIGURE 2.3 (a) TEM image of a bacterium *A. brasilense* Sp245 labeled with GNCs functionalized with O-specific antibodies. The bacteria were not washed before labeling, so the O-specific antigens of the LPS capsule are clearly visualized both on the bacteria surface and polar flagellum. (b) TEM image of the same bacteria after a special washing procedure and simultaneous labeling with a mixture of O-specific conjugates (50-nm Au–Ag nanocages + anti-LPS antibodies) and H-specific conjugates (15-nm GNPs + anti-PFL antibodies). The insets clearly indicate quite specific labeling of PFL and LPS antigens. The right upper inset also shows a small group of nanocage markers near the flagellum origin. This group corresponds to residual LPS antigens after washing. (Adapted from Khlebtsov, N.G., Bogatyrev, V.A., Dykman, L.A., Khlebtsov, B.N., Staroverov, S.A., Shirokov, A.A., Matora, L.Y., Khanadeev, V.A., Pylaev, T.E., Tsyganova, N.A., Terentyuk G.S., *Theranostics*, 3, 167–180, 2013.)

utilizing GNCs in combination with common CG nanospheres for a multiplexed immunoelectron microscopy visualization of at least two different antigen moieties. Figure 2.3 serves as an example of such multiplexed immunoelectron microscopy labeling [40]. Shown here is an *Azospirillum brasilense* Sp245 bacterium, which lives on its own in soil or in close associations with plants in the rhizosphere, thus promoting growth and increasing the yield of many plant species. To discriminate between H-antigens (the polar flagellum antigens [PFL]) and O-antigens of a lipopolysaccharide (LPS) capsule, we used two types of GNP conjugates. Namely, 15-nm GNPs were functionalized with antibodies against H-antigens, and 50-nm GNCs were functionalized with antibodies against O-antigens. In the case of native bacteria, their LPS capsule covers both the bacteria surface and polar flagellum (Figure 2.3b).

Recently, the popularity of visualization methods using GNPs and optical microscopy [41], in particular confocal laser microscopy, has also been on the rise. Confocal microscopy is a method for detecting microobjects with the aid of an optical system ensuring that light emission is recorded only when it comes from objects located in the system's focal point. This allows one to scan samples according to height and, ultimately, to create their three-dimensional (3D) images by superposition of scanograms. In this method, the use of GNPs and their antibody conjugates permits real-time detection of gold penetration into living (*e.g.*, cancer) cells at the level of a single particle and single molecule and even estimation of their number [42–47].

Confocal images can be obtained with, *e.g.*, detection of fluorescence emission (confocal fluorescence microscopy) or resonance elastic or two-photon (multiphoton) light scattering by plasmonic nanoparticles (confocal resonance-scattering or two-photon luminescence microscopy). These techniques are based on detecting microobjects with the optical microscope, in which the luminescence of an object is excited owing to simultaneous absorption of two (or more) photons, where the energy of each is lower than the energy needed for fluorescence excitation. The basic advantage of this method is the increase in contrast through a strong reduction in the background signal. Specifically, the use of two-photon luminescence of GNPs allows visualization of molecular markers of cancer on or inside cells [48–51], *Bacillus* spores [52], and the like. Figure 2.4a shows an example of combined bioimaging of a cancer cell with the help of adsorption, fluorescence, and luminescence plasmon resonance labels.

Dark-field microscopy remains to be one of the most popular bioimaging methods using GNPs [53]. It is based on light scattering by microscopic objects, including those whose sizes are smaller than the resolution limit for the light microscope (Figure 2.4b and c). In dark-field microscopy, the light entering the objective lens is solely that scattered by the object at side lighting (similarly to the Tyndall effect); therefore, the scattering object looks bright against a dark background. Compared with fluorescent labels, GNPs have considerable opportunity to reveal biospecific interactions with the help of dark-field microscopy [54,55] because the particle scattering cross-section is three to five orders of magnitude greater than the fluorescence cross-section for a single molecule. This principle was, for the first time, employed by Mostafa El-Sayed's group [56] for their new method of simple and reliable diagnosis of cancer with the use of GNPs. The method is based on the preferential binding of GNP conjugated with antibodies specific for tumor antigens to the surface of cancerous cells, as compared with binding to healthy cells. With dark-field microscopy, therefore, it is possible to "map out" a tumor with an accuracy of several cells (Figure 2.4b and c). Subsequently, GNRs [57], nanoshells [58], nanostars (GNSts) [59], nanocages [60], and nanoclusters [61] were used for these purposes.

The use of cell-assembled monolayers or island films, as well as nonspherical and/or composite particles, opens up fresh opportunities for increased sensitivity of detection of biomolecular binding on or near nanostructure surface. The principle of amplification

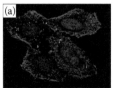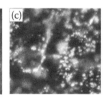

FIGURE 2.4 Confocal image of HeLa cells in the presence of GNPs (a). The nuclei are stained with Hoechst 33258 (blue), the actin cytoskeleton is labeled by Alexa Fluor 488 phalloidin (red), and unlabeled GNPs (green) were imaged by two-photon microscopy. Dark-field microscopic images of cancerous (b) and healthy (c) cells by using GNPs conjugated to antiepidermal growth factor antibodies. (Adapted from Maiorano, G., Sabella, S., Sorce, B., Brunetti, V., Malvindi, M.A., Cingolani, R., Pompa, P.P., *ACS Nano.*, 4, 7481–7491, 2010; El-Sayed, I.H., Huang, X.H., El-Sayed, M.A., *Nano Lett.*, 5, 829–834, 2005. With permission.)

of the biomolecular binding signal is based on strong local electromagnetic fields arising near nanoparticles with spiky surface sites or in narrow (on the order of 1 nm or smaller) gaps between two nanoparticles. This determines the increased plasmon-resonance sensitivity to the local dielectric environment [62] and the high scattering intensity as compared to equivolume spheres. Therefore, such nanostructures show considerable promise for use in dark-field-microscopy-assisted biomedical diagnostics [63–65].

In dark-field microscopy, GNPs are employed for the detection of microbial cells and their metabolites [66], bioimaging of cancerous cells [67–69] and revelation of receptors on their surface [70,71], study of endocytosis [72], and other purposes. In most biomedical applications, the effectiveness of conjugate labeling of cells is assessed at a qualitative level. One of the few exceptions is the work of Khanadeev *et al.* [73], in which a method was proposed for the quantitative estimation of the effectiveness of GNP labeling of cells, and its use was illustrated with the example of labeling of pig embryo kidney cells with gold nanoshell conjugates. Another approach was developed by Fu *et al.* [74].

Apart from the just listed means of recording biospecific interactions with the aid of various versions of optical microscopy and GNPs, the development of other state-of-the-art detection and bioimaging methods is currently active. These are referred to collectively as biophotonics methods [75,76]. Biophotonics combines all studies related to the interaction of light with biological cells and tissues. The most popular biophotonics methods include optical coherence tomography [77–79], x-ray imaging and computer and magnetic resonance tomography [80–84], photoacoustic microscopy [85,86] and tomography [87], fluorescence correlation microscopy [88], surface-enhanced Raman spectroscopy (SERS) [89], and other techniques. These methods also successfully use variously sized and shaped GNPs. In our opinion, biophotonics methods employing nonspherical gold particles may hold particular promise for bioimaging *in vivo* [26,90,91].

2.2.2 Analytical Diagnostic Methods

2.2.2.1 Homophase Techniques

Beginning in the 1980s, CG conjugates with recognizing biomacromolecules were coming into use in various analytical methods in clinical diagnostics. In 1980, Leuvering *et al.* [92] put forward a new immunoanalysis method that they called sol particle immunoassay (SPIA). This method is based on two principles: (1) the sol color and absorption spectrum change little when biopolymers are adsorbed on individual particles [93]; and (2) when particles move closer to each other by distances smaller than 0.1 of their diameter, the red color of the sol changes to purple or gray and the absorption spectrum broadens and red-shifts [94]. These changes in the absorption spectrum can easily be detected spectropfotometrically or visually [95–97] (Figure 2.5a and b).

The authors employed an optimized variation of SPIA (by using larger gold particles and monoclonal antibodies to various antigen sites) to detect human chorionic gonadotropin [98]. Subsequently, the assay was used for the immunoanalysis of *Shistosoma* [99] and *Rubella* [100] antigens; estimation of immunoglobulins [101,102], thrombin (by

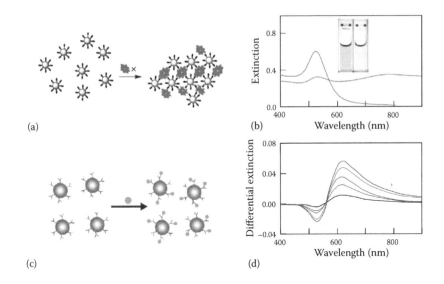

FIGURE 2.5 SPIA. (a) Scheme for the aggregation of conjugates as a result of binding by target molecules and (b) the corresponding changes in the spectra and in sol color. (c) Scheme for the formation of a secondary layer without conjugate aggregation and (d) the corresponding differential extinction spectra at 600 nm. (Adapted from Dykman, L.A., Khlebtsov, N.G., *Chem. Soc. Rev.*, 41, 2256–2282, 2012. With permission.)

using aptamers) [103], glucose [104], ATP [105], alpha-fetoprotein [106], C-reactive protein [107], lysozyme [108], fecal calprotectin [109], fibronectin [110], and selenoprotein P [111]; direct detection of cancerous cells [112] and antigens [113]; detection of *Leptospira* [114], bacteriophage [115], and influenza A virus [116]; detection of Alzheimer's disease markers [117]; determination of protease activity [118]; and other uses [119–123]. The simultaneous use of antibody conjugates of gold nanorods and nanospheres for the detection of tumor antigens was described by Liu *et al.* [124]. Wang *et al.* [125] provided data on the detection of hepatitis B virus in blood by using gold nanorods conjugated to specific antibodies.

All SPIA versions proved to be easy to implement and demonstrated high sensitivity and specificity. However, investigators came up against the fact than antigen–antibody reaction on sol particles do not necessarily lead to system destabilization (aggregation). Sometimes, despite the obvious complementarily of the pair, changes in solution color (and, correspondingly, in absorption spectra) were either absent or slight. Dykman *et al.* [126] proposed a model for the formation of a second protein layer on gold particles without loss of sol aggregative stability. The spectral changes arising from biopolymer adsorption on the surface of metallic particles are comparatively small [127] (Figure 2.5c and d). However, even such minor changes in absorption spectra, resulting from a change in the structure of the biopolymer layer (specifically, its average refractive index), near the GNP surface could be recorded and used for assay in biological application, as demonstrated by Englebienne *et al.* [128,129].

For increasing the sensitivity of the analytic reaction, new techniques for recording interaction are used, including photothermal spectroscopy [130], laser-based double beam absorption spectroscopy [131], hyper-Rayleigh scattering [132], differential light-scattering spectroscopy [133], and dynamic light scattering [134]. In addition, the vibrational spectroscopy methods—surface-enhanced infrared (IR) absorption spectroscopy [135–137] and SERS [138–141]—have been proposed for use in recording SPIA results.

Strategies for protein detection *via* SERS currently exploit the formation of randomly generated hot spots at the interfaces of metal colloidal nanoparticles, which are clustered together by intrusive chemical or physical processes in the presence of the target biomolecule. Matteini *et al.* [142] proposed a new approach based on selective and quantitative gathering of protein molecules at regular hot spots generated on the corners of individual silver nanocubes in aqueous medium at physiological pH. Advanced electron microscopy analyses and computational simulations outlined a unique strategy relying on a site-selective mechanism with superior Raman signal enhancement, which paves the way for highly controlled and reproducible routine SERS detection of proteins.

The ability of gold particles interacting with proteins to aggregate with a solution color change served as a basis for the development of a method for the colorimetric determination of proteins [143]. A new version gold aggregation assay using microplates and enzyme-linked immunosorbent assay (ELISA)-reader, with CG-conjugated trypsin as a specific agent for proteins, was advanced by Dykman *et al.* [144].

A new version of gold aggregation assay was advanced by Mirkin *et al.* [145] for the colorimetric detection of DNA. Currently, the colorimetric detection of DNA includes two strategies: (1) the use of GNPs conjugated to thiol-modifed ssDNA [145–149] or aptamers [150] and (2) the use of unmodified GNPs [151–153]. The first strategy is based on the aggregation of conjugates of 10–30-nm GNPs with thiol-modified ssDNA probes after the addition of target polynucleotides to the system. It used two types of probes complementary to two terminal target sites. Hybridization of the targets and probes leads to the formation of GNP aggregates, which is accompanied by a change in the absorption spectrum of the solution and can readily be detected visually, photometrically [154], or by the method of dynamic light scattering [148,155]. Within the limits of the first strategy, Sato *et al.* [156] used a diagnostic system based on the aggregation of GNPs modified with probes of one type, with DNA targets added to the solution under high ionic strength condition. Contrary to their data, Baptista *et al.* [147,157] devised a detection method based on the increased conjugate stability after the addition of complementary targets even at high ionic strength (2 M NaCl), and they observed aggregation for noncomplementary targets. The apparent contradictions between the two approaches were explained by Song *et al.* [158] as the difference in surface functionalization density.

The second strategy [152] is based on the fact that at high ionic strength, ssDNA protects unmodified GNPs against aggregation, whereas the formation of duplex during hybridization cannot stabilize the system. This approach was employed to detect hepatitis C virus RNA [159]. Recently, Xia *et al.* [160] described a new version of the second strategy that uses ssDNA, unmodified GNPs, and a cationic polyelectrolyte. The same approach proved suitable for the detection of a wide range of targets, including peptides, amino acids, pesticides, antibiotics, and heavy metals. As distinct from techniques employing usual GNPs, He *et al.* [153] proposed a method for detecting HIV-1 U5 virus DNA by using nanorods stabilized with CTAB and light scattering with a detection limit of 100 pM. In an optimized version using absorption spectroscopy [161], the detection limit was lowered to 0.1 pM. Pylaev *et al.* [162] demonstrated the use of CTAB-coated positively charged GNPs in combination with spectroscopic and dynamic light scattering methods and provided also a critical discussion of reported detection limits for light scattering and absorption techniques.

The above-listed versions of the method of sol particle aggregation caused by the hybridization reaction have been used for the detection of the DNA of *Mycobacterium* [147,163,164], *Staphylococcus* [165], *Streptococcus* [166], *Chlamydiae* [167], *Serratia* [168], *Bacillus* [169], *Salmonella* [170], and *Acinetobacter* [171] in clinical samples.

2.2.2.2 Dot Blot Immunoassay

At the early stages of immunoassay development, preference was given to liquid-phase techniques in which bound antibodies were precipitated or unbound antigen was removed by adsorption with dextran-coated activated charcoal. Currently, the most popular techniques are the solid-phase ones (first used for protein radioimmunoassay) because they permit the analysis to be considerably simplified and the background signal to be reduced. The most widespread solid-phase carriers are polystyrene plates and nitrocellulose membranes.

The solid-phase immunoassays are based on adsorption of antigens onto a solid substrate, followed by binding of adsorbed target molecules with biospecific labels. In the membrane version, the solid-phase immunoassay can be called "dot-immunoassay" as usually a drop of analyte is deposited into center of a 5 × 5-mm delineate square and the reaction outcome looks like a colored dot. The simplicity of analyses and the saving of antigens and reagents allow one to implement the solid-phase immunoassays in laboratory, field, or even domestic circumstances to detect proteins (Western blotting), DNA (Southern blotting), or RNA (Northern blotting).

Membrane immunoassays (dot, slot and blot assays) commonly employ radioactive isotopes (^{125}I, ^{14}C, ^{3}H) and enzymes (peroxydase, alkaline phosphatase, *etc.*) as labels. In 1984, four independent reports were published [172–175] in which CG was proposed for use as a label in solid-phase immunoassay. The use of GNP conjugates in solid-phase assay is based on the fact that the intense red coloration of a gold-containing marker allows the results of a reaction run on a solid carrier to be determined visually [176,177]. Immunogold methods in dot-blot assay outperform other techniques (*e.g.*, enzyme immunoassay) in sensitivity (Table 2.1), rapidity, and low cost [178–180]. After an appropriate immunochemical reaction is run, the sizes of GNPs can be increased by enhancement with salts of silver [181] or gold (autometallography) [182], considerably increasing the method sensitivity. An optimized solid-phase assay using a densitometry system afforded a linear detection range from 1 pM to 1 µM [183], with detection limit of 100 aM, which was lowered to 100 zM by silver enhancement. The use of state-of-the-art instrumental detection methods, such as photothermal deflection of the laser beam, caused by heating of the local environment near absorbing particles by heating laser

TABLE 2.1 Sensitivity Limits for Immunodot/Blot Methods
on Nitrocellulose Filters by Using Various Labels

Label	Sensitivity Limit (pg of protein/fraction)
^{125}I	5
Horseradish peroxidase	10
Alkaline phosphatase	1
Colloidal gold	1
Colloidal gold + silver	0.1
Fluorescein isothiocyanate	1000

Source: Bio-Rad Bulletin 1310, Bio-Rad, Richmond, 1987.

impulses [184], also ensures a very broad detection range (up to three orders of magnitude to the extent of several individual particles in a dot spot).

In specific staining, a membrane with applied material under study is incubated in a solution of antibodies (or other biospecific probes) labeled with CG [185]. As probes, "gold" dot or blot assay uses immunoglobulins, Fab and scFv antibody fragments, staphylococcal protein A, lectins, enzymes, avidin, aptamers, and other probing molecules. Sometimes, several labels are used simultaneously (*e.g.*, CG and peroxidase or alkaline phosphatase) for detection of multiple antigens on a membrane [186,187].

CG in membrane assay have been used for the diagnosis of parasitic [188–192], viral [194–197], and fungal [199,199] diseases; tuberculosis [200–202]; melioidosis [203]; syphilis [204]; brucellosis [205]; shigellosis [206]; *Escherichia coli* infections [207]; salmonellosis [208,209]; yersiniosis [210,211] and early pregnancy [212]; blood group determination [213]; dot blot hybridization [214]; detection of diphtheria toxin [215], ferritin [216], thrombin [217], β-amyloid peptide [218], tumor-associated antigens [219], and antibiotics [220]; diagnosis of myocardial infarction [221] and hepatitis B [222]; and other purposes.

The immunodot assay is one of the simplest methods for analyzing membrane-immobilized antigens. In some cases, it permits quantitative determination of antigen. Most commonly, the immunodot assay is employed to study soluble antigens [223]. However, there have been several reports in which corpuscular antigens (whole bacterial cells) were used as a research object in dot assays with enzyme labels [224]. Bogatyrev *et al.* [225,226] were the first to perform a dot assay of whole bacterial cells, with the reaction products being visualized with immunogold markers ("cell gold immunoblotting") to serotype the nitrogen-fixing soil microorganisms of the genus *Azospirillum*. Subsequently, this method was applied for the rapid diagnosis of enteric infections [227] and for the study of surface physicochemical properties of microorganisms [228]. Gas *et al.* [229] used a dot assay with GNPs to detect whole cells of the toxic phytoplankton *Alexandrium minutum*.

Khlebtsov *et al.* [230,231] first presented experimental results for the use of gold nanoshells (GNSs) as biospecific labels in dot assay. To demonstrate the method principles, three types of GNSs were examined that had silica core diameters of 100, 140, and 180 nm and a gold shell thickness of about 15 nm. The biospecific pair was normal rabbit serum (target molecules) and sheep antirabbit immunoglobulins (recognizing molecules). When the authors used a standard protocol for a nitrocellulose-membrane

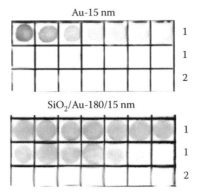

FIGURE 2.6 Dot assay of normal rabbit serum (1) by using suspensions of conjugates of 15-nm GNPs and SiO_2/Au nanoshells (180 nm core diameter) with sheep antirabbit antibodies. The amount of IgG in the first square of the top row is 1 μg and decreases from left to right in accordance with twofold dilutions. The lower rows (2) correspond to the application of 10 μg of bovine serum albumin to each square as a negative control. The detected analyte quantity is 15 ng for 15-nm GNPs and 0.2 ng for nanoshells. (Adapted from Khlebtsov, B.N., Khanadeev, V.A., Bogatyrev, V.A., Dykman, L.A., Khlebtsov, N.G., *Nanotechnol. Russ.*, 3, 442–455, 2008. With permission.)

dot assay, with 15-nm CG as labels, the minimum detectable quantity of rabbit IgG was 15 ng. Replacing CG conjugates with GNSs increased the assay sensitivity to 0.2 ng for 180/15-nm gold nanoshells and to 0.4 ng for 100/15-nm and 140/15-nm nanoshells (Figure 2.6). Such a noticeable increase in sensitivity with nanoshells, as compared with nanospheres, can be explained by the different optical properties of the particles [232].

By using multicolor composite GNPs (Ag nanocubes, Au–Ag alloy nanoparticles, and Au–Ag nanocages), Panfilova *et al.* [233] developed a multiplexed variant of immunodot assay. As in usual immunodot assay, the multiplexed variant is based on the staining of analyte drops on a nitrocellulose membrane strip by using multicolor nanoparticles conjugated with biospecific probing molecules. Depending on the Ag–Au conversion ratio, the particle plasmon resonance was tuned from 450 to 700 nm and the suspension color changed from yellow to blue (Figure 2.7).

The particles of yellow, red, and blue suspensions were functionalized with chicken, rat, and mouse IgG molecular probes, respectively. The multiplex capability of the assay is illustrated by a proof-of-concept experiment on simultaneous one-step determination of target molecules (rabbit antichicken, antirat, and antimouse antibodies) with a mixture of fabricated conjugates. Under naked eye examination, no cross-colored spots or nonspecific bioconjugate adsorption was observed, and the low detection limit was about 20 fmol.

A very promising direction is the use of CG to analyze large arrays of antigens in micromatrices (immunochips) [234,235]. These enable an analyte to be detected in 384 samples simultaneously at a concentration of 60–70 ng/l or, with account taken of the microliter amounts of sample and detecting immunogold marker, with a detection limit of lower than 1 pg. One area with prospects is the development of commercially available,

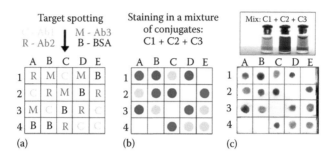

FIGURE 2.7 Scheme for the multiplexed dot immunoassay (a, b) and its experimental verification (c). At the first step (panel a), antichicken rabbit antibodies Ab1 were spotted in squares A2, B3, C1, D4, and E3; antirat rabbit antibodies Ab2 were spotted in squares A1, B2, C4, D3, and E2; antimouse rabbit antibodies Ab3 were spotted in squares A3, B1, C2, and D1; and for negative control, BSA was spotted in squares A4, B4, C3, D2, and E1. The concentration of all analytes was 100 µg/mL. After staining in a mixture of conjugates (C1 + C2 + C3), the expected spot colors are shown in panel b. The experimental panel c confirms the expected assay results. (Adapted from Panfilova, E., Shirokov, A., Khlebtsov, B., Matora, L., Khlebtsov, N., *Nano Res.*, 5, 124–134, 2012. With permission.)

handheld, sensitive readers for evaluation of quantitative results and the integration into systems designed to optimize the performance of the overall assay for samples under examination [236,237].

2.2.2.3 Immunochromatographic Assays

In 1990, several companies began to manufacture immunochromatographic test systems for instrument-free hand held diagnostics. Owing to high specificity and sensitivity, these strip tests have found a wide utility in the detection of narcotics and toxins and in screening for highly dangerous infections and urogenital diseases [238–245]. Methods have been developed for the diagnosis of tuberculosis [246,247], helicobacteriosis [248], staphylococcal infection [249,250], hepatitis B [251], shigellosis [252,253], diphtheria [254], pseudorabies [255], botulism [256], chlamydiosis [257], *E. coli* infections [258], prostatitis [259,260], and early pregnancy [261]; for DNA hybridization [262]; for the detection of pesticides [263,264], aflatoxin [265,266], fumonisin [267], hexoestrol [268], and antibiotics [269–271] in environmental constituents; and other purposes.

The immunochromatographic assay is based on eluent movement along the membrane (lateral diffusion), giving rise to specific immune complexes at different membrane sites; the complexes are visualized as colored bands. As labels, these systems use enzymes, colored latexes, but mostly GNPs [272–274] and quantum dots [235].

The sample being examined migrates along the test strip due to capillary forces. If the sample contains the sought-for substance or immunochemically related compounds, there occurs a reaction with colloidal-gold-labeled specific antibodies at the instant the sample passes through the absorbing device. The reaction is accompanied by the formation of antigen–antibody complex. The colloidal preparation enters into a competitive binding reaction, with the antigen immobilized in the test zone (as a rule, the detection of low-molecular-weight compounds employs conjugates of haptens with protein carrier for immobilization). If the antigen concentration in the sample exceeds the threshold level, the conjugate does not possess free valences for interaction in the test zone and the colored band corresponding to the formation of the complex is not revealed. When the sample does not contain the sought-for substance or when the concentration of that substance is lower than the threshold level, the antigen immobilized in the test zone of the strip reacts with the antibodies on the surface of CG, which leads to the development of a colored band.

When the liquid front moves on, the gold particles with immobilized antibodies that have not reacted with the antigen in the strip test zone bound to antispecific antibodies in the control zone of the test strip. The appearance of a colored band in the control zone confirms that the test was done correctly and that the system's components are diagnostically active. A negative test result—the appearance of two colored bands (in the test zone and in the control zone)—indicates that the antigen is absent from the sample or that its concentration is lower than the threshold level. A positive test result—the appearance of one colored band in the control zone—indicates that antigen concentration exceeds the threshold level (Figure 2.8).

Studies have shown that such assay systems are highly stable, their results are reproducible, and they correlate with alternative methods. Densitometric characterization of the dissimilarity degree for the bands detected yield values ranging from 5% to 8%, allowing reliable visual determination of the analysis results. These assays are very simple and convenient to use.

In summary, being effective diagnostic tools, rapid tests allow qualitative and quantitative determination, in a matter of minutes, of antigens, antibodies, hormones, and other diagnostically important substances in humans and animals. Rapid tests are highly sensitive and accurate, as they can detect more than 100 diseases (including

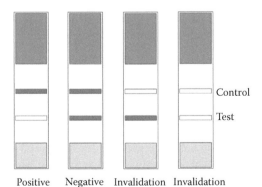

Positive   Negative   Invalidation   Invalidation

FIGURE 2.8 Results of an immunochromatographic assay: positive, negative, and invalid determination because of the absent coloration in the control zone. (Adapted from Chen, L., Wang, Z., Ferreri, M., Su, J., Han, B., *J. Agric. Food Chem.*, 57, 4674–4679, 2009. With permission.)

tuberculosis, syphilis, gonorrhea, chlamydiosis, various types of viral hepatitis, *etc.*) and the whole gamut of narcotic substances used, with the reliability of detection being high. An important advantage of these tests is their use in diagnostics *in vitro*, which does not require a patient's presence.

However, immunochromatographic test strips are not devoid of weak points, related to reliability, sensitivity, and cost-effectiveness. Reliability and sensitivity depend, first, on the quality of monoclonal antibodies used in a test and, second, on the antigen concentration in a biomaterial. The quality of monoclonal antibodies depends on the methods of their preparation, purification, and fixation on a carrier. The antigen concentration depends on the disease state and the biomaterial quantity. For increasing the analysis sensitivity, it has been proposed to employ the silver enhancement procedure [275] or GNRs [276] or GNSts [277] as labels. Several studies demonstrate the novel application of an artificial nonimmunoglobulin structures (aptamers, DARPins) as the new line of a visible detector using a rapid diagnostic test with characteristics that have the potential to be superior to those that utilize antibody-based tests [278,279]. In addition, semiquantitative and quantitative instrumental formats of immunochromatographic analysis have been developed that use special readers for recording the intensity of a label's signal in the test zone of a test strip [280–283].

2.2.2.4 Plasmonic Biosensors

Currently, gold and silver nanoparticles and their composites are widely utilized as effective optical detectors of biospecific interactions [284]. In particular, the resonant optical properties of nanometer-sized metallic particles have been successfully used for the development of biochips and biosensors. There are many types of sensors, *viz.* colorimetric, refractometric, electrochemical, piezoelectric, and certain others [285, 286]. Such devices are of much interest in biology (determination of nucleic acid, protein, and metabolite content), medicine (screening of drugs, analysis of antibodies and antigens, and diagnosis of infection), and chemistry (rapid environmental monitoring and assays of solutions and disperse systems). Of particular significance is the detection of specified nucleic acid (gene) sequences and the construction of new materials, which is based on the formation of 3D ordered structures during hybridization in solutions of complementary oligonucleotides that are covalently attached to metallic nanoparticles [287].

The detection of biospecific interaction that is based on a change in the optical properties of the nanoparticle-carrier system can be assigned to biosensoric. The biosensor is constituted either by the system itself in its entirety or by an individual marker particle (an elementary sensor). Among the localized plasmon resonance biosensor, CG occupies a special place because it can serve both as a label in a nanosensing device and as a tool in molecular biological studies, which is used *in vitro*, *in situ*, and *in vivo*.

Biospecific interactions have been studied in systems in which GNPs are represented as ordered structures: self-assembling (thin films) [288] or as part of polymer matrices [289]. Such structures have been actively used for the detection of biomolecules and infectious agents, development of DNA chips, and other purposes. In this case, investigators directly realize the possibility in principle of using the sharp enhancement of the optical signal from the probe (GNPs conjugated to biospecific macromolecules) resulting from the strengthening of the exciting local field in the aggregate formed from gold nanoclusters. Currently, biosensors are built with novel, unique technologies, including monolayer self-assembly of metallic particles [290–293], nanolithography [294], vacuum evaporation [295], and others. It is of fundamental importance to note that the optical response of nanoparticles or their aggregates (especially ordered ones) is substantially influenced by the particle size and shape [296], interparticle distance [297,298], and the local dielectric environment [299,300], which enables the "tuning" of sensors to be controlled. These properties of metallic clusters served as a basis for creation of new promising (localized) surface plasmon resonance biosensing systems [(L)SPR biosensors] based on the transformation of biospecific interactions into an optical signal. The basic principles behind the design of such systems and their use in practice have been considered in numerous books and reviews [301–334].

In experimental work with SPR biosensors, three stages can be identified [302]: (1) one of the reagents (target-recognizing molecules) is covalently attached to the sensor surface; (2) the other reagent (target molecules) is added at a definite concentration to the sensor surface along with the flow of the buffer. The process of complex formation is then recorded; and (3) the sensor is regenerated (dissociation of the formed complexes).

As this takes place, the following conditions should be met:

- Reagent immobilization on the substrate should not lead a critical change in the conformation of native molecules;
- The relatively small difference between the refractive indices of most biological macromolecules forces one to use a high local concentration of binding sites on the sensor surface (10–100 µM);
- The reagent being added should be vigorously agitated to achieve effective binding to the immobilized molecules, and unbound reagent should be promptly removed from the sensor surface to avoid nonspecific sorption.

Apart from that, the sensitivity, stability, and resolution of a sensor depend directly on the characteristics of the optical system being used for recording. The most popular sensor system of this type is Biacore [335,336]. The measurement principle of the planar, prismatic, or mirror biosensors is similar to the principle of the attenuated total internal reflection, traditionally employed to measure the thickness and refractive index of ultra-thin organic films on metallic (reflecting) surfaces [284]. The excitation of the plasmon resonance in a planar gold layer occurs when polarized light is incident on the surface at a certain angle. At the metal/dielectric interface, electromagnetic fields are excited that run along the interface and are localized near it at the cost of an exponential decrease in amplitude perpendicularly to the dielectric with a typical attenuation length of up to 200 nm (the effect of total internal reflection; Figure 2.9). The reflection coefficient at a certain angle and light wavelength depends on the dielectric properties of the thin layer at the interface, which are ultimately determined by the concentration of target molecules in the layer [337].

Various types of biosensors using GNPs have been developed for the immunodiagnostics of tick-borne encephalitis [338], the papilloma [339] and HIV [340] viruses,

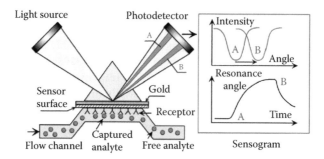

FIGURE 2.9 Typical setup for analyte detection in a BIAcoret-type SPR biosensor. The instrument detects changes in the local refractive index near a thin gold layer coated by a sensor surface with probing molecules. SPR is observed as a minimum in the reflected light intensity at an angle dependent on the mass of captured analyte. The minimum SPR angle shifts from A to B when the analyte binds to the sensor surface. The sensogram is a plot of resonance angle versus time that allows for real-time monitoring of an association/dissociation cycle. (Adapted from Cooper, M.A., *Anal. Bioanal. Chem.*, 373, 834–842, 2003. With permission.)

and Alzheimer's disease [341,342]; the detection of organophosphorus substances and pesticides [343], antibiotics [344], allergens [345], cytokines [346], carbohydrates [347], immunoglobulins [348], and thrombin [349]; the detection of tumor [350] and bacterial [351] cells; the detection of brain cell activity [352], cancer antigens [353,354], and chemical pollutants [355]; and other purposes.

GNP-based biosensors are used not only in immunoassay [356–358] but also for the supersensitive detection of nucleotide sequences [144,286,359,360]. In their pioneering works, Raschke *et al.* [361] and McFarland *et al.* [362] obtained record-high sensitivity of such sensors in the zeptomolar range, and they showed the possibility of detecting spectra of resonance scattering from individual particles. This opened up the way to the recording of intermolecular interactions at the level of individual molecules [314,363]. To make the response stronger, investigators often use avidin–biotin, barnase–barstar, and other biospecific molecular systems [364]. In addition, GNPs are applied in other analytical methods (various versions of chromatography, electrophoresis, and mass spectrometry) [365].

SPR and LSPR biosensors were compared in side-by-side experiments by Yonzon *et al.* [366] for concanavilin A binding to monosaccharides and by Svedendahl *et al.* [367] for biotin–streptavidin binding. It was found that both techniques demonstrate similar performance. As the bulk refractive index sensitivity is known to be higher for SPR, the earlier similarity was attributed to the long decay length of propagating plasmons, as compared to localized ones. The overall comparison of SPR and LSPR sensors can be found in [314,318].

The future development of low-cost SPR and LSPR biomedical sensors needs increasing the detection sensitivity and creating substrates that can operate in biological fluids and can be easy to functionalize with probing molecules, to clean, and to reuse [368,369].

2.3 GNPs in Therapy

2.3.1 Plasmonic Photothermal Therapy

Photothermal damaging of cells is currently one of the promising research avenues in the treatment of both cancer and infectious diseases. The essence of the phenomenon is as follows: GNPs have an absorption maximum in the visible or near-IR region and get very hot when irradiated with corresponding light. If, in this case, they are located inside or around the target cells (which can be achieved by conjugating gold particles to antibodies or other molecules), these cells die [370,371]. Progress in the application of plasmonic AuNPs for laser hyperthermia of tumor tissues is due to the multiplicative effects of increased local absorption of laser radiation by plasmonic GNPs and their targeted delivery [372].

The thermal treatment of cancerous cells has been used in tumor therapy since the eighteenth century, employing both local heating (with microwave, ultrasonic, and radio radiation) and general hyperthermia (heating to 41°C–47°C for 1 h) [373]. For local heating to 70°C, the heating time may be reduced to 3–4 min. Local and general hyperthermia leads to irreversible injury to the cells, which is caused by disruption of cell membrane permeability and protein denaturation. Naturally, the process also injures healthy tissues, which imposes serious limitations on the use of this method.

The revolution in thermal cancer therapy is associated with the use of laser radiation, which enabled controlled and limited injury to tumor tissues to be achieved [374]. Combining laser radiation with fiber-optic waveguides produced excellent results and was named interstitial laser hyperthermia [375]. The weak point of laser therapy is its low selectivity, related to the need for use of high-powered laser for effective stimulation of tumor cell death [376]. A photothermal therapy version was also proposed in which photothermal agents are used to achieve selective heating of the local environment [377]. Selective photothermal therapy is based on the principle of selective photothermolysis of a biological tissue containing a chromophore—a natural or artificial substance with a high coefficient of light emission absorption.

GNPs were first used as agents for photothermal therapy in 2003 [378,379]. Subsequently, it was proposed to call this method plasmonic photothermal therapy (PPTT) [372]. Pitsillides *et al.* [380] first described a new method for selective damaging of target cells that is based on the use of 20- and 30-nm gold nanospheres irradiated with 20-ns laser pulses (532 nm) to create local heating. For pulse photothermia in a model experiment, these authors were the first to employ a sandwich technology for labeling T lymphocytes with GNP conjugates.

A particularly promising method is the use of GNPs for the photothermal therapy of chemotherapy-resistant cancers [381]. As distinct from photosensitizers (see below), the unique properties of GNPs are determined by the long-term preservation of optical properties in cells under certain conditions. Successive irradiation with several laser pulses allows control of cell inactivation by a nontraumatic means. The simultaneous use of the scattering and absorbing properties of GNPs permits PPTT to be controlled by optical tomography [58]. Figure 2.10 shows an example of successful therapy of an implanted tumor in a model experiment with mice [382].

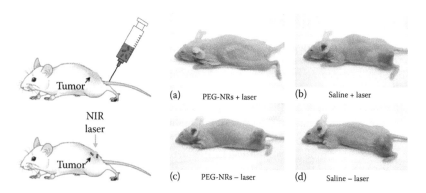

FIGURE 2.10 Scheme and the results of an experiment on the photothermal destruction of an implanted tumor in a mouse (2–3 weeks after injection of human cancer cells MDA-MB-435). Laser irradiation (a, b, 810 nm, 2 W/cm², 5 min) was performed at 72 h after injection of gold nanorods functionalized with PEG (a, c, 20 mg Au/kg) or of buffer (b, d). It can be seen that the tumor continued developing after particle-free irradiation (control b), as it did after did after particle or buffer administration without irradiation (control a and d) and that complete destruction was obtained only in experiment (a). (Adapted from von Maltzahn, G., Park, J.-H., Agrawal, A., Bandaru, N.K., Das, S.K., Sailor, M.J., Bhatia, S.N., *Cancer Res.*, 69, 3892–3900, 2009.)

The further development of PPTT and its acceptance in actual clinical practice [383] depends on success in solving many problems, with the most important ones being (1) the choice of nanoparticles with optimal optical properties, (2) the enhancement of nanoparticle accumulation in tumor and the lowering of total potential toxicity, and (3) the development of methods for the delivery of optical radiation to the targets and the search for alternative radiation sources combining high permeability with a possibility of heating GNPs.

The first requirement is determined by the concordance of the spectral position of the absorption plasmon resonance peak with the spectral window for biological tissues in the 700–900 nm near-IR region [384]. Khlebtsov *et al.* [297] made a resumptive theoretical analysis of the photothermal effectiveness of GNPs, depending on their size, shape, structure, and aggregation extent. They showed that although gold nanospheres themselves are ineffective in the near-IR range, aggregates formed from such particles can be very effective at sufficiently short interparticle distance (shorter than 1/10th the diameter). Such clusters are formed both on the surface of and inside cells [385]. Experimental data indicating an enhancement of PPTT through clusterization have been presented in the works [386–388]. Specifically, Huang *et al.* [386] demonstrated that small aggregates composed of 30-nm particles enable cancerous cells to be destroyed at a radiation power 20-fold lower than that in the particle-free control.

The gold nanoshell and nanorod parameters optimal for PPTT have also been defined [296,289]. By now, there have been quite a few publications dealing with the application in PPTT of gold nanorods [57,390–392], nanoshells [378,393–398], and gold-silver nanocages [90,399,400]. Experimental data comparing the heating efficiencies of nanorods, nanoshells, nanocages, and nanostars have been reported in some works [401–404].

In connection with the problem of particle parameter optimization, one should be aware of three matters of principle. First, self-absorption is not the sole parameter determining the effectiveness of PPTT [405]. Rapid heating of nanoparticles or aggregates gives rise to vapor bubbles [406], which can cause cavitation damage to cells irradiated with visible [386] or near-IR [407] light. The effectiveness of vapor bubble formation increases substantially when nanoparticle aggregates are formed [378,384]. Possibly, it is this effect, and not enhanced absorption, that is responsible for greater damage to cells, with all other factors being the same [404]. Finally, particle irradiation with high-power resonant nanosecond IR pusses can lead to particle destruction as early as after the first pulse (see, *e.g.*, Refs. [408,409] and references therein to earlier publications). In a series of investigations, Lapotko *et al.* ([405,410] and references therein) paid their attention to the fact that the heating of GNPs and their destruction can sharply decrease the photothermal effectiveness of "cold" particles, which are tuned to the laser wavelength. The use of femtosecond pulses provides no solution to the problem because of the low energy supplied, and for this reason, it is necessary to exert close control over the preservation of nanoparticles' properties for the chosen irradiation mode.

We now shift to consider the second question, associated with targeted nanoparticle delivery to a tumor. This question has two important aspects: increasing the particle concentration in the target and decreasing the side effects caused by GNP accumulation in other organs, primarily in the liver and spleen. Usually, two delivery strategies are employed. One is based on the conjugation of GNPs to poly(ethylene glycol) (PEG), and the other, on the conjugation with antibodies developed to specific marker proteins of tumor cells. PEG is used to enhance the bioavailability and stability of nanoparticles, ultimately prolonging the time of their circulation in the blood stream. Citrate-coated gold nanospheres, CTAB-coated nanorods, and nanoshells have low stability in saline solutions. When nanoparticles are conjugates to PEG, their stability is considerably improved and salt aggregation is prevented.

In vivo PEGylated nanoparticles preferentially accumulate in tumor tissue owing to the increased permeability of the tumor vessels [411] and are retained in it owing to the decreased lymph outflow ("the enhanced permeability and retention effect"). In addition, PEGylated nanoparticles are less accessible to the immune system (stealth properties). This method of delivery is called passive, as distinct from the active version, which uses antibodies [412,413] (Figure 2.11) or other targeting moieties. The active method of delivery is more reliable and effective, and it uses antibodies to specific tumor markers, most often to epidermal growth factor receptor (EGFR) and its varieties (*e.g.*, Her2) [392,414,415], tumor necrosis factor (TNF) [416]. Particular promise is offered by the simultaneous use of GNP–antibody conjugates for both diagnosis and PPTT (methods of what is known as theranostics) [417,418]. In addition to antibodies, active delivery may also use folic acid, which serves as a ligand for the numerous folate receptors of tumor cells [390,419,420], and hormones [421].

The effectiveness of targeted nanoparticle delivery to tumors has again become a subject for detailed study and discussion in a recent paper by Huang *et al.* [422]. In experiments with liposomes labeled with anti-Her2 [423] and GNPs labeled with

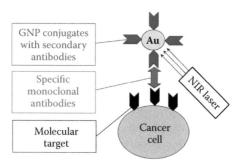

FIGURE 2.11   Scheme for plasmon photothermal therapy employing active delivery of GNPs to cancer cells.

transferrin [424], the functionalization was shown to improve the nanoparticle penetration into cells but produces no appreciable increase in particle accumulation in the tumor. Huang *et al.* [422] examined the biodistribution and localization of gold nanorods labeled with three types of probing molecules, including (1) a scFv to target EGFR, (2) an amino terminal fragment peptide that recognized the urokinase plasminogen activator receptor, and (3) a cyclic RGD peptide that recognizes the $a_v\beta_3$ integrin receptor. The authors showed that when injected intravenously, all three ligands induce insignificant increases in nanoparticle accumulation in cell models and in tumors but greatly affect extracellular distribution and intracellular localization. The authors concluded that for PPTT, the direct administration of particles to the tumor can be more effective than intravenous injection. This conclusion is in line with data by Terentyuk *et al.* [425], who also used direct intratumoral administration for successful PPTT of implanted tumor in rats.

The final important question in current PPTT concerns effective delivery of radiation to a biological target. Because the absorption of biological tissue chromophores in visible range is lower than that in the near-IR range by two orders of magnitude [370], the use of IR radiation radically decreases the nontarget heat load and enhances the penetration of radiation into the tissue interior. Nevertheless, the depth of penetration usually does not exceed 5–10 mm [76,388], so it is necessary to look for alternative solutions. One approach consists in using pulse (nanoseconds) irradiation modes in preference to continuous ones, which enables irradiation power to be enhanced without increasing side effects. Another approach involves the use of fiber-optic devices for endoscopic or intratissue delivery of radiation. The strong and weak points of such an approach are evident. Finally, for hyperthermia, it is possible to use radiations with greater depths of penetration, *e.g.*, radiofrequency [426–430] or nonthermal air plasma [431].

GNPs conjugated to antibiotics and antibodies have also been used as photothermal agents for selective damage to protozoa, bacteria [432–436], and bacterial biofilm [437,438]. Information on certain issues in the use of PPTT can be found in several books and reviews [372,439–450].

In summary, gold nanostructures with a plasmon resonance offer considerable promise for selective PPTT of cancer and other diseases. Without doubt, several questions await further study, including the stability and biocompatibility of nanoparticle bioconjugates, their chemical interaction in physiological environments, the period of circulation in blood, penetration into the tumor, interaction with the immune system, and nanoparticle excretion. We expect that the success of the initial stages of nanoparticle use for selective PPTT can be effectively enhanced at the clinical stage, provided that further studies are made on the optimal procedural parameters. In particular, one can mention the efforts of Feldmann's group [451] related to thermoplasmonics—a field that is not well understood yet and that is nowadays is a trend for hyperthermia and delivery upon light-to-heat conversion.

2.3.2 Photodynamic Therapy with the Use of GNPs

The photodynamic method of treatment of oncological diseases and certain skin or infectious diseases is based on the application of light-sensitizing agents called photosensitizers (including dyes) and, as a rule, of visible light at a specific wavelength [452–454]. Most often, sensitizers are administered intravenously, but they can also be contact type or per oral. The substances used for photodynamic therapy (PDT) can selectively accumulate in tumors or other target tissues (cells). The affected tissues are irradiated with laser light at a wavelength corresponding to the peak of dye absorption. In this case, apart from the usual heat emission through absorption [455], an essential role is played by another mechanism, related to the photochemical generation of singlet oxygen and the formation of highly active radicals, which induce necrosis and apoptosis in tumor cells. PDT also disrupts the nutrition of the tumor and leads to its death through damaging its microvessels. The major shortcoming of PDT is that the photosensitizer remains in the organism for a long time, leaving patient tissues highly sensitive to light. On the other hand, the effectiveness of use of dyes for selective tissue heating [455] is low because of the small cross-section of chromophore absorption.

It is well known [456] that metallic nanoparticles are effective fluorescence quenchers. However, it has been shown recently [457] that fluorescence intensity can be enhanced by a plasmon particle if the molecules are placed at an optimal distance from the metal. In principle, this idea can be used to improve the effectiveness of PDT.

Several investigators have proposed methods for the delivery of drugs as part of polyelectrolyte capsules on GNP, which decompose when acted on by laser radiation and delivery the drug to the targets [458,459], or by using nanoparticles surrounded by a layer of polymeric nanogel [460,461]. Apart from that, the composition of nanoconjugates includes photoactive substances [462,463], peptides (*e.g.*, CALLNN), and proteins (*e.g.*, transferrin), which facilitate intracellular penetration [423,464,465]. Recently, Bardhan *et al.* [466] suggested the use of composite nanoparticles, including, in addition to gold nanoshells, magnetic particles, a photodynamic dye, PEG, and antibodies. Finally, according to the data of Kuo *et al.* [467,468], nanoparticles conjugated to photodynamic dyes can demonstrate a synergetic antimicrobial effect, although the absence of such an effect has also been reported [469]. For more detailed information about the recent progress in PPTT, the readers are referred to an excellent review by Abadeer and Murphy [371].

2.3.3 GNPs as a Therapeutic Agent

In addition to being used in diagnostics and cell photothermolysis, GNPs have been increasingly applied directly for therapeutic purposes [470,471]. In 1997, Abraham and Himmel [472] first reported success in treating rheumatoid arthritis in humans by using CG. In 2008, Abraham published a great body of data from a decade of clinical trials of Aurasol—an oral preparation for the treatment of severe rheumatoid arthritis [473]. Tsai *et al.* [474] described positive results obtained when rats with collagen-induced arthritis were intraarticularly injected with CG. The authors explain the positive effect by on enhancement of antiangiogenic activity resulting from the binding of GNPs to vascular endothelium growth factor and, consequently, a decrease in macrophage infiltration and in inflammation. Similar results were obtained by Brown *et al.* [475,476], who subcutaneously injected rats with collagen- and pristane-induced arthritis with GNPs.

A series of papers by a research team from Maryland University have described the use of a CG vector for the delivery of TNF to solid tumors in rats [477–480]. When injected intravenously, GNPs conjugated to TNF accumulate rapidly in tumor cells and are not detected in cells of the liver, spleen, and other healthy animal organs. The accumulation of GNPs in tumors is proven by a recordable change in tumor color because it takes on a bright red-purple color (characteristic of CG and its aggregates), coincident with the peak of the tumor-specific activity of TNF (Figure 2.12). The CG–TNF vector was less toxic and more effective in tumor reduction than was the native TNF, because the maximal antitumor reaction was attained at lower drug doses. A medicinal preparation based

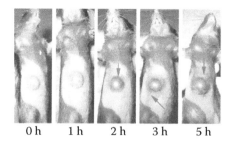

0 h   1 h   2 h   3 h   5 h

FIGURE 2.12 Accumulation of the GNP–TNF conjugate in the tumor after 1–5 h. Diseased mice were intravenously injected with 15 µg of the GNP–TNF vector. The belly images were obtained at the indicated times and show changes in tumor color within 5 h. The red arrow show vector accumulation in the tumor, and the blue arrows mark the accumulation in the tissues around the tumor. (Adapted from Dykman, L.A., Khlebtsov, N.G., *Chem. Soc. Rev.*, 41, 2256–2282, 2012. With permission.)

on the GNP–TNF conjugate, which is called AurImmune and intended for intravenous injection, was checked in clinical trials [481] (see also Ref. [482] for other nanomedicines in preclinical and clinical trials).

In experiments *in vitro* and *in vivo*, Bhattacharya *et al.* [483] and Mukherjee *et al.* [484] have demonstrated the antiangiogenic properties of GNPs, showing that they interact with heparin-binding glycoproteins, including vascular permeability factor/vascular endothelial growth factor and basic fibroblast growth factor. These substances mediate angiogenesis, including that in tumor tissues, and inhibit their activity at the cost of a change in molecular conformation [485]. Because intense angiogenesis (the formation of new vessels in organs or tissues) is a major factor in the pathogenesis of tumor growth, the presence of antiangiogenic properties in GNPs makes them potentially promising in oncotherapy [486–488]. The same research team has shown that GNPs enhance the apoptosis of programmed-death-resistant cells of chronic lymphocytic leukemia [489] and inhibit the proliferation of multiple myeloma cells [490].

In 2011, Wang *et al.* [491] revealed that PEG-coated GNRs have an unusual property: they can induce tumor cell death by accumulating in mitochondria and subsequently damaging them. Unexpectedly, for normal and stem cells, such an effect is either absent or less pronounced.

2.4 GNPs as Drug Carriers

2.4.1 Targeted Delivery of Anticancer Drugs

One of the most promising aspects of GNP use in medicine, currently under intense investigation, is targeted drug delivery [492–501]. The most popular objects for targeted delivery are antitumor preparations [502] and antibiotics [503].

GNPs have been conjugated to a variety of antitumor substances, including paclitaxel [477], methotrexate [504], daunorubicin [505], gemcitabine [506], 6-mercaptopurine [507], dodecylcysteine [508], sulfonamide [509], 5-fluorouracil [510], platinum complexes [511–513], Kahalalide F [514], tamoxifen [515], herceptin [516], β-lapachon [517], doxorubicin [518], prospidin [519], docetaxel [520], chloroquine [521], cetuximab [522], metelimumab [523], indole-3-carbinol [524], and other preparations (Table 2.2).

Conjugation was done both by simple physical adsorption of preparations on GNPs and by using alkanethyol linkers. The action of the conjugates was evaluated both *in vitro* (primarily), with tumor cell cultures, and *in vivo*, with mice bearing implanted tumors of various nature and localization (Lewis lung carcinoma, pancreatic adenocarcinoma, *etc.*).

For the creation of a delivery system, target molecules (*e.g.*, cetuximab) were used along with the active substance so as to ensure better anchoring and penetration of the complex into the target cells [506]. It has also been suggested that multimodal delivery systems be used [525], in which GNPs are loaded with several drugs (both hydrophilic and hydrophobic) and with auxiliary substances (target molecules, PDT dyes, aptamers, *etc.* [526–529]; Figure 2.13). Most researchers have noted the high effectiveness of antitumor preparations conjugated to GNPs [530].

TABLE 2.2 Antitumor Substances Conjugated with GNPs

Drugs	Particles	Methods of Functionalization	Auxiliary Substances	Cell Lines or Animals
Paclitaxel	Gold nanospheres, 26 nm	Paclitaxel-SH	PEG-SH TNF	MC-38; C57/BL6 mice implanted with B16/F10 melanoma cells
Methotrexate	Gold nanospheres, 13 nm	Physical adsorption	–	LL2, ML-1, MBT-2, TSGH 8301, TCC-SUP, J82, PC-3, HeLa
Daunorubicin	Gold nanospheres, 5 nm, 16 nm	Mercaptopropionic acid as a linker	–	K562/ADM
Gemcitabine	Gold nanospheres, 5 nm	Physical adsorption	Cetuximab (monoclonal antibodies)	PANC-1, AsPC-1, MIA Paca2
6-Mercaptopurine	Gold nanospheres, 5 nm	Physical adsorption	–	K-562
Dodecylcysteine	Gold nanospheres, 3–6 nm	Physical adsorption	–	EAC
5-Fluorouracil	Gold nanospheres, 2 nm	Thiol ligand	–	MCF-7
Pt(IV) pro drug	Gold nanospheres, 13 nm	Amide linkages	DNA	HeLa, U2OS, PC3
Cisplatin	Gold nanospheres, 5 nm	PEG-SH as a linker	Folic acid, PEG-SH	OV-167, OVCAR-5, HUVEC, OSE
Oxaliplatin	Gold nanospheres, 30 nm	PEG-SH as a linker	PEG-SH	A549, HCT116, HCT15, HT29, RKO

(Continued)

TABLE 2.2 (CONTINUED) Antitumor Substances Conjugated with GNPs

Drugs	Particles	Methods of Functionalization	Auxiliary Substances	Cell Lines or Animals
Kahalalide F	Gold nanospheres, 20 nm, 40 nm	Physical adsorption	–	HeLa
Tamoxifen	Gold nanospheres, 25 nm	PEG-SH as a linker	PEG-SH	MDA-MB-231, MCF-7, HSC-3
Herceptin	Gold nanorods	Mercaptoundecanoic acid as a linker	–	BT474, SKBR3, MCF-7
β-Lapachon	Gold nanospheres, 25 nm	Physical adsorption	Cyclodextrin as a drug pocket, anti-EGFR, PEG-SH	MCF-7
Doxorubicin	Gold nanospheres, 12 nm	Physical adsorption	Folate-modified PEG	KB
Prospidin	Gold nanospheres, 50 nm	Physical adsorption	–	HeLa
Docetaxel	Gold nanospheres, 2.5 nm	Bromododecanethiol	–	MCF-7, HCT15
Chloroquine	Gold nanospheres, 7 nm	Mercaptoundecanoic acid as a linker	–	MCF-7
Cetuximab	Gold nanospheres, 25 nm	PEG-SH as a linker	p-SCN-Bz-DOTA	A549 tumor xenograft mouse
Metelimumab	Gold nanospheres, 50, 80, 100, 200, 400 nm; gold nanorods; gold nanostars; gold nanocubes	Mercaptopropionic acid as a linker	Methotrexate, folic acid	MDA-MB-231
Indole-3-carbinol	Gold nanospheres, 3 nm	Direct synthesis	–	EAC, DAL

Note: The following designations are used in the table: PEG, poly(ethylene glycol); TNF, tumor necrosis factor; EGFR, epidermal growth factor receptor.

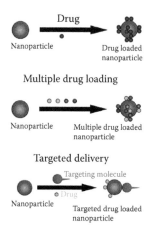

FIGURE 2.13 Schemes for different versions of drug delivery systems. (Adapted from http://mayoresearch.mayo.edu/mayo/research/dev_lab/nanogold.cfm.)

2.4.2 Delivery of other Substances and Genes

Besides antitumor substances, other objects used for the delivery of GNPs are antibiotics and other antibacterial agents. Gu *et al.* [531] demonstrated the possibility of obtaining a stable vancomycin–CG complex and its effectiveness toward various (including vancomycin-resistant) enteropathogenic strains of *E. coli*, *Enterococcus faecium*, and *Enterococcus faecalis*. Similar results were presented by Rosemary *et al.* [532]: a complex formed between ciprofloxacin and gold nanoshells showed high antibacterial activity against *E. coli*. Selvaraj and Alagar [533] reported that a conjugate of the antileukemic drug 5-fluorouracil to CG exhibited noticeable antibacterial and antifungal activities against *Micrococcus luteus*, *Staphylococcus aureus*, *Pseudomonas aeruginosa*, *E. coli*, *Aspergillus fumigatus*, and *Aspergillus niger*. Noteworthy is the fact that in all those cases, the drug–GNP complexes were stable, which could be judged by the optical spectra of the conjugates.

In contrast, Saha *et al.* [534] (antibiotics: ampicillin, streptomycin, and kanamycin; bacteria: *E. coli*, *M. luteus*, and *S. aureus*) and Grace and Pandian [535,536] (amino-glycoside antibiotics: gentamicin, neomycin, and streptomycin; quinolone antibiotics: ciprofloxacin, gatifloxacin, and norfloxacin; bacteria: *E. coli*, *M. luteus*, *S. aureus*, and *P. aeruginosa*) failed to prepare stable complexes with GNPs. Nevertheless, those authors showed that depending on the antibiotic used, the increase in the activity of an antibiotic–CG mixture, as compared to that of the nature drug, ranged from 12% to 40%. From these data, it was concluded that the antibacterial activity of the antibiotics is enhanced at the cost of GNPs. However, the question as to the mechanisms responsible for such possible enhancement remained unclarified, which was noted by the authors themselves. Burygin *et al.* [539] experimentally proved that free gentamicin and its mixture with GNPs do not significantly differ in antimicrobial activity in assays on solid and in liquid nutrient media. They proposed that a necessary condition for enhancement of antibacterial activity is the preparation of stable conjugates of nanoparticles coated with antibiotic molecules. Specifically, Rai *et al.* [538] suggested the use of the antibiotic cefaclor directly in the synthesis of GNPs. As a result, they prepared a stable conjugate that had a high antibacterial activity against *E. coli* and *S. aureus*. GNPs have been also conjugated to a variety of antibiotics, including aminoglycosides [539–541], aminopenicillins [542,543], ansamycins [544], cephalosporins [545], and other groups.

Other drugs conjugated to GNPs are referred to much more rarely in the literature. However, some of those works deserve mention. Nie *et al.* [546] demonstrated a high antioxidant activity of GNPs complexed with tocoferol and suggested potential applications of the complex. Bowman *et al.* [547] provided data to show that a conjugate of GNPs with the preparation TAK-779 exhibited more pronounced activity against HIV than did a nature preparation at the cost of the high local concentration. Joshi *et al.* [548] and Cho *et al.* [549] described a procedure for oral and intranasal administration of colloidal-gold-conjugated insulin to diabetic rats, showing a significant decrease in blood sugar comparable with that obtained by subcutaneous insulin injection. Ehsan *et al.* [550] and Shilo *et al.* [551] reported the binding of insulin to GNPs and its application or efficiency in subcutaneous delivery for the therapeutic treatment of diabetes mellitus. Finally, Chamberland *et al.* [552] reported a therapeutic effect of the antirheumatism drug Etanercept conjugated to gold nanorods. Similar results were obtained by Huang *et al.* [553], who subcutaneously injected rats with galectin-1–nanogold complex, and Gomes *et al.* [554] with methotrexate conjugated to GNPs. GNPs have been also conjugated to a variety of other drugs, *e.g.*, silymarin [555], octreotide [556], therapeutic peptides [557], *etc.*

In conclusion, it is necessary to mention gene therapy, which can be seen as an ideal strategy for the treatment of genetic and acquired diseases [558]. The term *gene therapy* is used in reference to a medical approach based on the administration, for therapeutic purposes, of gene constructs to cells and the organism [559]. The desired effect is achieved either as a result of expression of the introduced gene or through partial or complete suppression of the function of a damaged or overexpressing gene. There have also been recent attempts at correcting the structure and function of an improperly functioning ("bad") gene. In such a case, too, GNPs can serve as an effective means of delivery of genetic material to the cytoplasm and the cell nucleus [560–564].

Small interfering RNA (siRNA) is an effective method for regulating the expression of proteins, even "undruggable" ones that are nearly impossible to target through traditional small molecule therapeutics. Delivery to the cell and then to the cytosol is the primary requirement for realization of therapeutic potential of siRNA. Effective delivery of siRNA at the organismic and cellular levels, coupled with the low immunogenic response of GNPs, will facilitate the clinical translation of siRNA treatments for genetic diseases [565,566].

2.5 Concluding Remarks

Owing to the success of the rapid development of technologies for the chemical synthesis of GNPs during the past decade, investigators currently have at their disposal an enormous diversity of available particles with the required parameters in respect of size, shape, structure, and optical properties. Moreover, the question is now on having the desired properties, which is followed by the development of a procedure for the synthesis of a theoretical nanostructure.

From the standpoint of medical application, much significance was held by the development of effective technologies for the functionalization of GNPs with molecules belonging to various classes, which ensure nanoparticle stabilization *in vivo* and targeted interaction with biological targets. At this juncture, the best stabilizers are thioloated derivatives of PEG and other molecules. Specifically, PEG-coated particles can circulate in the blood stream for longer times and are less susceptible to the attack of the cellular components of the immune system. However, the creation of conjugates "stealth" to the immune system, which could bind to biological targets in an effective and target-oriented way, remains an unresolved problem.

It is now generally recognized that GNP conjugates are excellent labels for bioimaging, which can be realized by various biophotonics technologies, including dark-field resonance scattering microscopy, confocal laser microscopy, various version of two-photon scattering and GNPs' own luminescence, optical coherence and acoustic tomography, and so on.

GNP conjugates have found numerous applications in analytical research based on both state-of-the-art instrumental methods (SERS, surface enhanced infrared absorption spectroscopy, laser-induced scattering around a NanoAbsorber, *etc.*) and simple solid-phase or homophase techniques (dot assay, immunochromatography, *etc.*). The following two examples are illustrative: (1) by using GNP–antibody conjugates, it is possible to detect a prostate-specific antigen with a sensitivity that is millionfold greater than that in the ELISA [567]. (2) The sharp dependence of the color of the system on interparticle distances enables mutant DNAs to be detected visually in a test known as the Northwestern spot test [146]. Along with the literature examples of clinical diagnostics of cancer, Alzheimer's disease, AIDS, hepatitis, tuberculosis, diabetes, and other diseases, new diagnostic application of GNPs should be expected. Progress in this direction will be determined by success achieved in increasing the sensitivity of analytical tests with retention of the simplicity of detection. The limitations of homophasic methods with visual detection are due to the need to use a large number (on the order of 10^{10} [454]) of nanoparticles. Even at the minimal ratio between target molecules and particles (1:1), the detection limit will be on the order of 0.01 pM, which is considerably (millionfold) higher than the quantity of the target molecules that needs to be detected, *e.g.*, in typical biopsy samples [454]. Thus, sensitivity can be improved either by enhancing the signal (polymerase chain reaction, autometallography, *etc.*) or by using sensitive instrumental methods. For instance, single-particle instrumental methods [362] have a single-molecule detection limit that is attainable in principle. Specifically, the SERS is a trend technique used to detect very low concentrations of solutes (see, *e.g.*, the recent review and reports by Liz-Marzán and coworkers [568–571] regarding SERS detection of biological molecules). However, the topical problem is to create multiplex sensitive tests that do not require equipment and can be performed by the end user under nonlaboratory conditions. An example of the prototype of such devices is Pro Strips, which can simultaneously detect five threads: anthrax, ricin toxin, botulinum toxin, *Yersinia pestis* (plague), and staphylococcal enterotoxin B. The physical basis of the new tests may by associated with the dependence of the plasmon resonance wavelength on the local dielectric environment or on the interparticle distance.

GNP-assisted PPTT of cancer, first described in 2003, is now in the stage of clinical trials [370]. The actual clinical success of this technology will depend on how quickly it will be possible to solve several topical problems: (1) development of effective methods for the delivery of radiation to tumors inside the organism by using fiber-optic technologies or nonoptical heating methods, (2) improvement of the methods of conjugate delivery to tumors and enhancement of the contrast and accumulation uniformity, and (3) development of methods for controlling the process of photothermolysis *in situ*.

GNP-aided targeted delivery of DNA, antigens, and drugs is one of the most promising directions in biomedicine. Specifically, research conducted by Warren Chan's teams at Toronto University [572] has shown the size-dependent possibility of delivery of GNPs complexed with herceptin to cancer cells with much greater effectiveness than that obtained with a pure preparation. The recent critical reexamination of the concept of PPTT, based on the intravenous targeted delivery of GNPs conjugated with molecular probes to the receptor of cancer [421], indicates that there is a pressing need to continue research in this direction.

References

1. Hermanson G.T. *Bioconjugate Techniques*. San Diego: Academic Press. 1996.

2. Thiele H., Hoppe K., Moll G. Über das kolloide Gold // *Kolloid-Z. Z. Polymere.* 1962. Bd. 185. S. 45–52.

3. Chow M.K., Zukoski C.F. Gold sols formation mechanisms: Role of colloidal stability // *J. Colloid Interface Sci.* 1994. V. 165. P. 97–109.

4. Zhou J., Ralston J., Sedev R., Beattie D.A. Functionalized gold nanoparticles: Synthesis, structure and colloid stability // *J. Colloid Interface Sci.* 2008. V. 331. P. 251–262.

5. Krpetić Ž., Nativo P., Porta F., Brust M. A multidentate peptide for stabilization and facile bioconjugation of gold nanoparticles // *Bioconjugate Chem.* 2009. V. 20. P. 619–624.

6. Sapsford K.E., Algar W.R., Berti L., Gemmill K.B., Casey B.J., Oh E., Stewart M.H., Medintz I.L. Functionalizing nanoparticles with biological molecules: Developing chemistries that facilitate nanotechnology // *Chem. Rev.* 2013. V. 113. P. 1904–2074.

7. Festag G., Klenz U., Henkel T., Csáki A., Fritzsche W. Biofunctionalization of metallic nanoparticles and microarrays for biomolecular detection // In: *Biofunctionalization of Nanomaterials* / Ed. Kumar C.S.S.R. Weinheim: Wiley. 2005. P. 150–182.

8. Shenton W., Davis S.A., Mann S. Directed self-assembly of nanoparticles into macroscopic materials using antibody-antigen recognition // *Adv. Mater.* 1999. V. 11. P. 449–452.

9. Dubois L.H., Nuzzo R.G. Synthesis, structure, and properties of model organic surfaces // *Ann. Rev. Phys. Chem.* 1992. V. 43. P. 437–467.

10. Lowe C.R. Nanobiotechnology: The fabrication and applications of chemical and biological nanostructures // *Curr. Opin. Struct. Biol.* 2000. V. 10. P. 428–434.

11. Ulman A. Formation and structure of self-assembled monolayers // *Chem. Rev.* 1996. V. 96. P. 1533–1554.

12. Mirkin C.A., Letsinger R.L., Mucic R.C., Storhoff J.J. A DNA-based method for rationally assembling nanoparticles into macroscopic materials // *Nature.* 1996. V. 382. P. 607–609.

13. Letsinger R.L., Elghanian R., Viswanadham G., Mirkin C.A. Use of a steroid cyclic disulfide anchor in constructing gold nanoparticle-oligonucleotide conjugates // *Bioconjug. Chem.* 2000. V. 11. P. 289–291.

14. Li Z., Jin R.S., Mirkin C.A., Letsinger R.L. Multiple thiol-anchor capped DNA-gold nanoparticle conjugates // *Nucleic Acids Res.* 2002. V. 30. P. 1558–1562.

15. Walton I.D., Norton S.M., Balasingham A., He L., Oviso D.F., Gupta D., Raju P.A., Natan M.J., Freeman R.G. Particles for multiplexed analysis in solution: Detection and identification of striped metallic particles using optical microscopy // *Anal. Chem.* 2002. V. 74. P. 2240–2247.

16. Ghosh S.S., Kao P.M., McCue A.W., Chappelle H.L. Use of maleimide-thiol coupling chemistry for efficient syntheses of oligonucleotide-enzyme conjugate hybridization probes // *Bioconjug. Chem.* 1990. V. 1. P. 71–76.

17. Cheng Y., Samia A.C., Li J., Kenney M.E., Resnick A., Burda C. Delivery and efficacy of a cancer drug as a function of the bond to the gold nanoparticle surface // *Langmuir.* 2010. V. 26. P. 2248–2255.

18. Sperling R.A., Parak W.J. Surface modification, functionalization and bioconjugation of colloidal inorganic nanoparticles // *Phil. Trans. R. Soc. A.* 2010. V. 368. P. 1333–1383.

19. Mout R., Moyano D.F., Rana S., Rotello V.M. Surface functionalization of nanoparticles for nanomedicine // *Chem. Soc. Rev.* 2012. V. 41. P. 2539–2544.

20. Biju V. Chemical modifications and bioconjugate reactions of nanomaterials for sensing, imaging, drug delivery and therapy // *Chem. Soc. Rev.* 2014. V. 43. P. 744–764.

21. Nicol J.R, Dixon D., Coulter J.A. Gold nanoparticle surface functionalization: A necessary requirement in the development of novel nanotherapeutics // *Nanomedicine (Lond.).* 2015. V. 10. P. 1315–1326.

22. Oh J.-H., Park D.H., Joo J.H., Lee J.-S. Recent advances in chemical functionalization of nanoparticles with biomolecules for analytical applications // *Anal. Bioanal. Chem.* 2015. V. 407. P. 8627–8645.

23. Zhou W., Gao X., Liu D., Chen X. Gold nanoparticles for *in vitro* diagnostics // *Chem. Rev.* 2015. V. 115. P. 10575–10636.

24. Jazayeri M.H., Amani H., Pourfatollah A.A., Pazoki-Toroudi H., Moghadam B.S. Various methods of gold nanoparticles (GNPs) conjugation to antibodies // *Sens. Bio-Sensing Res.* 2016. V. 9. P. 17–22.

25. Agasti S.S., Rana S., Park M.H., Kim C.K., You C.C., Rotello V.M. Nanoparticles for detection and diagnosis // *Adv. Drug Deliv. Rev.* 2010. V. 62. P. 316–328.

26. Hahn M.A., Singh A.K., Sharma P., Brown S.C., Moudgil B.M. Nanoparticles as contrast agents for *in vivo* bioimaging: Current status and future perspectives // *Anal. Bioanal. Chem.* 2011. V. 399. P. 3–27.

27. Amiry-Moghaddam M., Ottersen O.P. Immunogold cytochemistry in neuroscience // *Nat. Neurosci.* 2013. V. 16. P. 798–804.

28. Dykman L.A., Khlebtsov N.G. Gold nanoparticles in biomedical applications: Recent advances and perspectives // *Chem. Soc. Rev.* 2012. V. 41. P. 2256–2282.

29. Faulk W., Taylor G. An immunocolloid method for the electron microscope // *Immunochemistry.* 1971. V. 8. P. 1081–1083.

30. Horisberger M., Rosset J., Bauer H. Colloidal gold granules as markers for cell surface receptors in the scanning electron microscopy // *Experimentia.* 1975. V. 31. P. 1147–1151.

31. *Colloidal Gold: Principles, Methods, and Applications* / Ed. Hayat M.A. San Diego: Acad. Press. 1989.

32. Jürgens L., Nichtl A., Werner U. Electron density imaging of protein films on gold-particle surfaces with transmission electron microscopy // *Cytometry.* 1999. V. 37. P. 87–92.

33. Schröfel A., Cmarko D., Bártová E., Raška I. Gold nanoparticles for high resolution imaging in modern immunocytochemistry // In: *Intracellular Delivery II* / Eds. Prokop A., Iwasaki Y., Harada A. Dordrecht: Springer. 2014. P. 189–206.

34. Ho K.-C., Tsai P.-J., Lin Y.-S., Chen Y.-C. Using biofunctionalized nanoparticles to probe pathogenic bacteria // *Anal. Chem.* 2004. V. 76. P. 7162–7168.

35. Bunin V.D., Ignatov O.V., Gulii O.I., Voloshin A.G., Dykman L.A., O'Neil D., Ivnitskii D. Investigation of electrophysical properties of Listeria monocytogenes cells during the interaction with monoclonal antibodies // *Biophysics.* 2005. V. 50. P. 299–302.

36. Apicella M.A., Post D.M.B., Fowler A.C., Jones B.D., Rasmussen J.A., Hunt J.R., Imagawa S., Choudhury B., Inzana T.J., Maier T.M., Frank D.W., Zahrt T.C., Chaloner K., Jennings M.P., McLendon M.K., Gibson B.W. Identification, characterization and immunogenicity of an O-antigen capsular polysaccharide of *Francisella tularensis* // *PLoS One.* 2010. V. 5. e11060.

37. Drygin Yu.F., Blintsov A.N., Osipov A.P., Grigorenko V.G., Andreeva I.P., Uskov A.I., Varitsev Yu.A., Anisimov B.V., Novikov V.K., Atabekov J.G. High-sensitivity express immunochromatographic method for detection of plant infection by tobacco mosaic virus // *Biochemistry (Moscow).* 2009. V. 74. P. 986–983.

38. Naja G., Hrapovic S., Male K., Bouvrette P., Luong J.H.T. Rapid detection of microorganisms with nanoparticles and electron microscopy // *Microsc. Res. Tech.* 2008. V. 71. P. 742–748.

39. Phillips R.L., Miranda O.R., You C.C., Rotello V.M., Bunz U.H. Rapid and efficient identification of bacteria using gold-nanoparticle-poly(*para*-phenyleneethynylene) constructs // *Angew. Chem. Int. Ed.* 2008. V. 47. P. 2590–2594.

40. Khlebtsov N.G., Bogatyrev V.A., Dykman L.A., Khlebtsov B.N., Staroverov S.A., Shirokov A.A., Matora L.Y., Khanadeev V.A., Pylaev T.E., Tsyganova N.A., Terentyuk G.S. Analytical and theranostic applications of gold nanoparticles and multifunctional nanocomposites // *Theranostics.* 2013. V. 3. P. 167–180.

41. Wang G., Stender A.S., Sun W., Fang N. Optical imaging of non-fluorescent nanoparticle probes in live cells // *Analyst.* 2010. V. 135. P. 215–221.

42. Sokolov K., Follen M., Aaron J., Pavlova I., Malpica A., Lotan R., Richards-Kortum R. Real-time vital optical imaging of precancer using anti-epidermal growth factor receptor antibodies conjugated to gold nanoparticles // *Cancer Res.* 2003. V. 63. P. 1999–2004.

43. Kho K.W., Kah J.C.Y., Lee C.G.L., Sheppard C.J.R., Shen Z.X., Soo K.C., Olivo M.C. Applications of gold nanoparticles in the early detection of oral cancer // *J. Mech. Med. Biol.* 2007. V. 7. P. 1–17.

44. Tsai S.-W., Chen Y.-Y., Liaw J.-W. Compound cellular imaging of laser scanning confocal microscopy by using gold nanoparticles and dyes // *Sensors.* 2008. V. 8. P. 2306–2316.

45. Klein S., Petersen S., Taylor U., Rath D., Barcikowski S. Quantitative visualization of colloidal and intracellular gold nanoparticles by confocal microscopy // *J. Biomed. Opt.* 2010. V. 15. 036015.

46. Ng V.W.K., Berti R., Lesage F., Kakkar A. Gold: A versatile tool for in vivo imaging // *J. Mater. Chem. B.* 2013. V. 1. P. 9–25.

47. Leduc C., Si S., Gautier J.J., Gao Z., Shibu E.S., Gautreau A., Giannone G., Cognet L., Lounis B. Single-molecule imaging in live cell using gold nanoparticles // *Methods Cell Biol.* 2015. V. 125. P. 13–27.

48. Wang H., Huff T.B., Zweifel D.A., He W., Low P.S., Wei A., Cheng J.-X. *In vitro* and *in vivo* two-photon luminescence imaging of single gold nanorods // *Proc. Natl. Acad. Sci. USA.* 2005. V. 102. P. 15752–15756.

49. Durr N.J., Larson T., Smith D.K., Korgel B.A., Sokolov K., Ben-Yakar A. Two-photon luminescence imaging of cancer cells using molecularly targeted gold nanorods // *Nano Lett.* 2007. V. 7. P. 941–945.

50. Park J., Estrada A., Sharp K., Sang K., Schwartz J.A., Smith D.K., Coleman C., Payne J.D., Korgel B.A., Dunn A.K., Tunnell J.W. Two-photon-induced photoluminescence imaging of tumors using near-infrared excited gold nanoshells // *Opt. Express.* 2008. V. 16. P. 1590–1599.

51. Maiorano G., Sabella S., Sorce B., Brunetti V., Malvindi M.A., Cingolani R., Pompa P.P. Effects of cell culture media on the dynamic formation of protein–nanoparticle complexes and influence on the cellular response // *ACS Nano.* 2010. V. 4. P. 7481–7491.

52. He W., Henne W.A., Wei Q., Zhao Y., Doorneweerd D.D., Cheng J.X., Low P.S., Wei A. Two-photon luminescence imaging of *Bacillus* spores using peptide-functionalized gold nanorods // *Nano Res.* 2008. V. 1. P. 450–456.

53. Liu M., Chao J., Deng S., Wang K., Li K., Fan C. Dark-field microscopy in imaging of plasmon resonant nanoparticles // *Colloids Surf. B.* 2014. V. 124. P. 111–117.

54. Bogatyrev V.A., Dykman L.A., Alekseeva A.V., Khlebtsov B.N., Novikova A.P., Khlebtsov N.G. Observation of time-dependent single-particle light scattering from gold nanorods and nanospheres by using unpolarized dark-field microscopy // *Proc. SPIE.* 2006. V. 1664. P. 1–10.

55. Hu M., Novo C., Funston A., Wang H., Staleva H., Zou S., Mulvaney P., Xia Y., Hartland G.V. Dark-field microscopy studies of single metal nanoparticles: Understanding the factors that influence the linewidth of the localized surface plasmon resonance // *J. Mater. Chem.* 2008. V. 18. P. 1949–1960.

56. El-Sayed I.H., Huang X.H., El-Sayed M.A. Surface plasmon resonance scattering and absorption of anti-EGFR antibody conjugated gold nanoparticles in cancer diagnostics: Applications in oral cancer // *Nano Lett.* 2005. V. 5. P. 829–834.

57. Huang X., El-Sayed I.H., Qian W., El-Sayed M.A. Cancer cell imaging and photothermal therapy in the near-infrared region by using gold nanorods // *J. Am. Chem. Soc.* 2006. V. 128. P. 2115–2120.

58. Loo C., Hirsch L., Lee M., Chang E., West J., Halas N., Drezek R. Gold nanoshell bioconjugates for molecular imaging in living cells // *Opt. Lett.* 2005. V. 30. P. 1012–1014.

59. Aaron J., de la Rosa E., Travis K., Harrison N., Burt J., José-Yakamán M., Sokolov K. Polarization microscopy with stellated gold nanoparticles for robust monitoring of molecular assemblies and single biomolecules // *Opt. Express.* 2008. V. 16. P. 2153–2167.

60. Au L., Zhang Q., Cobley C.M., Gidding M., Schwartz A.G., Chen J., Xia Y. Quantifying the cellular uptake of antibody-conjugated Au nanocages by two-photon microscopy and inductively coupled plasma mass spectrometry // *ACS Nano.* 2010. V. 4. P. 35–42.

61. Shang L., Nienhaus G.U. Gold nanoclusters as novel optical probes for *in vitro* and *in vivo* fluorescence imaging // *Biophys. Rev.* 2012. V. 4. P. 313–322.

62. Khlebtsov N.G., Trachuk L.A., Mel'nikov A.G. The effect of the size, shape, and structure of metal nanoparticles on the dependence of their optical properties on the refractive index of a disperse medium // *Opt. Spectrosc.* 2005. V. 98. P. 77–83.

63. Mohamed M.B., Volkov V., Link S., El-Sayed M.A. The "lightning" gold nanorods: Fluorescence enhancement of over a million compared to the gold metal // *Chem. Phys. Lett.* 2000. V. 317. P. 517–523.

64. Schultz D.A. Plasmon resonant particles for biological detection // *Curr. Opin. Biotechnol.* 2003. V. 14. P. 13–22.

65. Sönnichsen C., Alivisatos A.P. Gold nanorods as novel nonbleaching plasmon-based orientation sensors for polarized single-particle microscopy // *Nano Lett.* 2005. V. 5. P. 301–304.

66. York J., Spetzler D., Hornung T., Ishmukhametov R., Martin J., Frasch W.D. Abundance of *Escherichia coli* F1-ATPase molecules observed to rotate via single-molecule microscopy with gold nanorod probes // *J. Bioenerg. Biomembr.* 2007. V. 39. P. 435–439.

67. He H., Xie C., Ren J. Nonbleaching fluorescence of gold nanoparticles and its applications in cancer cell imaging // *Anal. Chem.* 2008. V. 80. P. 5951–5957.

68. Huang Y.-F., Lin Y.-W., Lin Z.-H., Chang H.-T. Aptamer-modified gold nanoparticles for targeting breast cancer cells through light scattering // *J. Nanopart. Res.* 2009. V. 11. P. 775–783.

69. Hu R., Yong K.T., Roy I., Ding H., He S., Prasad P.N. Metallic nanostructures as localized plasmon resonance enhanced scattering probes for multiplex dark field targeted imaging of cancer cells // *J. Phys. Chem. C.* 2009. V. 113. P. 2676–2684.

70. Curry A.C., Crow M., Wax A. Molecular imaging of epidermal growth factor receptor in live cells with refractive index sensitivity using dark-field microspectroscopy and immunotargeted nanoparticles // *J. Biomed. Opt.* 2008. V. 13. 014022.

71. Bickford L., Chang J., Fu K., Sun J., Hu Y., Gobin A., Yu T.K., Drezek R. Evaluation of immunotargeted gold nanoshells as rapid diagnostic imaging agents for HER2-overexpressing breast cancer cells: A time-based analysis // *Nanobiotechnology*. 2008. V. 4. P. 1–8.

72. Wang S.-H., Lee C.-W., Chiou A., Wei P.-K. Size-dependent endocytosis of gold nanoparticles studied by three-dimensional mapping of plasmonic scattering images // *J. Nanobiotechnol.* 2010. V. 8. 33.

73. Khanadeev V.A., Khlebtsov B.N., Staroverov S.A., Vidyasheva I.V., Skaptsov A.A., Ilineva E.S., Bogatyrev V.A., Dykman L.A., Khlebtsov N.G. Quantitative cell bio-imaging using gold nanoshell conjugates and phage antibodies // *J. Biophotonics.* 2011. V. 4. P. 74–83.

74. Fu K., Sun J., Bickford L.R., Lin A.W.H., Halas N.J., Yu T.-K., Drezek R.A. Measurement of immunotargeted plasmonic nanoparticles' cellular binding: A key factor in optimizing diagnostic efficacy // *Nanotechnology.* 2008. V. 19. 045103.

75. Khlebtsov N.G. Optics and biophotonics of nanoparticles with a plasmon resonance // *Quantum Electron.* 2008. V. 38. P. 504–529.

76. *Handbook of Photonics for Biomedical Science* / Ed. Tuchin V.V. Boca Raton: CRC Press. 2010.

77. Gobin A.M., Lee M.H., Halas N.J., James W.D., Drezek R.A., West J.L. Near-infrared resonant nanoshells for combined optical imaging and photothermal cancer therapy // *Nano Lett.* 2007. V. 7. P. 1929–1934.

78. Adler D.C., Huang S.-W., Huber R., Fujimoto J.G. Photothermal detection of gold nanoparticles using phase-sensitive optical coherence tomography // *Opt. Express.* 2008. V. 16. P. 4376–4393.

79. Zagaynova E.V., Shirmanova M.V., Kirillin M.Y., Khlebtsov B.N., Orlova A.G., Balalaeva I.V., Sirotkina M.A., Bugrova M.L., Agrba P.D., Kamensky V.A. Contrasting properties of gold nanoparticles for optical coherence tomography: Phantom, *in vivo* studies and Monte Carlo simulation // *Phys. Med. Biol.* 2008. V. 53.P. 4995–5009.

80. Kim D., Park S., Lee J.H., Jeong Y.Y., Jon S. Antibiofouling polymer-coated gold nanoparticles as a contrast agent for *in vivo* x-ray computed tomography imaging // *J. Am. Chem. Soc.* 2007. V. 129. P. 7661–7665.

81. Patra C.R., Jing Y., Xu Y.-H., Bhattacharya R., Mukhopadhyay D., Glockner J.F., Wang J.-P., Mukherjee P. A core-shell nanomaterial with endogenous therapeutic and diagnostic functions // *Cancer Nano.* 2010. V. 1. P. 13–18.

82. Xi D., Dong S., Meng X., Lu Q., Meng L., Ye J. Gold nanoparticles as computerized tomography (CT) contrast agents // *RSC Adv.* 2012. V. 2. P. 12515–12524.

83. Ahn S., Jung S.Y., Lee S.J. Gold nanoparticle contrast agents in advanced x-ray imaging technologies // *Molecules.* 2013. V. 18. P. 5858–5890.

84. Cole L.E., Ross R.D., Tilley J.M.R., Vargo-Gogola T., Roeder R.K. Gold nanoparticles as contrast agents in x-ray imaging and computed tomography // *Nanomedicine (Lond.).* 2015. V. 10. P. 321–341.

85. Zharov V., Galanzha E., Shashkov E., Khlebtsov N., Tuchin V. *In vivo* photoacoustic flow cytometry for monitoring circulating cells and contrast agents // *Opt. Lett.* 2006. V. 31. P. 3623–3625.

86. Mallidi S., Larson T., Aaron J., Sokolov K., Emelianov S. Molecular specific optoacoustic imaging with plasmonic nanoparticles // *Opt. Express*. 2007. V. 15. P. 6583–6588.

87. Li C., Wang L.V. Photoacoustic tomography and sensing in biomedicine // *Phys. Med. Biol.* 2009. V. 54. P. R59–R97.

88. Chen J., Irudayaraj J. Quantitative investigation of compartmentalized dynamics of ErbB2 targeting gold nanorods in live cells by single molecule spectroscopy // *ACS Nano.* 2009. V. 3. P. 4071–4079.

89. Henry A.-I., Sharma B., Cardinal M.F., Kurouski D., Van Duyne R.P. Surface-enhanced Raman spectroscopy biosensing: *In vivo* diagnostics and multimodal imaging // *Anal. Chem.* 2016. V. 88. P. 6638–6647.

90. Chen J., Claus C., Laforest R., Zhang Q., Yang M., Gidding M., Welch M., Xia Y. Gold nanocages as photothermal transducers for cancer treatment // *Small.* 2010. V. 7. P. 811–817.

91. Hutter E., Boridy S., Labrecque S., Lalancette-Hébert M., Kriz J., Winnik F.M., Maysinger D. Microglial response to gold nanoparticles // *ACS Nano.* 2010. V. 4. P. 2595–2606.

92. Leuvering J.H.W., Thal P.J.H.M., van der Waart M., Schuurs A.H.W.M. Sol particle immunoassay (SPIA) // *J. Immunoassay.* 1980. V. 1. P. 77–91.

93. Khlebtsov N.G. Optical models for conjugates of gold and silver nanoparticles with biomacromolecules // *J. Quant. Spectrosc. Radiat. Transfer.* 2004. V. 89. P. 143–152.

94. Khlebtsov N.G., Melnikov A.G., Dykman L.A., Bogatyrev V.A. Optical properties and biomedical applications of nanostructures based on gold and silver bioconjugates // In: *Photopolarimetry in Remote Sensing* / Eds. Videen G., Yatskiv Ya.S., Mishchenko M.I. Dordrecht: Kluwer Acad. Publ. 2004. P. 265–308.

95. Wu S.H., Wu Y.S., Chen C.H. Colorimetric sensitivity of gold nanoparticles: Minimizing interparticular repulsion as a general approach // *Anal. Chem.* 2008. V. 80. P. 6560–6566.

96. Tsai C.S., Yu T.B., Chen C.T. Gold nanoparticle-based competitive colorimetric assay for detection of protein-protein interactions // *Chem. Commun. (Camb).* 2005. V. 34. P. 4273–4275.

97. Zhang Y., McKelvie I.D., Cattrall R.W., Kolev S.D. Colorimetric detection based on localised surface plasmon resonance of gold nanoparticles: Merits, inherent shortcomings and future prospects // *Talanta.* 2016. V. 152. P. 410–422.

98. Leuvering J.H.W., Coverde B.C., Thal P.J.H.M., Schuurs A.H.W.M. A homogeneous sol particle immunoassay for human chorionic gonadotrophin using monoclonal antibodies // *J. Immunol. Methods.* 1983. V. 60. P. 9–23.

99. Deelder A.M., Dozy M.H. Applicability of sol particle immunoassay (SPIA) for detection of *Schistosoma mansoni* circulating antigens // *Acta Leiden.* 1982. V. 48. P. 17–22.

100. Wielaard F., Denissen A., van der Veen L., Rutjes I. A sol-particle immunoassay for determination of anti-rubella antibodies. Development and clinical validation // *J. Virol. Meth.* 1987. V. 17. P. 149–158.

101. Zeisler R., Stone S.F., Viscidi R.P., Cerny E.H. Sol particle immunoassays using colloidal gold and neutron activation // *J. Radioanal. Nucl. Chem.* 1993. V. 167. P. 445–452.

102. Gasparyan V.K. Hen egg immunoglobulin Y in colloidal gold agglutination assay: Comparison with rabbit immunoglobulin G // *J. Clin. Lab. Anal.* 2005. V. 19. P. 124–127.

103. Pavlov V., Xiao Y., Shlyahovsky B., Willner I. Aptamer-functionalized Au nanoparticles for the amplified optical detection of thrombin // *J. Am. Chem. Soc.* 2004. V. 126. P. 11768–11769.

104. Aslan K., Lakowicz J.R., Geddes C.D. Nanogold-plasmon-resonance-based glucose sensing // *Anal. Biochem.* 2004. V. 330. P. 145–155.

105. Liao Y.-J., Shiang Y.-C., Chen L.-Y., Hsu C.-L., Huang C.-C., Chang H.-T. Detection of adenosine triphosphate through polymerization-induced aggregation of actin-conjugated gold/silver nanorods // *Nanotechnology.* 2013. V. 24. 444003.

106. Zhu J., Yu Z., Li J.-j., Zhao J.-w. Coagulation induced attenuation of plasmonic absorption of Au nanorods: Application in ultra sensitive detection of alpha-fetoprotein // *Sens. Actuators B.* 2013. V. 188. P. 318–325.

107. Byun J.Y., Shin Y.B., Li T., Park J.H., Kim D.M., Choi D.H., Kim M.G. The use of an engineered single chain variable fragment in a localized surface plasmon resonance method for analysis of the C-reactive protein // *Chem. Commun.* 2013. V. 49. P. 9497–9499.

108. Truong P.L., Choi S.P., Sim S.J. Amplification of resonant Rayleigh light scattering response using immunogold colloids for detection of lysozyme // *Small.* 2013. V. 9. P. 3485–3492.

109. Inoue K., Aomatsu T., Yoden A., Okuhira T., Kaji E., Tamai H. Usefulness of a novel and rapid assay system for fecal calprotectin in pediatric patients with inflammatory bowel diseases // *J. Gastroenterol. Hepatol.* 2014, V. 29. P. 1406–1412.

110. Nekouian R., Khalife N.J., Salehi Z. Anti human fibronectin–gold nanoparticle complex, a potential nanobiosensor tool for detection of fibronectin in ECM of cultured cells // *Plasmonics.* 2014. V. 9. P. 1417–1423.

111. Tanaka M., Saito Y., Misu H., Kato S., Kita Y., Takeshita Y., Kanamori T., Nagano T., Nakagen M., Urabe T., Takamura T., Kaneko S., Takahashi K., Matsuyama N. Development of a sol particle homogeneous immunoassay for measuring full-length selenoprotein P in human serum // *J. Clin. Lab. Anal.* 2016. V. 30. P. 114–122.

112. Medley C.D., Smith J.E., Tang Z., Wu Y., Bamrungsap S., Tan W. Gold nanoparticle-based colorimetric assay for the direct detection of cancerous cells // *Anal. Chem.* 2008. V. 80. P. 1067–1072.

113. Zhang K., Shen X. Cancer antigen 125 detection using the plasmon resonance scattering properties of gold nanorods // *Analyst.* 2013. V. 138. P. 1828–1834.

114. Chirathaworn C., Chantaramalai T., Sereemaspun A., Kongthong N., Suwancharoen D. Detection of *Leptospira* in urine using anti-*Leptospira*-coated gold nanoparticles // *Comp. Immunol. Microbiol. Infect. Dis.* 2011. V 34. P. 31–34.

115. Lesniewski A., Los M., Jonsson-Niedziółka M., Krajewska A., Szot K., Los J.M., Niedziolka-Jonsson J. Antibody modified gold nanoparticles for fast and selective, colorimetric T7 bacteriophage detection // *Bioconjug. Chem.* 2014. V. 25. P. 644–648.

116. Liu Y., Zhang L., Wei W., Zhao H., Zhou Z., Zhang Y., Liu S. Colorimetric detection of influenza A virus using antibody-functionalized gold nanoparticles // *Analyst.* 2015. V. 140. P. 3989–3995.

117. Neely A., Perry C., Varisli B., Singh A.K., Arbneshi T., Senapati D., Kalluri J.R., Ray P.C. Ultrasensitive and highly selective detection of Alzheimer's disease biomarker using two-photon Rayleigh scattering properties of gold nanoparticle // *ACS Nano.* 2009. V. 3. P. 2834–2840.

118. Guarise C., Pasquato L., De Filippis V., Scrimin P. Gold nanoparticles-based protease assay // *Proc. Natl. Acad. Sci. USA.* 2006. V. 103. P. 3978–3982.

119. Huang H., Liu F., Huang S., Yuan S., Liao B., Yi S., Zeng Y., Chu P.K. Sensitive and simultaneous detection of different disease markers using multiplexed gold nanorods // *Anal. Chim. Acta.* 2012. V. 755. P. 108–114.

120. Andresen H., Mager M., Grießner M., Charchar P., Todorova N., Bell N., Theocharidis G., Bertazzo S., Yarovsky I., Stevens M.M. Single-step homogeneous immunoassays utilizing epitope-tagged gold nanoparticles: On the mechanism, feasibility, and limitations // *Chem. Mater.* 2014. V. 26. P. 4696–4704.

121. Liu H., Rong P., Jia H., Yang J., Dong B., Dong Q., Yang C., Hu P., Wang W., Liu H., Liu D. A wash-free homogeneous colorimetric immunoassay method // *Theranostics.* 2016. V. 6. P. 54–64.

122. Kashid S.B., Tak R.D., Raut R.W. Antibody tagged gold nanoparticles as scattering probes for the pico molar detection of the proteins in blood serum using nanoparticle tracking analyzer // *Colloids Surf. B.* 2015. V. 133. P. 208–213.

123. Yeo E.L.L., Chua A.J.S., Parthasarathy K., Yeo H.Y., Ng M.L., Kah J.C.Y. Understanding aggregation-based assays: Nature of protein corona and number of epitopes on antigen matters // *RSC Adv.* 2015. V. 5. P. 14982–14993.

124. Liu X., Dai Q., Austin L., Coutts J., Knowles G., Zou J., Chen H., Huo Q. A one-step homogeneous immunoassay for cancer biomarker detection using gold nanoparticle probes coupled with dynamic light scattering // *J. Am. Chem. Soc.* 2008. V. 130. P. 2780–2782.

125. Wang X., Li Y., Wang H., Fu Q., Peng J., Wang Y., Du J., Zhou Y., Zhan L. Gold nanorod-based localized surface plasmon resonance biosensor for sensitive detection of hepatitis B virus in buffer, blood serum and plasma // *Biosens. Bioelectron.* 2010. V. 26. P. 404–410.

126. Dykman L.A., Krasnov Ya.M, Bogatyrev V.A., Khlebtsov N.G. Quantitative immunoassay method based on extinction spectra of colloidal gold bioconjugates // *Proc. SPIE.* 2001. V. 4241. P. 37–41.

127. Khlebtsov N.G., Bogatyrev V.A., Khlebtsov B.N., Dykman L.A., Englebienne P. A multilayer model for gold nanoparticle bioconjugates: Application to study of gelatin and human IgG adsorption using extinction and light scattering spectra and the dynamic light scattering method // *Colloid J.* 2003. V. 65. P. 622–635.

128. Englebienne P. Use of colloidal gold surface plasmon resonance peak shift to infer affinity constants from the interactions between protein antigens and antibodies specific for single or multiple epitopes // *Analyst*. 1998. V. 123. P. 1599–1603.

129. Englebienne P., van Hoonacker A., Verhas M., Khlebtsov N.G. Advances in high-throughput screening: Biomolecular interaction monitoring in real-time with colloidal metal nanoparticles // *Comb. Chem. High Throughput Screen*. 2003. V. 6. P. 777–787.

130. Sakashita H., Tomita A., Umeda Y., Narukawa H., Kishioka H., Kitamori T., Sawada T. Homogeneous immunoassay using photothermal beam deflection spectroscopy // *Anal. Chem*. 1995. V. 67. P. 1278–1282.

131. Thanh N.T.K., Rees J.H., Rosenzweig Z. Laser-based double beam absorption detection for aggregation immunoassays using gold nanoparticles // *Anal. Bioanal. Chem*. 2002. V. 374. P. 1174–1178.

132. Zhang C.X., Zhang Y., Wang X., Tang Z.M., Lu Z.H. Hyper-Rayleigh scattering of protein-modified gold nanoparticles // *Anal. Biochem*. 2003. V. 320. P. 136–140.

133. Khlebtsov N.G., Bogatyrev V.A., Melnikov A.G., Dykman L.A., Khlebtsov B.N., Krasnov Y.M. Differential light-scattering spectroscopy: A new approach to studying of colloidal gold nanosensors // *J. Quant. Spectrosc. Radiat. Transfer*. 2004. V. 89. P. 133–142.

134. Huo Q. Protein complexes/aggregates as potential cancer biomarkers revealed by a nanoparticle aggregation immunoassay // *Colloids Surf. B*. 2010. V. 78. P. 259–265.

135. Kamnev A.A., Dykman L.A., Tarantilis P.A., Polissiou M.G. Spectroimmunochemistry using colloidal gold bioconjugates // *Biosci. Rep*. 2002. V. 22. P. 541–547.

136. Aroca R.F., Ross D.J., Domingo C. Surface-enhanced infrared spectroscopy // *Appl. Spectrosc*. 2004. V. 58. P. 324A–338A.

137. López-Lorente Á.I., Mizaikoff B. Recent advances on the characterization of nanoparticles using infrared spectroscopy // *TrAC*. 2016. V. 84. P. 97–106.

138. Porter M.D., Lipert R.J., Siperko L.M., Wang G., Narayanan R. SERS as a bioassay platform: Fundamentals, design, and applications // *Chem. Soc. Rev*. 2008. V. 37. P. 1001–1011.

139. Kneipp K., Kneipp H., Kneipp J. Surface-enhanced Raman scattering in local optical fields of silver and gold nanoaggregates—From single-molecule Raman spectroscopy to ultrasensitive probing in live cells // *Acc. Chem. Res*. 2006. V. 39. P. 443–450.

140. Ding S.-Y., Yi J., Li J.-F., Ren B., Wu D.-Y., Panneerselvam R., Tian Z.-Q. Nanostructure-based plasmon-enhanced Raman spectroscopy for surface analysis of materials // *Nat. Rev. Mater*. 2016. V. 1. 16021.

141. Fazio B., D'Andrea C., Foti A., Messina E., Irrera A., Donato M.G., Villari V., Micali N., Maragò O.M., Gucciardi P.G. SERS detection of biomolecules at physiological pH via aggregation of gold nanorods mediated by optical forces and plasmonic heating // *Sci. Rep*. 2016. V. 6. 26952.

142. Matteini P., Cottat M., Tavanti F., Panfilova E., Scuderi M., Nicotra G., Menziani M.C., Khlebtsov N., de Angelis M., Pini R. Site-selective surface-enhanced Raman detection of proteins // *ACS Nano*. 2017. V. 11. P. 518–526.

143. Stoschek C.M. Protein assay sensitive at nanogram levels // *Anal. Biochem.* 1987. V. 160. P. 301–305.

144. Dykman L.A., Bogatyrev V.A., Khlebtsov B.N., Khlebtsov N.G. A protein assay based on colloidal gold conjugates with trypsin // *Anal. Biochem.* 2005. V. 341. P. 16–21.

145. Mirkin C.A., Letsinger R.L., Mucic R.C., Storhoff J.J. A DNA-based method for rationally assembling nanoparticles into macroscopic materials // *Nature.* 1996. V. 382. P. 607–609.

146. Elghanian R., Storhoff J.J., Mucic R.C., Letsinger R.L., Mirkin C.A. Selective colorimetric detection of polynucleotides based on the distance-dependent optical properties of gold nanoparticles // *Science.* 1997. V. 277. P. 1078–1081.

147. Sato K., Onoguchi M., Sato Y., Hosokawa K., Maeda M. Non-cross-linking gold nanoparticle aggregation for sensitive detection of single-nucleotide polymorphisms: Optimization of the particle diameter // *Anal. Biochem.* 2006. V. 350. P. 162–164.

148. Baptista P.V., Koziol-Montewka M., Paluch-Oles J., Doria G., Franco R. Gold-nanoparticle-probe-based assay for rapid and direct detection of *Mycobacterium tuberculosis* DNA in clinical samples // *Clin. Chem.* 2006. V. 52. P. 1433–1434.

149. Dai Q., Liu X., Coutts J., Austin L., Huo Q. A one-step highly sensitive method for DNA detection using dynamic light scattering // *J. Am. Chem. Soc.* 2008. V. 130. P. 8138–8139.

150. Zhang J., Wang L., Pan D., Song S., Boey F.Y.C., Zhang H., Fan C. Visual cocaine detection with gold nanoparticles and rationally engineered aptamer structures // *Small.* 2008. V. 8. P. 1196–1200.

151. Wang L.H., Liu X.F., Hu X.F., Song S.P., Fan C.F. Unmodified gold nanoparticles as a colorimetric probe for potassium DNA aptamers // *Chem. Commun. (Camb.).* 2006. No. 36. P. 3780–3782.

152. Li H.X., Rothberg L. Colorimetric detection of DNA sequences based on electrostatic interactions with unmodified gold nanoparticles // *Proc. Natl. Acad. Sci. USA.* 2004. V. 101. P. 14036–14039.

153. He W., Huang C.Z., Li Y.F., Xie J.P., Yang R.G., Zhou P.F., Wang J. One-step label-free optical genosensing system for sequence-specific DNA related to the human immunodeficiency virus based on the measurements of light scattering signals of gold nanorods // *Anal. Chem.* 2008. V. 80. P. 8424–8430.

154. Storhoff J.J., Elghanian R., Mucic R.C., Mirkin C.A., Letsinger R.L. One-pot colorimetric differentiation of poly-nucleotides with single base imperfections using gold nano-particle probes // *J. Am. Chem. Soc.* 1998. V. 120. P. 1959–1964.

155. Witten K.G., Bretschneider J.C., Eckert T., Richtering W., Simon U. Assembly of DNA-functionalized gold nanoparticles studied by UV/Vis-spectroscopy and dynamic light scattering // *Phys. Chem. Chem. Phys.* 2008. V. 10. P. 1870–1875.

156. Sato K., Hosokawa K., Maeda M. Rapid aggregation of gold nanoparticles induced by non-cross-linking DNA hybridization // *J. Am. Chem. Soc.* 2003. V. 125. P. 8102–8103.

157. Doria G., Franco R., Baptista P. Nanodiagnostics: Fast colorimetric method for single nucleotide polymorphism/mutation detection // *IET Nanobiotechnol.* 2007. V. 1. P. 53–57.

158. Song J., Li Z., Cheng Y., Liu C. Self-aggregation of oligonucleotide-functionalized gold nanoparticles and its applications for highly sensitive detection of DNA // *Chem. Commun. (Camb.).* 2010. V. 46. P. 5548–5550.

159. Shawky S.M., Bald D., Azzazy H.M.E. Direct detection of unamplified hepatitis C virus RNA using unmodified gold nanoparticles // *Clin. Biochem.* 2010. V. 43. P. 1163–1168.

160. Xia F., Zuo X., Yang R., Xiao Y., Kang D., Vallée-Bélisle A., Gong X., Yuen J.D., Hsu B.B., Heeger A.J., Plaxco K.W. Colorimetric detection of DNA, small molecules, proteins, and ions using unmodified gold nanoparticles and conjugated polyelectrolytes // *Proc. Natl. Acad. Sci. USA.* 2010. V. 107. P. 10837–10841.

161. Ma Z., Tian L., Wang T., Wang C. Optical DNA detection based on gold nanorods aggregation // *Anal. Chim. Acta.* 2010. V. 673. P. 179–184.

162. Pylaev T.E., Khanadeev V.A., Khlebtsov B.N., Dykman L.A., Bogatyrev V.A., Khlebtsov N.G. Colorimetric and dynamic light scattering detection of DNA sequences by using positively charged gold nanospheres: A comparative study with gold nanorods // *Nanotechnology.* 2011. V. 22. 285501.

163. Soo P.C., Horng Y.T., Chang K.C., Wang J.Y., Hsueh P.R., Chuang C.Y., Lu C.C., Lai H.C. A simple gold nanoparticle probes assay for identification of *Mycobacterium tuberculosis* and *Mycobacterium tuberculosis* complex from clinical specimens // *Mol. Cell Probes.* 2009. V. 23. P. 240–246.

164. Liandris E., Gazouli M., Andreadou M., Comor M., Abazovic N., Sechi L.A, Ikonomopoulos J. Direct detection of unamplified DNA from pathogenic mycobacteria using DNA-derivatized gold nanoparticles // *J. Microbiol. Methods.* 2009. V. 78. P. 260–264.

165. Storhoff J.J., Marla S.S., Bao P., Hagenow S., Mehta H., Lucas A., Garimella V., Patno T., Buckingham W., Cork W., Müller U.R. Gold nanoparticle-based detection of genomic DNA targets on microarrays using a novel optical detection system // *Biosens. Bioelectron.* 2004. V. 19. P. 875–883.

166. Storhoff J.J., Lucas A.D., Garimella V., Bao Y.P., Müller U.R. Homogeneous detection of unamplified genomic DNA sequences based on colorimetric scatter of gold nanoparticle probes // *Nat. Biotechnol.* 2004. V. 22. P. 883–887.

167. Parab H.J., Jung C., Lee J.H., Park H.G. A gold nanorod-based optical DNA biosensor for the diagnosis of pathogens // *Biosens. Bioelectron.* 2010. V. 26. P. 667–673.

168. Wang X., Li Y., Wang J., Wang Q., Xu L., Du J., Yan S., Zhou Y., Fu Q., Wang Y., Zhan L. A broad-range method to detect genomic DNA of multiple pathogenic bacteria based on the aggregation strategy of gold nanorods // *Analyst.* 2012. V. 137. P. 4267–4273.

169. Deng H., Zhang X., Kumar A., Zou G., Zhang X., Liang X.-J. Long genomic DNA amplicons adsorption onto unmodified gold nanoparticles for colorimetric detection of *Bacillus anthracis* // *Chem. Commun. (Camb).* 2013. V. 49. P. 51–53.

170. Kalidasan K., Neo J.L., Uttamchandani M. Direct visual detection of *Salmonella* genomic DNA using gold nanoparticles // *Mol. BioSyst.* 2013. V. 9. P. 618–621.

171. Khalil M.A.F., Azzazy H.M.E., Attia A.S., Hashem A.G.M. A sensitive colorimetric assay for identification of *Acinetobacter baumannii* using unmodified gold nanoparticles // *J. Appl. Microbiol.* 2014. V. 117. P. 465–471.

172. Brada D., Roth J. "Golden blot"—Detection of polyclonal and monoclonal antibodies bound to antigens on nitrocellulose by protein A–gold complexes // *Anal. Biochem.* 1984. V. 142. P. 79–83.

173. Moeremans M., Daneles G., van Dijck A., Langanger G., De Mey J. Sensitive visualization of antigen-antibody reactions in dot and blot immune overlay assays with immunogold and immunogold/silver staining // *J. Immunol. Methods.* 1984. V. 74. P. 353–360.

174. Surek B., Latzko E. Visualization of antigenic proteins blotted onto nitrocellulose using the immuno-gold-staining (IGS)-method // *Biochem. Biophys. Res. Commun.* 1984. V. 121. P. 284–289.

175. Hsu Y.-H. Immunogold for detection of antigen on nitrocellulose paper // *Anal. Biochem.* 1984. V. 142. P. 221–225.

176. Daneels G., Moeremans M., De Raeymaeker M., De Mey J. Sequential immunostaining (gold/silver) and complete protein staining (AuroDye) on western blots // *J. Immunol. Methods.* 1986. V. 89. P. 89–91.

177. Everts M., Saini V., Leddon J.L., Kok R.J., Stoff-Khalili M., Preuss M.A., Millican C.L., Perkins G., Brown J.M., Bagaria H., Nikles D.E., Johnson D.T., Zharov V.P., Curiel D.T. Covalently linked Au nanoparticles to a viral vector: Potential for combined photothermal and gene cancer therapy // *Nano Lett.* 2006. V. 6. P. 587–591.

178. Western blotting detection systems: How do you choose? / *Bio-Rad Bulletin* 1310. Richmond: Bio-Rad. 1987.

179. Edwards P., Wilson T. Choose your labels // *Laboratory Pract.* 1987. V. 36. P. 13–17.

180. Goldman A., Harper S., Speicher D.W. Detection of proteins on blot membranes // *Curr. Protoc. Protein Sci.* 2016. V. 86. P. 10.8.1–10.8.11.

181. Danscher G. Localization of gold in biological tissue. A photochemical method for light and electron microscopy // *Histochemistry.* 1981. V. 71. P. 81–88.

182. Ma Z., Sui S.-F. Naked-eye sensitive detection of immunoglobulin G by enlargement of Au nanoparticles *in vitro* // *Angew. Chem. Int. Ed.* 2002. V. 41. P. 2176–2179.

183. Hou S.-Y., Chen H.-K., Cheng H.-C., Huang, C.-Y. Development of zeptomole and attomolar detection sensitivity of biotin-peptide using a dot-blot goldnanoparticle immunoassay // *Anal. Chem.* 2007. V. 79. P. 980–985.

184. Blab G.A., Cognet L., Berciaud S., Alexandre I., Husar D., Remacle J., Lounis B. Optical readout of gold nanoparticle-based DNA microarrays without silver enhancement // *Biophys. J.* 2006. V. 90. P. L13–L15.

185. Dykman L.A., Bogatyrev V.A. Colloidal gold in solid-phase assays. A review // *Biochemistry* (Moscow). 1997. V. 62. P. 350–356.

186. Steffen W., Linck R.W. Multiple immunoblot: A sensitive technique to stain proteins and detect multiple antigens on a single two-dimensional replica // *Electrophoresis.* 1989. V. 10. P. 714–718.

187. Poltavchenko A.G., Zaitsev B.N., Ersh A.V., Korneev D.V., Taranov O.S., Filatov P.V., Nechitaylo O.V. The selection and optimization of the detection system for self-contained multiplexed dot-immunoassay // *J. Immunoassay Immunochem.* 2016. V. 37. P. 540–554.

188. Petchclai B., Hiranras S., Potha U. Gold immunoblot analysis of IgM-specific antibody in the diagnosis of human leptospirosis // *Am. J. Trop. Med. Hyg.* 1991. V. 45. P. 672–675.

189. Scott J.M., Shreffler W.G., Ghalib H.W., el Asad A., Siddig M., Badaro R., Reed S.G. A rapid and simple diagnostic test for active visceral leishmaniasis // *Am. J. Trop. Med. Hyg.* 1991. V. 44. P. 272–277.

190. Liu Y.S., Du W.P., Wu Z.X. Dot-immunogold-silver staining in the diagnosis of cysticercosis // *Int. J. Parasitol.* 1996. V. 26. P. 127–129.

191. Liu Y.S., Du W.P., Wu Y.M., Chen Y.G., Zheng K.Y., Shi J.M., Hu X.Z., Li G.Y., You C.F., Wu Z.X. Application of dot-immunogold-silver staining in the diagnosis of clonorchiasis // *J. Trop. Med. Hyg.* 1995. V. 98. P. 151–154.

192. Thiruppathiraja C., Kamatchiammal S., Adaikkappan P., Alagar M. An advanced dual labeled gold nanoparticles probe to detect Cryptosporidium parvum using rapid immuno-dot blot assay // *Biosens. Bioelectron.* 2011. V. 26. P. 4624–4627.

193. Feodorova V.A., Polyanina T.I., Zaytsev S.S., Laskavy V.N., Ulianova O.V., Dykman L.A. Development of dot-ELISA for the screening rapid diagnosis of chlamydial infection in sheep // *Veterinariya.* 2015. No. 8. P. 21–24 (in Russian).

194. Chu F., Ji Q., Yan R.-M. Study on using colloidal gold immuno-dot assay to detect special antibody of hemorrhagic fever renal syndrome // *Chinese J. Integr. Tradit. West. Med.* 2001. V. 21. P. 504–506 (in Chinese).

195. Dar V.S., Ghosh S., Broor S. Rapid detection of rotavirus by using colloidal gold particles labeled with monoclonal antibody // *J. Virol. Methods.* 1994. V. 47. P. 51–58.

196. Fernandez D., Valle I., Llamos R., Guerra M., Sorell L., Gavilondo J. Rapid detection of rotavirus in feces using a dipstick system with monoclonal-antibodies and colloidal gold as marker // *J. Virol. Methods.* 1994. V. 48. P. 315–323.

197. Yee J.L., Jennings M.B., Carlson J.R., Lerche N.W. A simple, rapid immunoassay for the detection of simian immunodeficiency virus antibodies // *Lab. Anim. Sci.* 1991. V. 41. P. 119–122.

198. Reboli A.C. Diagnosis of invasive candidiasis by a dot immunobinding assay for candida antigen-detection // *J. Clin. Microbiol.* 1993. V. 31. P. 518–523.

199. Poulain D., Mackenzie D.W., van Cutsem J. Monoclonal antibody-gold silver staining dot assay for the detection of antigenaemia in candidosis // *Mycoses.* 1991. V. 34. P. 221–226.

200. Vera-Cabrera L., Rendon A., Diaz-Rodriguez M., Handzel V., Laszlo A. Dot blot assay for detection of antidiacyltrehalose antibodies in tuberculous patients // *Clin. Diagn. Lab. Immunol.* 1999. V. 6. P. 686–689.

201. Staroverov S.A., Vidyasheva I.V., Fomin A.S., Vasilenko O.A., Malinin M.L., Gabalov K.P., Bogatyrev V.A., Dykman L.A. Development of immunogold diagnostic systems for identification of the causative agent of tuberculosis *in situ* // *Russ. Vet. J.* 2011. No. 2. P. 29–32 (in Russian).

202. Hou D.,Wang X., Huang Y. Diagnostic value of gold standard method detecting of serum anti-TB antibody on TB // *Chin. Med. Factory Mine.* 2001. V. 14. P. 413–414.

203. Kunakorn M., Petchclai B., Khupulsup K., Naigowit P. Gold blot for detection of immunoglobulin M (IgM)- and IgG-specific antibodies for rapid serodiagnosis of melioidosis // *J. Clin. Microbiol.* 1991. V. 29. P. 2065–2067.

204. Huang Q., Lan X., Tong T., Wu X., Chen M., Feng X., Liu R., Tang Y., Zhu Z. Dot-immunogold filtration assay as a screening test for syphilis // *J. Clin. Microbiol.* 1996. V. 34. P. 2011–2013.

205. Zagoskina T.Y., Markov E.Y., Kalinovskiĭ A.I., Golubinskiĭ E.P. Use of specific antibodies, labeled with colloidal gold particles, for the detection of *Brucella* antigens using dot-immunoassay // *Zh. Mikrobiol. Epidemiol. Immunobiol.* 1998. No. 6. P. 64–69 (in Russian).

206. Lazarchik V.A., Titov L.P., Vorobyova T.N., Ermakova T.S., Vrublevskaya O.N., Vlasik N.V. Test-system based on antibacterial antibody conjugates with colloidal gold for revelation of shigells antigens in biological liquids // *Proc. Natl. Acad. Sci. Belarus Ser. Med. Sci.* 2005. No. 3. P. 44–47 (in Russian).

207. Kamma S., Tang L., Leung K., Ashton E., Newman N., Suresh M.R. A rapid two dot filter assay for the detection of *E. coli* O157 in water samples // *J. Immunol. Methods.* 2008. V. 336. P. 159–165.

208. Fang S.B., Tseng W.Y., Lee H.C., Tsai C.K., Huang J.T., Hou S.Y. Identification of *Salmonella* using colony-print and detection with antibody-coated gold nanoparticles // *J. Microbiol. Methods.* 2009. V. 77. P. 225–228.

209. Pandey S.K., Suri C.R., Chaudhry M., Tiwari R.P., Rishi P. A gold nanoparticles based immuno-bioprobe for detection of Vi capsular polysaccharide of Salmonella enterica serovar Typhi // *Mol. BioSyst.* 2012. V. 8. P. 1853–1860.

210. Khadzhu A., Ivaschenko S.V., Fomin A.S., Scherbakov A.A., Staroverov S.A., Dykman L.A. Usage of hyperimmune serum obtained for the DMSO-antigen of intestinal yersiniosis microbe in the indirect dot-immunoanalysis with colloidal gold conjugate // *Sci. Rev.* 2015. No. 5. P. 30–34 (in Russian).

211. Noskova O.A., Zagoskina T.Y., Subycheva E.N., Markov E.Y., Popova Y.O., Gridneva L.G., Mikhailov E.P. Application of dot-immunoassay for detection of plague agent antigens in the field samples // *Probl. Osobo Opas. Infekc.* 2014. No. 4. P. 69–71 (in Russian).

212. Xu Z. Immunogold dot assay for diagnosis of early pregnancy // *J. Chin. Med. Assoc.* 1992. V. 72. P. 216–218 (in Chinese).

213. Matsuzawa S., Kimura H., Itoh Y., Wang H., Nakagawa T. A rapid dot-blot method for species identification of bloodstains // *J. Forensic Sci.* 1993. V. 38. P. 448–454.

214. Cremers A.F., Jansen in de Wal N., Wiegant J., Dirks R.W., Weisbeek P., van der Ploeg M., Landegent J.E. Non-radioactive *in situ* hybridization. A comparison of several immunocytochemical detection systems using reflection-contrast and electron microscopy // *Histochemistry.* 1987. V. 86. P. 609–615.

215. Kolodkina V.L., Denisevich T.N., Dykman L.A., Vrublevskaya O.N. Preparation of gold-labeled antibody probe and its use in dot-immunogold assay for detection of diphtheria toxin // *Med. J.* 2009. No. 2. P. 66–69 (in Russian).

216. Staroverov S.A., Volkov A.A., Fomin A.S., Laskavuy V.N., Mezhennyy P.V., Kozlov S.V., Larionov S.V., Fedorov M.V., Dykman L.A., Guliy O.I. The usage of phage mini-antibodies as a means of detecting ferritin concentration in animal blood serum // *J. Immunoassay Immunochem.* 2015. V. 36. P. 100–110.

217. Wang Y.L., Li D., Ren W., Liu Z.J., Dong S.J. Wang, E.K. Ultrasensitive colorimetric detection of protein by aptamer–Au nanoparticles conjugates based on a dot-blot assay // *Chem. Commun.* 2008. No. 22. P. 2520–2522.

218. Wang C., Liu D., Wang Z. Gold nanoparticle based dot-blot immunoassay for sensitively detecting Alzheimer's disease related β-amyloid peptide // *Chem. Commun.* 2012. V. 48. P. 8392–8394.

219. Tavernaro I., Hartmann S., Sommer L., Hausmann H., Rohner C., Ruehl M., Hoffmann-Roedere A., Schlecht S. Synthesis of tumor-associated MUC1-glycopeptides and their multivalent presentation by functionalized gold colloids // *Org. Biomol. Chem.* 2015. V. 13. P. 81–97.

220. Sui J., Lin H., Xu Y., Cao L. Enhancement of dot-immunogold filtration assay (DIGFA) by activation of nitrocellulose membranes with secondary antibody // *Food Anal. Methods.* 2011. V. 4. P. 245–250.

221. Guo H., Zhang J., Yang D., Xiao P., He N. Protein array for assist diagnosis of acute myocardial infarction // *Colloids Surf. B.* 2005. V. 40. P. 195–198.

222. Xi D., Luo X., Ning Q., Lu Q., Yao K., Liu Z. The detection of HBV DNA with gold nanoparticle gene probes // *J. Nanjing Med. Univ.* 2007. V. 21. P. 207–212.

223. Starodub N.F., Artyukh V.P., Nazarenko V.I., Kolomiets L.I. Protein immunoblot and immunodot in biochemical studies // *Ukr. Biokhim. Zh.* 1987. V. 59. P. 108–120 (in Russian).

224. Fenoll A., Jado I., Vicioso D., Casal J. Dot blot assay for the serotyping of pneumococci // *J. Clin. Microbiol.* 1997. V. 35. P. 764–766.

225. Bogatyrev V.A., Dykman L.A., Matora L.Yu., Schwartsburd B.I. Serotyping of azospirillae using a colloidal gold solid-phase immunoassay // *Microbiology.* 1991. V. 60. P. 366–370.

226. Bogatyrev V.A., Dykman L.A., Matora L.Yu., Schwartsburd B.I. The serotyping of *Azospirillum* Spp by cell gold immunoblotting // *FEMS Microbiol. Lett.* 1992. V. 96. P. 115–118.

227. Dykman L.A., Bogatyrev V.A. Use of the dot-immunogold assay for the rapid diagnosis of acute enteric infections // *FEMS Immunol. Med. Microbiol.* 2000. V. 27. P. 135–137.

228. Matveev V.Y., Bogatyrev V.A., Dykman L.A., Matora L.Y., Schwartsburd B.I. Cell surface physico-chemical properties of the R- and S-variants of *Azospillum brasilense* strain // *Microbiology.* 1992. V. 61. P. 454–459.

229. Gas F., Pinto L., Baus B., Gaufres L., Crassous M.P., Compere C., Quéméneur E. Monoclonal antibody against the surface of *Alexandrium minutum* used in a whole-cell ELISA. // *Harmful Algae.* 2009. V. 8. P. 538–545.

230. Khlebtsov B.N., Dykman L.A., Bogatyrev V.A., Zharov V., Khlebtsov N.G. A solid-phase dot assay using silica/gold nanoshells // *Nanoscale Res. Lett.* 2007. V. 2. P. 6–11.

231. Khlebtsov B.N., Khanadeev V.A., Bogatyrev V.A., Dykman L.A., Khlebtsov N.G. Use of gold nanoshells in solid-phase immunoassay // *Nanotechnol. Russ.* 2008. V. 3. P. 442–455.

232. Khlebtsov B.N., Khlebtsov N.G. Enhanced solid-phase immunoassay using gold nanoshells: Effect of nanoparticle optical properties // *Nanotechnology*. 2008. V. 19. 435703.

233. Panfilova E., Shirokov A., Khlebtsov B., Matora L., Khlebtsov N. Multiplexed dot immunoassay using Ag nanocubes, Au/Ag alloy nanoparticles, and Au/Ag nanoboxes // *Nano Res.* 2012. V. 5. P. 124–134.

234. Han A., Dufva M., Belleville E., Christensen C.B.V. Detection of analyte binding to microarrays using gold nanoparticle labels and a desktop scanner // *Lab. Chip.* 2003. V. 3. P. 329–332.

235. Duan L., Wang Y., Li S.S.-c., Wan Z., Zhai J. Rapid and simultaneous detection of human hepatitis B virus and hepatitis C virus antibodies based on a protein chip assay using nano-gold immunological amplification and silver staining method // *BMC Infect. Dis.* 2005. V. 5. 53.

236. Goryacheva I.Y., Lenain P., De Saeger S. Nanosized labels for rapid immunotests // *Trends Analyt. Chem.* 2013. V. 46. P. 30–43.

237. Yang H.-W., Tsai R.-Y., Chen J.-P., Ju S.-P., Liao J.-F., Wei K.-C., Zou W.-L., Hua M.-Y. Fabrication of a nanogold-dot array for rapid and sensitive detection of vascular endothelial growth factor in human serum // *ACS Appl. Mater. Interfaces*. 2016. V. 8. P. 30845–30852.

238. Peruski A.H., Peruski L.F. Immunological methods for detection and identification of infectious disease and biological warfare agents // *Clin. Diagn. Lab. Immunol.* 2003. V. 10. P. 506–513.

239. Long G.W., O'Brien T. Antibody-based systems for the detection of *Bacillus anthracis* in environmental samples // *J. Appl. Microbiol.* 1999. V. 87. P. 214.

240. Bird C.B., Miller R.L., Miller B.M. Reveal for *Salmonella* test system // *J. AOAC Int.* 1999. V. 82. P. 625–633.

241. Wu S.J., Paxton H., Hanson B., Kung C.G., Chen T.B., Rossi C., Vaughn D.W., Murphy G.S., Hayes C.G. Comparison of two rapid diagnostic assays for detection of immunoglobulin M antibodies to dengue virus // *Clin. Diagn. Lab. Immunol.* 2000. V. 7. P. 106–110.

242. Engler K.H., Efstratiou A., Norn D., Kozlov R.S., Selga I., Glushkevich T.G., Tam M., Melnikov V.G., Mazurova I.K., Kim V.E., Tseneva G.Y., Titov L.P., George R.C. Immunochromatographic strip test for rapid detection of diphtheria toxin: Description and multicenter evaluation in areas of low and high prevalence of diphtheria // *J. Clin. Microbiol.* 2002. V. 40. P. 80–83.

243. Shyu R.H., Shyu H.F., Liu H.W., Tang S.S. Colloidal gold-based immunochromatographic assay for detection of ricin // *Toxicon*. 2002. V. 40. P. 255–258.

244. Chanteau S., Rahalison L., Ralafiarisoa L., Foulon J., Ratsitorahina M., Ratsifasoamanana L., Carniel E., Nato F. Development and testing of a rapid diagnostic test for bubonic and pneumonic plague // *Lancet*. 2003. V. 361. P. 211–216.

245. Saidi A., Mirzaei M., Zeinali S. Using antibody coated gold nanoparticles as fluorescence quenchers for simultaneous determination of aflatoxins (B1, B2) by soft modeling method // *Chemometr. Intell. Lab.* 2013. V. 127. P. 29–34.

246. Grobusch M.P., Schormann D., Schwenke S., Teichmann D., Klein E. Rapid immunochromatographic assay for diagnosis of tuberculosis // *J. Clin. Microbiol.* 1998. V. 36. P. 3443.

247. Koo H.C., Park Y.H., Ahn J., Waters W.R., Palmer M.V., Hamilton M.J., Barrington G., Mosaad A.A., Park K.T., Jung W.K., Hwang I.Y., Cho S.-N., Shin S.J., Davis W.C. Use of rMPB70 protein and ESAT-6 peptide as antigens for comparison of the enzyme-linked immunosorbent immunochromatographic, and latex bead agglutination assays for serodiagnosis of bovine tuberculosis // *J. Clin. Microbiol.* 2005. V. 43. P. 4498–4506.

248. Treepongkaruna S., Nopchinda S., Taweewongsounton A., Atisook K., Pienvichit P., Vithayasai N., Simakachorn N., Aanpreung P. A rapid serologic test and immunoblotting for the detection of *Helicobacter pylori* infection in children // *J. Trop. Pediatr.* 2006. V. 52. P. 267–271.

249. Huang S.-H. Gold nanoparticle-based immunochromatographic assay for the detection of *Staphylococcus aureus* // *Sens. Actuator B Chem.* 2007. V. 127. P. 335–340.

250. Wiriyachaiporn S., Howarth P.H., Bruce K.D., Dailey L.A. Evaluation of a rapid lateral flow immunoassay for *Staphylococcus aureus* detection in respiratory samples // *Diagn. Microbiol. Infect. Dis.* 2013. V. 75. P. 28–36.

251. Chelobanov B.P., Afinogenova G.N., Cheshenko I.O., Sharova T.V., Zyrianova A.V., Veliev S.N. Development of an immunochromatography tests for rapid diagnostics of viral hepatitis type B patients // *Bull. SB RAMS.* 2007. No. 5. P. 83–87 (in Russian).

252. Taneja N., Nato F., Dartevelle S., Sire J.M., Garin B., Thi Phuong L.N., Diep T.T., Shako J.C., Bimet F., Filliol I., Muyembe J.-J., Ungeheuer M.N., Ottone C., Sansonetti P., Germani Y. Dipstick test for rapid diagnosis of *Shigella dysenteriae* 1 in bacterial cultures and its potential use on stool samples // *PLoS One.* 2011. V. 6. e24830.

253. Duran C., Nato F., Dartevelle S., Thi Phuong L.N., Taneja N., Ungeheuer M.N., Soza G., Anderson L., Benadof D., Zamorano A., Diep T.T., Nguyen T.Q., Nguyen V.H., Ottone C., Bégaud E., Pahil S., Prado V., Sansonetti P., Germani Y. Rapid diagnosis of diarrhea caused by Shigella sonnei using dipsticks; comparison of rectal swabs, direct stool and stool culture // *PLoS One.* 2013. V. 8. e80267.

254. Lyubavina I.A., Valyakina T.I., Grishin E.V. Monoclonal antibodies labeled with colloidal gold for immunochromatographic express analysis of diphtheria toxin // *Russ. J. Bioorganic Chem.* 2011. V. 37. P. 326–332.

255. Joon Tam Y., Mohd Lila M.A., Bahaman A.R. Development of solid-based paper strips for rapid diagnosis of Pseudorabies infection // *Trop. Biomed.* 2004. V. 21. P. 121–134.

256. Han S.-M., Cho J.-H., Cho I.-H., Paek E.-H., Oh H.-B., Kim B.-S., Ryu C., Lee K., Kim Y.-K., Paek S.-H. Plastic enzyme-linked immunosorbent assays (ELISA)-on-a-chip biosensor for botulinum neurotoxin A // *Anal. Chim. Acta.* 2007. V. 587. P. 1–8.

257. Feodorova V.A., Polyanina T.I., Zaytsev S.S., Ulianova O.V., Laskavy V.N., Dykman L.A. Development of immunochromatography test strip for rapid diagnosis of ovine enzootic abortion // *Veterinaria i kormlenie*. 2015. No. 5. P. 16–19 (in Russian).

258. Cui X., Huang Y., Wang J., Zhang L., Rong Y., Lai W., Chen T. A remarkable sensitivity enhancement in a gold nanoparticle-based lateral flow immunoassay for the detection of *Escherichia coli* O157:H7 // *RSC Adv*. 2015. V. 5. P. 45092–45097.

259. Fernández-Sánchez C., McNeil C.J., Rawson K., Nilsson O., Leung H.Y., Gnanapragasam V. One-step immunostrip test for the simultaneous detection of free and total prostate specific antigen in serum // *J. Immunol. Methods*. 2005. V. 307. P. 1–12.

260. Andreeva I.P., Grigorenko V.G., Egorov A.M., Osipov A.P. Quantitative lateral flow immunoassay for total prostate specific antigen in serum // *Anal. Lett*. 2015. V.49. P. 579–588.

261. Tanaka R., Yuhi T., Nagatani N., Endo T., Kerman K., Takamura Y., Tamiya E. A novel enhancement assay for immunochromatographic test strips using gold nanoparticles // *Anal. Bioanal. Chem*. 2006. V. 385. P. 1414–1420.

262. Glynou K., Ioannou P.C., Christopoulos T.K., Syriopoulou V. Oligonucleotide-functionalized gold nanoparticles as probes in a dry-reagent strip biosensor for DNA analysis by hybridization // *Anal. Chem*. 2003. V. 75. P. 4155–4160.

263. Zhou P., Lu Y., Zhu J., Hong J., Li B., Zhou J., Gong D., Montoya A. Nanocolloidal gold-based immunoassay for the detection of the *N*-methylcarbamate pesticide carbofuran // *J. Agric. Food Chem*. 2004. V. 52. P. 4355–4359.

264. Zhang C., Zhang Y., Wang S. Development of multianalyte flow-through and lateral-flow assays using gold particles and horseradish peroxidase as tracers for the rapid determination of carbaryl and endosulfan in agricultural products // *J. Agric. Food Chem*. 2006. V. 54. P. 2502–2507.

265. Xiulan S., Xiaolian Z., Jian T., Zhou J., Chu F.S. Preparation of gold-labeled antibody probe and its use in immunochromatography assay for detection of aflatoxin B1 // *Int. J. Food Microbiol*. 2005. V. 99. P. 185–194.

266. Zhang D., Li P., Zhang Q., Zhang W. Ultrasensitive nanogold probe-based immunochromatographic assay for simultaneous detection of total aflatoxins in peanuts // *Biosens. Bioelectron*. 2011. V. 26. P. 2877–2882.

267. Wang S., Quan Y., Lee N., Kennedy I.R. Rapid determination of fumonisin B1 in food samples by enzyme-linked immunosorbent assay and colloidal gold immunoassay // *J. Agric. Food Chem*. 2006. V. 54. P. 2491–2495.

268. Huo T., Peng C., Xu C., Liu L. Immumochromatographic assay for determination of hexoestrol residues // *Eur. Food Res. Technol*. 2007. V. 225. P. 743–747.

269. Zhao Y., Zhang G., Liu Q., Teng M., Yang J., Wang J. Development of a lateral flow colloidal gold immunoassay strip for the rapid detection of enrofloxacin residues // *J. Agric. Food Chem*. 2008. V. 56. P. 12138–12142.

270. Chen L., Wang Z., Ferreri M., Su J., Han B. Cephalexin residue detection in milk and beef by ELISA and colloidal gold based one-step strip assay // *J. Agric. Food Chem*. 2009. V. 57. P. 4674–4679.

271. Byzova N.A., Smirnova N.I., Zherdev A.V., Eremin S.A., Shanin I.A., Lei H.T., Sun Y., Dzantiev B.B. Rapid immunochromatographic assay for ofloxacin in animal original foodstuffs using native antisera labeled by colloidal gold // *Talanta*. 2014. V. 119. P. 125–132.

272. Cho J.-H., Paek S.-H. Semiquantitative, bar-code version of immunochromatographic assay system for human serum albumin as model analyte // *Biotechnol. Bioeng*. 2001. V. 75. P. 725–732.

273. Wang S., Zhang C., Wang J., Zhang Y. Development of colloidal gold-based flow-through and lateral-flow immunoassays for the rapid detection of the insecticide carbaryl // *Anal. Chim. Acta*. 2005. V. 546. P. 161–166.

274. Bahadır E.B., Sezgintürk M.K. Lateral flow assays: Principles, designs and labels // *Trends Anal. Chem*. 2016. V. 82. P. 286–306.

275. Gupta S., Huda S., Kilpatrick P.K., Velev O.D. Characterization and optimization of gold nanoparticle-based silver-enhanced immunoassays // *Anal. Chem*. 2007. V. 79. P. 3810–3820.

276. Venkataramasubramani M., Tang L. Development of gold nanorod lateral flow test for quantitative multi-analyte detection // *IFMBE Proc*. 2009. V. 24. P. 199–202.

277. Ren W., Huang Z., Xu Y., Li Y., Ji Y., Su B. Urchin-like gold nanoparticle-based immunochromatographic strip test for rapid detection of fumonisin B1 in grains // *Anal. Bioanal. Chem*. 2015. V. 407. P. 7341–7348.

278. Wang L., Ma W., Chen W., Liu L., Ma W., Zhu Y., Xu L., Kuang H., Xu C. An aptamer-based chromatographic strip assay for sensitive toxin semi-quantitative detection // *Biosens. Bioelectron*. 2011. V. 26. P. 3059–3062.

279. Nangola S., Thongkum W., Saoin S., Ansari A.A., Tayapiwatana C. An application of capsid-specific artificial ankyrin repeat protein produced in *E. coli* for immunochromatographic assay as a surrogate for antibody // *Appl. Microbiol. Biotechnol*. 2014. V. 98. P. 6095–6103.

280. Dzantiev B.B., Byzova N.A., Urusov A.E., Zherdev A.V. Immunochromatographic methods in food analysis // *Trends Anal. Chem*. 2014. V. 55. P. 81–93.

281. Quesada-González D., Merkoçi A. Nanoparticle-based lateral flow biosensors // *Biosens. Bioelectron*. 2015. V. 73. P. 47–63.

282. Mak W.C., Beni V., Turner A.P.F. Lateral-flow technology: From visual to instrumental // *Trends Anal. Chem*. 2016. V. 79. P. 297–305.

283. Huang X., Aguilar Z.P., Xu H., Lai W., Xiong Y. Membrane-based lateral flow immunochromatographic strip with nanoparticles as reporters for detection: A review // *Biosens. Bioelectron*. 2016. V. 75. P. 166–180.

284. Schalkhammer Th. Metal nano clusters as transducers for bioaffinity interactions // *Chem. Monthly*. 1998. V. 129. P. 1067–1092.

285. *Biosensors and Biodetection* / Eds. Rasooly A., Herold K.E. New York: Humana Press, 2009.

286. Li Y., Schluesener H.J., Xu S. Gold nanoparticle-based biosensors // *Gold Bull*. 2010. V. 43. P. 29–41.

287. Peng H.I., Miller B.L. Recent advancements in optical DNA biosensors: Exploiting the plasmonic effects of metal nanoparticles // *Analyst*. 2011. V. 136. P. 436–447.

288. Musick M.D., Keating C.D., Lyon L.A., Botsko S.L., Peña D.J., Holliway W.D., McEvoy T.M., Richardson J.N., Natan M.J. Metal films prepared by stepwise assembly // *Chem. Mater.* 2000. V. 12. P. 2869–2881.

289. Shipway A.N., Katz E., Willner I. Nanoparticle arrays on surfaces for electronic, optical, and sensor applications // *ChemPhysChem.* 2000. V. 1. P. 18–52.

290. Grabar K.C., Freeman R.G., Hommer M.B., Natan M.J. Preparation and characterization of Au colloid monolayers // *Anal. Chem.* 1995. V. 67. P. 735–743.

291. Ulman A. Formation and structure of self-assembled monolayers // *Chem. Rev.* 1996. V. 96. P. 1533–1554.

292. Nath N., Chilkoti A. A colorimetric gold nanoparticle sensor to interrogate biomolecular interactions in real time on a surface // *Anal. Chem.* 2002. V. 74. P. 504–509.

293. Prasad B.L.V., Stoeva S.I., Sorensen C.M., Klabunde K.J. Digestive-ripening agents for gold nanoparticles: Alternatives to thiols // *Chem. Mater.* 2003. V. 15. P. 935–943.

294. Haynes C.L., Van Duyne R.P. Nanosphere lithography: A versatile nanofabrication tool for studies of size-dependent nanoparticle optics // *J. Phys. Chem. B.* 2001. V. 105. P. 5599–5611.

295. Lyon L.A., Musick M.D., Natan M.J. Colloidal Au-enhanced surface-plasmon resonance immunosensing // *Anal. Chem.* 1998. V. 70. P. 5177–5183.

296. Miller M.M., Lazarides A.A. Sensitivity of metal nanoparticle surface plasmon resonance to the dielectric environment // *J. Phys. Chem. B.* 2005. V.109. P. 21556–21565.

297. Khlebtsov B.N., Melnikov A.G., Zharov V.P., Khlebtsov N.G. Absorption and scattering of light by a dimer of metal nanospheres: Comparison of dipole and multipole approaches // *Nanotechnology.* 2006. V. 17. P. 1437–1445.

298. Jain P.K., El-Sayed M.A. Plasmonic coupling in noble metal nanostructures // *Chem. Phys. Lett.* 2010. V. 487. P. 153–164.

299. Templeton A.C., Pietron J.J., Murray R.W., Mulvaney P. Solvent refractive index and core charge influences on the surface plasmon absorbance of alkanethiolate monolayer-protected gold clusters // *J. Phys. Chem. B.* 2000. V. 104. P. 564–570.

300. Khlebtsov N.G., Dykman L.A., Bogatyrev V.A., Khlebtsov B.N. Two-layer model of colloidal gold bioconjugates and its application to the optimization of nanosensors // *Colloid J.* 2003. V. 65. P. 508–517.

301. Penn S.G., He L., Natan M.J. Nanoparticles for bioanalysis // *Curr. Opin. Chem. Biol.* 2003. V. 7. P. 609–615.

302. Schuk P. Use of surface plasmon resonance to probe the equilibrium and dynamic aspects of interactions between biological macromolecules // *Annu. Rev. Biophys. Biomol. Struct.* 1997. V. 26. P. 541–566.

303. Homola J., Yee S.S., Gauglitz G. Surface plasmon resonance sensors: Review // *Sens. Actuator B* Chem. 1999. V. 54. P. 3–15.

304. Mullett W.M., Lai E.P.C., Yeung J.M. Surface plasmon resonance-based immunoassays // *Methods.* 2000. V. 22. P. 77–91.

305. Niemeyer C.M. Nanoparticles, proteins, and nucleic acids: Biotechnology meets materials science // *Angew. Chem. Int. Ed.* 2001. V. 40. P. 4128–4158.

306. Jain K.K. Nanodiagnostics: Application of nanotechnology in molecular diagnostics // *Expert Rev. Mol. Diagn.* 2003. V. 3. P. 153–161.
307. Parak W.J., Gerion D., Pellegrino T., Zanchet D., Micheel C., Williams S.C., Boudreau R., Le Gros M.A., Larabell C.A., Alivisatos A.P. Biological applications of colloidal nanocrystals // *Nanotechnology.* 2003. V. 14. P. R15–R27.
308. Riboh J.C., Haes A.J., McFarland A.D., Yonzon C.R., Van Duyne R.P. A nanoscale optical biosensor: Real-time immunoassay in physiological buffer enabled by improved nanoparticle adhesion // *J. Phys. Chem. B.* 2003. V. 107. P. 1772–1780.
309. Rosi N.L., Mirkin C.A. Nanostructures in biodiagnostics // *Chem. Rev.* 2005. V. 105. P. 1547–1562.
310. Stewart M.E., Anderton C.R., Thompson L.B., Maria J., Gray S.K., Rogers J.A., Nuzzo R.G. Nanostructured plasmonic sensors // *Chem. Rev.* 2008. V. 108. P. 494–521.
311. Shtykov S.N., Rusanova T.Y. Nanomaterials and nanotechnologies in chemical and biochemical sensors: Capabilities and applications // *Russ. J. Gen. Chem.* 2008. V. 7. P. 2521–2531.
312. Sepúlveda B., Angelomé P.C., Lechuga L.M., Liz-Marzán L.M. LSPR-based nanobiosensors // *Nano Today.* 2009. V. 4. P. 244–251.
313. Daghestani H.N., Day B.W. Theory and applications of surface plasmon resonance, resonant mirror, resonant waveguide grating, and dual polarization interferometry biosensors // *Sensors.* 2010. V. 10. P. 9630–9646.
314. Lee S.E., Lee L.P. Biomolecular plasmonics for quantitative biology and nanomedicine // *Curr. Opin. Biotechnol.* 2010. V. 21. P. 489–497.
315. Csáki A., Berg S., Jahr N., Leiterer C., Schneider T., Steinbrück A., Zopf D., Fritzsche W. // In: *Gold Nanoparticles: Properties, Characterization and Fabrication* / Ed. Chow P.E. New York: Nova Science Publisher. 2010. P. 245–261.
316. Stefan-van Staden R.-I., van Staden J.F., Balasoiu S.-C., Vasile O.-R. Micro- and nanosensors, recent developments and features: A minireview // *Anal. Lett.* 2010. V. 43. P. 1111–1118.
317. Abbas A., Linman M.J., Cheng Q. New trends in instrumental design for surface plasmon resonance-based biosensors // *Biosens. Bioelectron.* 2011. V. 26. P. 1815–1824.
318. Pérez-López B., Merkoçi A. Nanoparticles for the development of improved (bio) sensing systems // *Anal. Bioanal. Chem.* 2011. V. 399. P. 1577–1590.
319. Mayer K.M., Hafner J.H. Localized surface plasmon resonance sensors // *Chem. Rev.* 2011. V. 111. P. 3828–3857.
320. Zhao J., Bo B., Yin Y.-M. Gold nanoparticles-based biosensors for biomedical application // *Nano Life.* 2012. V. 2. 1230008.
321. Upadhyayula V.K.K. Functionalized gold nanoparticle supported sensory mechanisms applied in detection of chemical and biological threat agents: A review // *Anal. Chim. Acta.* 2012. V. 715. P. 1–18.
322. Saha K., Agasti S.S., Kim C., Li X., Rotello V.M. Gold nanoparticles in chemical and biological sensing // *Chem. Rev.* 2012. V. 112. P. 2739–2779.
323. Chen P.-C., Roy P., Chen L.-Y., Ravindranath R., Chang H.-T. Gold and silver nanomaterial-based optical sensing systems // *Part. Part. Syst. Charact.* 2014. V. 31. P. 917–942.

324. Cao J., Sun T., Grattan K.T.V. Gold nanorod-based localized surface plasmon reso-nance biosensors: A review // *Sens. Actuators B.* 2014. V. 195. P. 332–351.

325. Kedem O., Vaskevich A., Rubinstein I. Critical issues in localized plasmon sensing // *J. Phys. Chem. C.* 2014. V. 118. P. 8227–8244.

326. Howes P.D., Rana S., Stevens M.M. Plasmonic nanomaterials for biodiagnostics // *Chem. Soc. Rev.* 2014. V. 43. P. 3835–3853.

327. Howes P.D., Chandrawati R., Stevens M.M. Colloidal nanoparticles as advanced biological sensors // *Science.* 2014. V. 346. 1247390.

328. Szunerits S., Spadavecchia J., Boukherroub R. Surface plasmon resonance: Signal amplification using colloidal gold nanoparticles for enhanced sensitivity // *Rev. Anal. Chem.* 2014. V. 33. P. 153–164.

329. *Gold Nanoparticles in Analytical Chemistry* / Eds. Valcárcel M., López-Lorente Á.I. Amsterdam: Elsevier. 2014.

330. Dykman L.A., Khlebtsov N.G., Shchyogolev S.Y. Gold nanoparticles in analyti-cal methods // In: *Nanoobjects and Nanotechnology in Chemical Analysis* / Ed. Shtykov S.N. Moscow: Nauka. 2015. P. 42–74 (in Russian).

331. Strobbia P., Languirand E., Cullum B.M. Recent advances in plasmonic nano-structures for sensing: A review // *Opt. Eng.* 2015. V. 54. 100902.

332. Sotnikov D.V., Zherdev A.V., Dzantiev B.B. Detection of intermolecular interac-tions based on surface plasmon resonance registration // *Biochemistry (Moscow).* 2015. V. 80. P. 1820–1832.

333. Li M., Cushing S.K., Wu N. Plasmon-enhanced optical sensors: A review // *Analyst.* 2015. V. 140. P. 386–406.

334. Unser S., Bruzas I., He J., Sagle L. Localized surface plasmon resonance biosens-ing: Current challenges and approaches // *Sensors.* 2015. V. 15. P. 15684–15716.

335. Johne B., Hansen K., Mørk E., Holtlund J. Colloidal gold conjugated monoclonal antibodies, studied in the BIAcore biosensor and in the Nycocard immunoassay format // *J. Immunol. Methods.* 1995. V. 183. P. 167–174.

336. Jason-Moller L., Murphy M., Bruno J. Overview of Biacore systems and their applications // *Curr. Protoc. Protein Sci.* 2006. Ch. 19. Unit 19.13.

337. Cooper M.A. Label-free screening of bio-molecular interactions // *Anal. Bioanal. Chem.* 2003. V. 373. P. 834–842.

338. Brainina K., Kozitsina A., Beikin J. Electrochemical immunosensor for Forest-Spring encephalitis based on protein A labeled with colloidal gold // *Anal. Bioanal. Chem.* 2003. V. 376. P. 481–485.

339. Baek T.J., Park P.Y., Han K.N., Kwon H.T., Seong G.H. Development of a photodi-ode array biochip using a bipolar semiconductor and its application to detection of human papilloma virus // *Anal. Bioanal. Chem.* 2008. V. 390. P. 1373–1378.

340. Mahmoud K.A., Luong J.H. Impedance method for detecting HIV-1 protease and screening for its inhibitors using ferrocene-peptide conjugate/Au nanoparticle/single-walled carbon nanotube modified electrode // *Anal. Chem.* 2008. V. 80. P. 7056–7062.

341. Georganopoulou D.G., Chang L., Nam J.M., Thaxton C.S., Mufson E.J., Klein W.L., Mirkin C.A. Nanoparticle-based detection in cerebral spinal fluid of a sol-uble pathogenic biomarker for Alzheimer's disease // *Proc. Natl. Acad. Sci. USA.* 2005. V. 102. P. 2273–2276.

342. Haes A.J., Chang L., Klein W.L., Van Duyne R.P. Detection of a biomarker for Alzheimer's disease from synthetic and clinical samples using a nanoscale optical biosensor // *J. Am. Chem. Soc.* 2005. V. 127. P. 2264–2271.

343. Simonian A.L., Good T.A., Wang S.-S., Wild J.R. Nanoparticle-based optical biosensors for the direct detection of organophosphate chemical warfare agents and pesticides // *Anal. Chim. Acta.* 2005. V. 534. P. 69–77.

344. Boghaert E.R., Khandke K.M., Sridharan L., Dougher M., DiJoseph J.F., Kunz A., Hamann P.R., Moran J., Chaudhary I., Damle N.K. Determination of pharmacokinetic values of calicheamicin-antibody conjugates in mice by plasmon resonance analysis of small (5 μl) blood samples // *Cancer Chemother. Pharmacol.* 2008. V. 61. P. 1027–1035.

345. Maier I., Morgan M.R., Lindner W., Pittner F. Optical resonance-enhanced absorption-based near-field immunochip biosensor for allergen detection // *Anal. Chem.* 2008. V. 80. P. 2694–2703.

346. Huang T., Nallathamby P.D., Xu X.H. Photostable single-molecule nanoparticle optical biosensors for real-time sensing of single cytokine molecules and their binding reactions // *J. Am. Chem. Soc.* 2008. V. 130. P. 17095–17105.

347. Aslan K., Zhang J., Lakowicz J.R., Geddes C.D. Saccharide sensing using gold and silver nanoparticles—a review // *J. Fluoresc.* 2004. V. 14. P. 391–400.

348. Wang L., Jia X., Zhou Y., Xie Q., Yao S. Sandwich-type amperometric immunosensor for human immunoglobulin G using antibody-adsorbed Au/SiO2 nanoparticles // *Microchim. Acta.* 2010. V. 168. P. 245–251.

349. Kwon M.J., Lee J., Wark A.W., Lee H.J. Nanoparticle-enhanced surface plasmon resonance detection of proteins at attomolar concentrations: Comparing different nanoparticle shapes and sizes // *Anal. Chem.* 2012. V. 84. P. 1702–1707.

350. de la Escosura-Muñiz A., Sánchez-Espinel C., Díaz-Freitas B., González-Fernández A., Maltez-da Costa M., Merkoçi A. Rapid identification and quantification of tumor cells using an electrocatalytic method based on gold nanoparticles // *Anal. Chem.* 2009. V. 81. P. 10268–10274.

351. Eum N.-S., Yeom S.-H., Kwon D.-H., Kim H.-R., Kang S.-W. Enhancement of sensitivity using gold nanorods—Antibody conjugator for detection of *E. coli* O157:H7 // *Sens. Actuator B* Chem. 2010. V. 143. P. 784–788.

352. Zhang J., Atay T., Nurmikko A.V. Optical detection of brain cell activity using plasmonic gold nanoparticles // *Nano Lett.* 2009. V. 9. P. 519–524.

353. Hwang W.S., Truong P.L., Sim S.J. Size-dependent plasmonic responses of single gold nanoparticles for analysis of biorecognition // *Anal. Biochem.* 2012. V. 421. P. 213–218.

354. Devi R.V., Doble M., Verma R.S. Nanomaterials for early detection of cancer biomarker with special emphasis on gold nanoparticles in immunoassays/sensors // *Biosens. Bioelectron.* 2015. V. 68. P. 688–698.

355. Wang C., Yu C. Detection of chemical pollutants in water using gold nanoparticles as sensors: A review // *Rev. Anal. Chem.* 2013. V. 32. P. 1–14.

356. Adamczyk M., Johnson D.D., Mattingly P.G., Moore J.A., Pan Y. Immunoassay reagents for thyroid testing. 3. Determination of the solution binding affinities of a T4 monoclonal antibody Fab fragment for a library of thyroxine analogs using surface plasmon resonance // *Bioconjugate Chem.* 1998. V. 9. P. 23–32.

357. Adamczyk M., Moore J.A., Yu Z. Application of surface plasmon resonance toward studies of low-molecular weight antigen-antibody binding interactions // *Methods.* 2000. V. 20. P. 319–328.

358. Seo K.H., Brackett R.E., Hartman N.F., Campbell D.P. Development of a rapid response biosensor for detection of *Salmonella typhimurium* // *J. Food Prot.* 1999. V. 62. P. 431–437.

359. Bao P., Frutos A.G., Greef Ch., Lahiri J., Muller U., Peterson T.C., Warden L., Xie X. High-sensitivity detection of DNA hybridization on microarrays using resonance light scattering // *Anal. Chem.* 2002. V. 74. P. 1792–1797.

360. Piliarik M., Šípová H., Kvasnička P., Galler N., Krenn J.R., Homola J. High-resolution biosensor based on localized surface plasmons // *Opt. Express.* 2012. V. 20. P. 672–680.

361. Raschke G., Kowarik S., Franzl T., Sönnichsen C., Klar T.A., Feldmann J., Nichtl A., Kürzinger K. Biomolecular recognition based on single gold nanoparticle light scattering // *Nano Lett.* 2003. V. 3. P. 935–942.

362. McFarland A.D. Van Duyne R.P. Single silver nanoparticles as real-time optical sensors with zeptomole sensitivity // *Nano Lett.* 2003. V. 3. P. 1057–1062.

363. Mayer K.M., Hao F., Lee S., Nordlander P., Hafner J.H. A single molecule immunoassay by localized surface plasmon resonance // *Nanotechnology.* 2010. V. 21. 255503.

364. Deyev S.M., Lebedenko E.N. Modern technologies for creating synthetic antibodies for clinical application // *Acta Naturae.* 2009. V. 1. P. 32–50.

365. Wu C.-S., Liu F.-K., Ko F.-H. Potential role of gold nanoparticles for improved analytical methods: An introduction to characterizations and applications // *Anal. Bioanal. Chem.* 2011. V. 399. P. 103–118.

366. Yonzon C.R., Jeoung E., Zou S.L., Schatz G.C., Mrksich M., Van Duyne R.P. A comparative analysis of localized and propagating surface plasmon resonance sensors: The binding of concanavalin a to a monosaccharide functionalized self-assembled monolayer // *J. Am. Chem. Soc.* 2004. V. 126. P. 12669–12676.

367. Svedendahl M., Chen S., Dmitriev A., Käll M. Refractometric sensing using propagating versus localized surface plasmons: A direct comparison // *Nano Lett.* 2009. V. 9. P. 4428–4433.

368. Yuan Z., Hu C.-C., Chang H.-T., Lu C. Gold nanoparticles as sensitive optical probes // *Analyst.* 2016. V. 141. P. 1611–1626.

369. Syedmoradi L., Daneshpour M., Alvandipour M., Gomez F.A., Hajghassem H., Omidfar K. Point of care testing: The impact of nanotechnology // *Biosens. Bioelectron.* 2017. V. 87. P. 373–387.

370. Kennedy L.C., Bickford L.R., Lewinski N.A., Coughlin A.J., Hu Y., Day E.S., West J.L., Drezek R.A. A new era for cancer treatment: Gold-nanoparticle-mediated thermal therapies // *Small.* 2011. V. 7. P. 169–183.

371. Abadeer N.S., Murphy C.J. Recent progress in cancer thermal therapy using gold nanoparticles // *J. Phys. Chem. C.* 2016. V. 120. P. 4691–4716.

372. Bucharskaya A., Maslyakova G., Terentyuk G., Yakunin A., Avetisyan Y., Bibikova O., Tuchina E., Khlebtsov B., Khlebtsov N., Tuchin V. Towards effective photothermal/photodynamic treatment using plasmonic gold nanoparticles // *Int. J. Molec. Sci.* 2016. V. 17. 1295.

373. Huang X., Jain P.K., El-Sayed I.H., El-Sayed M.A. Plasmonic photothermal therapy (PPTT) using gold nanoparticles // *Lasers Med. Sci.* 2008. V. 23. P. 217–228.

374. Minton J.P., Carlton D.M., Dearman J.R., McKnight W.B., Ketcham A.S. An evaluation of the physical response of malignant tumor implants to pulsed laser radiation // *Surg. Gynaecol. Obstet.* 1965. V. 121. P. 538–544.

375. Masters A., Bown S.G. Interstitial laser hyperthermia // *Br. J. Cancer.* 1992. V. 8. P. 242–249.

376. Sultan R.A. Tumour ablation by laser in general surgery // *Lasers Med. Sci.* 1990. V. 5. P. 185–193.

377. Anderson R.R., Parrish J.A. Selective photothermolysis: Precise microsurgery by selective absorption of pulsed radiation // *Science.* 1983. V. 220. P. 524–527.

378. Hirsch L.R., Stafford R.J., Bankson J.A., Sershen S.R., Price R.E., Hazle J.D., Halas N.J., West J.L. Nanoshell-mediated near infrared thermal therapy of tumors under MR guidance // *Proc. Natl. Acad. Sci.* 2003. V. 100. P. 13549–13554.

379. Zharov V.P., Galitovsky V., Viegas M. Photothermal detection of local thermal effects during selective nanophotothermolysis // *Appl. Phys. Lett.* 2003. V. 83. P. 4897–4899.

380. Pitsillides C.M., Joe E.K., Wei X., Anderson R.R., Lin C.P. Selective cell targeting with light-absorbing microparticles and nanoparticles // *Biophys J.* 2003. V. 84. P. 4023–4032.

381. Carpin L.B., Bickford L.R., Agollah G., Yu T.K., Schiff R., Li Y., Drezek R.A. Immunoconjugated gold nanoshell-mediated photothermal ablation of trastuzumab-resistant breast cancer cells // *Breast Cancer Res. Treat.* 2011. V. 125. P. 27–34.

382. von Maltzahn G., Park J.-H., Agrawal A., Bandaru N.K., Das S.K., Sailor M.J., Bhatia S.N. Computationally guided photothermal tumor therapy using long-circulating gold nanorod antennas // *Cancer Res.* 2009. V. 69. P. 3892–3900.

383. URL: http://clinicaltrials.gov/ct2/show/NCT00848042.

384. Weissleder R. A clearer vision for *in vivo* imaging // *Nat. Biotechnol.* 2001. V. 19. P. 316–317.

385. Lapotko D., Lukianova-Hleb E., Oraevsky A. A clusterization of nanoparticles during their interaction with living cells // *Nanomedicine.* 2007. V. 2. P. 241–253.

386. Huang X., Qian W., El-Sayed I.H., El-Sayed M.A. The potential use of the enhanced nonlinear properties of gold nanospheres in photothermal cancer therapy // *Lasers Surg. Med.* 2007. V. 39. P. 747–753.

387. Lapotko D., Lukianova E., Potapnev M., Aleinikova O., Oraevsky A. Method of laser activated nanothermolysis for elimination of tumor cells // *Cancer Lett.* 2005. V. 239. P. 36–45.

388. Lukianova-Hleb E., Hu Y., Latterini L., Tarpani L., Lee S., Drezek R.A., Hafner J.H., Lapotko D.O. Plasmonic nanobubbles as transient vapor nanobubbles generated around plasmonic nanoparticles // *ACS Nano.* 2010. V. 4. P. 2109–2123.

389. Harris N., Ford M.J., Cortie M.B. Optimization of plasmonic heating by gold nanospheres and nanoshells // *J. Phys. Chem. B.* 2006. V. 110. P. 10701–10707.

390. Takahashi H., Niidome T., Nariai A., Niidome Y., Yamada S. Gold nanorod-sensitized cell death: Microscopic observation of single living cells irradiated by pulsed near-infrared laser light in the presence of gold nanorods // *Chem. Lett.* 2006. V. 35. P. 500–501.

391. Huff T.B., Tong L., Zhao Y., Hansen M.N., Cheng J.X., Wei A. Hyperthermic effects of gold nanorods on tumor cells // *Nanomedicine.* 2007. V. 2. P. 125–132.

392. Pissuwan D., Valenzuela S.M., Killingsworth M.C., Xu X., Cortie M.B. Targeted destruction of murine macrophage cells with bioconjugated gold nanorods // *J. Nanopart. Res.* 2007. V. 9. P. 1109–1124.

393. Loo C., Lowery A., Halas N.J., West J.L., Drezek R. Immunotargeted nanoshells for integrated cancer imaging and therapy // *Nano Lett.* 2005. V. 5. P. 709–711.

394. Stern J.M., Stanfield J., Lotan Y., Park S., Hsieh J.-T., Cadeddu J.A. Efficacy of laser-activated gold nanoshells in ablating prostate cancer cells *in vitro* // *J. Endourol.* 2007. V. 21. P. 939–942.

395. Diagaradjane P., Shetty A., Wang J.C., Elliott A.M., Schwartz J., Shentu S., Park H.C., Deorukhkar A., Stafford R.J., Cho S.H., Tunnell J.W., Hazle J.D., Krishnan S. Modulation of *in vivo* tumor radiation response via gold nanoshell-mediated vascular-focused hyperthermia: Characterizing an integrated antihypoxic and localized vascular disrupting targeting strategy // *Nano Lett.* 2008. V. 8. P. 1492–1500.

396. Waldman S.A., Fortina P., Surrey S., Hyslop T., Kricka L.J., Graves D.J. Opportunities for near-infrared thermal ablation of colorectal metastases by guanylyl cyclase C targeted gold nanoshells // *Future Oncol.* 2006. V. 2. P. 705–716.

397. Bernardi R.J., Lowery A.R., Thompson P.A., Blaney S.M., West J.L. Immunonanoshells for targeted photothermal ablation in medulloblastoma and glioma: An *in vitro* evaluation using human cell lines // *J. Neurooncol.* 2008. V. 86. P. 165–172.

398. Day E.S., Thompson P.A., Zhang L., Lewinski N.A., Ahmed N., Drezek R.A., Blaney S.M., West J.L. Nanoshell-mediated photothermal therapy improves survival in a murine glioma model // *J. Neurooncol.* 2011. V. 104. P. 55–63.

399. Hu M., Petrova H., Chen J., McLellan J.M., Siekkinen A.R., Marquez M., Li X., Xia Y., Hartland G.V. Ultrafast laser studies of the photothermal properties of gold nanocages // *J. Phys. Chem. B.* 2006. V. 110. P. 1520–1524.

400. Chen J., Wang D., Xi J., Au L., Siekkinen A., Warsen A., Li Z.-Y., Zhang H., Xia Y., Li X. Immuno gold nanocages with tailored optical properties for targeted photothermal destruction of cancer cells // *Nano Lett.* 2007. V. 7. P. 1318–1322.

401. Khlebtsov B.N., Khanadeev V.A., Maksimova I.L., Terentyuk G.S., Khlebtsov N.G. Silver nanocubes and gold nanocages: Fabrication and optical and photothermal properties // *Nanotechnol. Russ.* 2010. V. 5. P. 454–468.

402. Terentyuk G.S., Maslyakova G,N., Suleymanova L.V., Khlebtsov N.G., Khlebtsov B.N., Akchurin G.G., Maksimova I.L., Tuchin V.V. Laser induced tissue hyperthermia mediated by gold nanoparticles: Towards cancer phototherapy // *J. Biomed. Opt.* 2009. V. 14. 021016.

403. Cole J.R., Mirin N.A., Knight M.W., Goodrich G.P., Halas N.J. Photothermal efficiencies of nanoshells and nanorods for clinical therapeutic applications // *J. Phys. Chem. C.* 2009. V. 113. P. 12090–12094.

404. Chirico G., Pallavicini P., Collini M. Gold nanostars for superficial diseases: A promising tool for localized hyperthermia? // *Nanomedicine*. 2014. V. 9. P. 1–3.
405. Lapotko D. Therapy with gold nanoparticles and lasers: What really kills the cells? // *Nanomedicine*. 2009. V. 4. P. 253–256.
406. Hleb E., Hu Y., Drezek R., Hafner J., Lapotko D. Photothermal bubbles as optical scattering probes for imaging living cells // *Nanomedicine*. 2008. V. 3. P. 797–812.
407. Zharov V.P., Galitovskaya E.N., Johnson C., Kelly T. Synergistic enhancement of selective nanophotothermolysis with gold nanoclusters: Potential for cancer therapy // *Lasers Surg. Med.* 2005. V. 37. P. 219–226.
408. Akchurin G., Khlebtsov B., Akchurin G., Tuchin V., Zharov V., Khlebtsov N. Gold nanoshell photomodification under a single-nanosecond laser pulse accompanied by color-shifting and bubble formation phenomena // *Nanotechnology*. 2008. V. 19. 015701.
409. Hleb E.Y., Lapotko D.O. Photothermal properties of gold nanoparticles under exposure to high optical energies // *Nanotechnology*. 2008. V. 19. 355702.
410. Lukianova-Hleb E.Y., Anderson L.J., Lee S., Hafner J.H., Lapotko D.O. Hot plasmonic interactions: A new look at the photothermal efficacy of gold nanoparticles // *Phys. Chem. Chem. Phys.* 2010. V. 12. P. 12237–12244.
411. Jain R.K., Booth M.F. What brings pericytes to tumor vessels? // *J. Clin. Invest.* 2003. V. 112. P. 1134–1136.
412. Philips M.A., Gran M.L., Peppas N.A. Targeted nanodelivery of drugs and diagnostics // *Nano Today*. 2010. V. 5. P. 143–159.
413. Nie S.M. Understanding and overcoming major barriers in cancer nanomedicine // *Nanomedicine*. 2010. V. 5. P. 523–528.
414. El-Sayed I.H., Huang X., El-Sayed M.A. Selective laser photo-thermal therapy of epithelial carcinoma using anti-EGFR antibody conjugated gold nanoparticles // *Cancer Lett.* 2006. V. 239. P. 129–135.
415. Melancon M.P., Lu W., Yang Z., Zhang R., Cheng Z., Elliot A.M., Stafford J., Olson T., Zhang J.Z., Li C. *In vitro* and *in vivo* targeting of hollow gold nanoshells directed at epidermal growth factor receptor for photothermal ablation therapy // *Mol. Cancer Ther.* 2008. V. 7. P. 1730–1739.
416. Visaria R.K., Griffin R.J., Williams B.W., Ebbini E.S., Paciotti G.F., Song C.W., Bischof J.C. Enhancement of tumor thermal therapy using gold nanoparticle–assisted tumor necrosis factor-A delivery // *Mol. Cancer Ther.* 2006. V. 5. P. 1014–1020.
417. Larson T.A., Bankson J., Aaron J., Sokolov K. Hybrid plasmonic magnetic nanoparticles as molecular specific agents for MRI/optical imaging and photothermal therapy of cancer cells // *Nanotechnology*. 2007. V. 18. 325101.
418. Ke H., Wang J., Dai Z., Jin Y., Qu E., Xing Z., Guo C., Yue X., Liu J. Gold-nanoshelled microcapsules: A theranostic agent for ultrasound contrast imaging and photothermal therapy // *Angew. Chem. Int. Ed.* 2011. V. 50. P. 3017–3021.
419. Leamon C.P., Low P.S. Folate-mediated targeting: From diagnostics to drug and gene delivery // *Drug Discov. Today*. 2001. V. 6. P. 44–51.

420. Nayak S., Lee H., Chmielewski J., Lyon L.A. Folatemediated cell targeting and cytotoxicity using thermoresponsive microgels // *J. Am. Chem. Soc.* 2004. V. 126. P. 10258–10259.

421. Lu W., Xiong C., Zhang G., Huang Q., Zhang R., Zhang J.Z., Li C. Targeted photo thermal ablation of murine melanomas with melanocyte-stimulating hormone analog-conjugated hollow gold nanospheres // *Clin. Cancer Res.* 2009. V. 15. P. 876–886.

422. Huang X., Peng X., Wang Y., Wang Y., Shin D.M., El-Sayed M.A., Nie S. A reexamination of active and passive tumor targeting by using rod-shaped gold nanocrystals and covalently conjugated peptide ligands // *ACS Nano.* 2010. V. 4. P. 5887–5896.

423. Kirpotin D.B., Drummond D.C., Shao Y., Shalaby M.R., Hong K., Nielsen U.B., Marks J.D., Benz C.C., Park J.W. Antibody targeting of long-circulating lipidic nanoparticles does not increase tumour localisation but does increase internalization in animal models // *Cancer Res.* 2006. V. 66. P. 6732–6740.

424. Choi C.H., Alabi C.A., Webster P., Davis M.E. Mechanism of active targeting in solid tumors with transferrin-containing gold nanoparticles // *Proc. Natl. Acad. Sci. USA.* 2010. V. 107. P. 1235–1240.

425. Terentyuk G., Panfilova E., Khanadeev V., Chumakov D., Genina E., Bashkatov A., Tuchin V., Bucharskaya A., Maslyakova G., Khlebtsov N., Khlebtsov B. Gold nanorods with hematoporphyrin-loaded silica shell for dual-modality photodynamic and photothermal treatment of tumors in vivo // *Nano Res.* 2104. V. 7. P. 325–337.

426. Hainfeld J.F., Slatkin D.N., Smilowitz H.M. The use of gold nanoparticles to enhance radiotherapy in mice // *Phys. Med. Biol.* 2004. V. 49. P. N309–N315.

427. Gannon C.J., Patra C.R., Bhattacharya R., Mukherjee P., Curley S.A. Intracellular gold nanoparticles enhance non-invasive radiofrequency thermal destruction of human gastrointestinal cancer cells // *J. Nanobiotechnology.* 2008. V. 6. 2.

428. Cardinal J., Klune J.R., Chory E., Jeyabalan G., Kanzius J.S., Nalesnik M., Geller D.A. Noninvasive radiofrequency ablation of cancer targeted by gold nanoparticles // *Surgery.* 2008. V. 144. P. 125–132.

429. Zhang X., Xing J.Z., Chen J., Ko L., Amanie J., Gulavita S., Pervez N., Yee D., Moore R., Roa W. Enhanced radiation sensitivity in prostate cancer by gold-nanoparticles // *Clin. Investig. Med.* 2008. V. 31. P. 160–167.

430. Liu X., Chen H.-J., Chen X., Alfadhl Y., Yu J., Wen D. Radiofrequency heating of nanomaterials for cancer treatment: Progress, controversies, and future development // *Appl. Phys. Rev.* 2015. V. 2. 011103.

431. Kim G.J., Park S.R., Kim G.C., Lee J.K. Targeted cancer treatment using anti-EGFR and -TFR antibody-conjugated gold nanoparticles stimulated by nonthermal air plasma // *Plasma Med.* 2011. V. 1. P. 45–54.

432. Pissuwan D., Valenzuela S.M., Miller C.M., Cortie M.B. A golden bullet? Selective targeting of *Toxoplasma gondii* tachyzoites using antibody-functionalized gold nanorods // *Nano Lett.* 2007. V. 7. P. 3808–3812.

433. Huang W.-C., Tsai P.-J., Chen Y.-C. Functional gold nanoparticles as phototermal agents for selective-killing of pathogenic bacteria // *Nanomedicine.* 2007. V. 2. P. 777–787.

434. Zharov V.P., Mercer K.E., Galitovskaya E.N., Smeltzery M.S. Photothermal nanotherapeutics and nanodiagnostics for selective killing of bacteria targeted with gold nanoparticles // *Biophys. J.* 2006. V. 90. P. 619–627.

435. Norman R.S., Stone J.W., Gole A., Murphy C.J., Sabo-Attwood T.L. Targeted photothermal lysis of the pathogenic bacteria, *Pseudomonas aeruginosa*, with gold nanorods // *Nano Lett.* 2008. V. 8. P. 302–306.

436. Ray P.C., Khan S.A., Singh A.K., Senapati D., Fan Z. Nanomaterials for targeted detection and photothermal killing of bacteria // *Chem. Soc. Rev.* 2012. V. 41. P. 3193–3209.

437. Jo W., Kim M.J. Influence of the photothermal effect of a gold nanorod cluster on biofilm disinfection // *Nanotechnology.* 2013. V. 24. 195104.

438. Castillo-Martínez J.C., Martínez-Castañón G.A., Martínez-Gutierrez F., Zavala-Alonso N.V., Patiño-Marín N., Niño-Martinez N., Zaragoza-Magaña V., Cabral-Romero C. Antibacterial and antibiofilm activities of the photothermal therapy using gold nanorods against seven different bacterial strains // *J. Nanomater.* 2015. V. 2015. 783671.

439. Pissuwan D., Valenzuela S.M., Cortie M.B. Therapeutic possibilities of plasmonically heated gold nanoparticles // *Trends Biotech.* 2006. V. 24. P. 62–67.

440. Allison R.R., Mota H.C., Bagnato V.S., Sibata C.H. Bio-nanotechnology and photodynamic therapy—State of the art review // *Photodiagn. Photodyn. Ther.* 2008. V. 5. P. 19–28.

441. Subramani K. Applications of nanotechnology in drug delivery systems for the treatment of cancer and diabetes // *Int. J. Nanotechnol.* 2006. V. 3. P. 557–580.

442. Cai W., Gao T., Hong H., Sun J. Applications of gold nanoparticles in cancer nanotechnology // *Nanotechnol. Sci. Appl.* 2008. V. 1. P. 17–32.

443. *Cancer Nanotechnology. Methods and Protocols* / Eds. Grobmyer S.R., Moudgil B.M. New York: Humana Press. 2010.

444. Baffou G., Quidant R. Thermo-plasmonics: Using metallic nanostructures as nano-sources of heat // *Laser Photonics Rev.* 2013. V. 7. P. 171–187.

445. Iancu C. Photothermal therapy of human cancers (PTT) using gold nanoparticles // *Biotechnol. Mol. Biol. Nanomed.* 2013. V. 1. P. 53–60.

446. Jaque D., Maestro L.M., del Rosal B., Haro-Gonzalez P., Benayas A., Plaza J.L., Rodríguez E.M., Solé J.G. Nanoparticles for photothermal therapies // *Nanoscale.* 2014. V. 6. P. 9494–9530.

447. Qiu J., Wei W.D. Surface plasmon-mediated photothermal chemistry // *J. Phys. Chem. C.* 2014. V. 118. P. 20735–20749.

448. Hwang S., Nam J., Jung S., Song J., Doh H., Kim S. Gold nanoparticle-mediated photothermal therapy: Current status and future perspective // *Nanomedicine (Lond.).* 2014. V. 9. P. 2003–2022.

449. Cheng L., Wang C., Feng L., Yang K., Liu Z. Functional nanomaterials for phototherapies of cancer // *Chem. Rev.* 2014. V. 114. P. 10869–10939.

450. Ahmad R., Fu J., He N., Li S. Advanced gold nanomaterials for photothermal therapy of cancer // *J. Nanosci. Nanotechnol.* 2016. V. 16. P. 67–80.

451. Sau T.K., Rogach A.L., Jäckel F., Klar T.A., Feldmann J. Properties and applications of colloidal nonspherical noble metal nanoparticles // *Adv. Mater.* 2010. V. 22. P. 1805–1825.

452. Wilson B.C. Photodynamic therapy/diagnostics // In: *Handbook of Photonics for Biomedical Science* / Ed. Tuchin V.V. Boca Raton: CRC Press. 2010. P. 649–686.

453. Gamaleia N.F., Shton I.O. Gold mining for PDT: Great expectations from tiny nanoparticles // *Photodiagnosis Photodyn. Ther.* 2015. V. 12. P. 221–231.

454. Yin R., Agrawal T., Khan U., Gupta G.K., Rai V., Huang Y.-Y., Hamblin M.R. Antimicrobial photodynamic inactivation in nanomedicine: Small light strides against bad bugs // *Nanomedicine (Lond.)*. 2015. V. 10. P. 2379–2404.

455. Wilson R. The use of gold nanoparticles in diagnostics and detection // *Chem. Soc. Rev.* 2008. V. 37. P. 2028–2045.

456. Lakowicz J.R., Ray K., Chowdhury M., Szmacinski H., Fu Y., Zhang J., Nowaczyk K. Plasmon-controlled fluorescence: A new paradigm in fluorescence spectroscopy // *Analyst*. 2008. V. 133. P. 1308–1346.

457. Bardhan R., Grady N.K., Cole J.R., Joshi A., Halas N.J. Fluorescence enhancement by Au nanostructures: Nanoshells and nanorods // *ACS Nano*. 2009. V. 3. P. 744–752.

458. Sershen S.R., Westcott S.L., Halas N.J., West J.L. Temperature-sensitive polymer-nanoshell composites for photothermally modulated drug delivery // *J. Biomed. Mater. Res.* 2000. V. 51. P. 293–298.

459. Radt B., Smith T.A., Caruso F. Optically addressable nanostructured capsule // *Adv. Mater.* 2004. V. 16. P. 2184–2189.

460. Shiotani A., Mori T., Niidome T., Niidome Y., Katayama Y. Stable incorporation of gold nanorods into *N*-isopropylacrylamide hydrogels and their rapid shrinkage induced by near-infrared laser irradiation // *Langmuir*. 2007. V. 23. P. 4012–4018.

461. Nakamura T., Tamura A., Murotani H., Oishi M., Jinji Y., Matsuishi K., Nagasaki Y. Large payloads of gold nanoparticles into the polyamine network core of stimuli-responsive PEGylated nanogels for selective and noninvasive cancer photothermal therapy // *Nanoscale*. 2010. V. 2. P. 739–746.

462. Thomas K.G., Kamat P.V. Chromophore-functionalized gold nanoparticles // *Acc. Chem. Res.* 2003. V. 36. P. 888–898.

463. Hu J., Tang Y., Elmenoufy A.H., Xu H., Cheng Z., Yang X. Nanocomposite-based photodynamic therapy strategies for deep tumor treatment // *Small*. 2015. V. 11. P. 5860–5887.

464. Chithrani B.D., Chan W.C. Elucidating the mechanism of cellular uptake and removal of protein-coated gold nanoparticles of different sizes and shapes // *Nano Lett*. 2007. V. 7. P. 1542–1550.

465. Liu S.Y., Liang Z.S., Gao F., Luo S.F., Lu G.Q. *In vitro* photothermal study of gold nanoshells functionalized with small targeting peptides to liver cancer cells // *J. Mater. Sci. Mater. Med*. 2010. V. 21. P. 665–674.

466. Bardhan R., Chen W., Bartels M., Perez-Torres C., Botero M.F., McAninch R.W., Contreras A., Schiff R., Pautler R.G., Halas N.J., Joshi A. Tracking of multimodal therapeutic nanocomplexes targeting breast cancer *in vivo* // *Nano Lett.* 2010. V. 10. P. 4920–4928.

467. Kuo W.-S., Chang C.-N., Chang Y.-T., Yeh C.-S. Antimicrobial gold nanorods with dual-modality photodynamic inactivation and hyperthermia // *Chem. Commun. (Camb.).* 2009. No. 32. P. 4853–4855.

468. Kuo W.-S., Chang C.-N., Chang Y.-T. Yang M.-H., Chien Y.-H., Chen S.-J., Yeh C.-S. Gold nanorods in photodynamic therapy, as hyperthermia agents, and in near-infrared optical imaging // *Angew. Chem. Int. Ed.* 2010. V. 49. P. 2711–2715.

469. Tuchina E.S., Tuchin V.V., Khlebtsov B.N., Khlebtsov N.G. Phototoxic effect of conjugates of plasmon-resonance nanoparticles with indocyanine green dye on Staphylococcus aureus induced by IR laser radiation // *Quantum Electron.* 2011. V. 41. P. 354–359.

470. Bhattacharyya S., Kudgus R.A., Bhattacharya R., Mukherjee P. Inorganic nanoparticles in cancer therapy // *Pharm. Res.* 2011. V. 28. P. 237–259.

471. Arvizo R.R., Bhattacharyya S., Kudgus R.A., Giri K., Bhattacharya R., Mukherjee P. Intrinsic therapeutic applications of noble metal nanoparticles: Past, present and future // *Chem. Soc. Rev.* 2012. V. 41. P. 2943–2970.

472. Abraham G.E., Himmel P.B. Management of rheumatoid arthritis: Rationale for the use of colloidal metallic gold // *J. Nutr. Med.* 1997. V. 7. P. 295–305.

473. Abraham G.E. Clinical applications of gold and silver nanocolloids // *Orig. Intern.* 2008. V. 15. P. 132–158.

474. Tsai C.Y., Shiau A.L., Chen S.Y., Chen Y.H., Cheng P.C., Chang M.Y., Chen D.H., Chou C.H., Wang C.R., Wu C.L. Amelioration of collagen-induced arthritis in rats by nanogold // *Arthritis Rheum.* 2007. V. 56. P. 544–554.

475. Brown C.L., Bushell G., Whitehouse M.W., Agrawal D.S., Tupe S.G., Paknikar K.M., Tiekink E.R.T. Nanogold-pharmaceutics // *Gold Bull.* 2007. V. 40. P. 245–250.

476. Brown C.L., Whitehouse M.W., Tiekink E.R.T., Bushell G.R. Colloidal metallic gold is not bio-inert // *Inflammopharmacology.* 2008. V. 16. P. 133–137.

477. Paciotti G.F., Myer L., Weinreich D., Goia D., Pavel N., McLaughlin R.E., Tamarkin L. Colloidal gold: A novel nanoparticle vector for tumor directed drug delivery // *Drug Deliv.* 2004. V. 11. P. 169–183.

478. Paciotti G.F., Kingston D.G.I., Tamarkin L. Colloidal gold nanoparticles: A novel nanoparticle platform for developing multifunctional tumor-targeted drug delivery vectors // *Drug Dev. Res.* 2006. V. 67. P. 47–54.

479. Farma J.M., Puhlmann M., Soriano P.A., Cox D., Paciotti G.F., Tamarkin L., Alexander H.R. Direct evidence for rapid and selective induction of tumor neovascular permeability by tumor necrosis factor and a novel derivative, colloidal gold bound tumor necrosis factor // *Int. J. Cancer.* 2007. V. 120. P. 2474–2480.

480. Stern S.T., Hall J.B., Yu L.L., Wood L.J., Paciotti G.F., Tamarkin L., Long S.E., McNeil S.E. Translational considerations for cancer nanomedicine // *J. Control. Release.* 2010. V. 146. P. 164–174.

481. Libutti S.K., Paciotti G.F., Byrnes A.A., Alexander H.R., Jr., Gannon W.E., Walker M., Seidel G.D., Yuldasheva N., Tamarkin L. Phase I and pharmacokinetic studies of CYT-6091, a novel PEGylated colloidal gold-rhTNF nanomedicine // *Clin. Cancer Res.* 2010. V. 16. P. 6139–6149.

482. Luque-Michel E., Imbuluzqueta E., Sebastián V., Blanco-Prieto M.J. Clinical advances of nanocarrier-based cancer therapy and diagnostics // *Expert Opin. Drug Deliv.* 2017. V. 14. P. 75–92.

483. Bhattacharya R., Mukherjee P., Xiong Z., Atala A., Soker S., Mukhopadhyay D. Gold nanoparticles inhibit VEGF165-induced proliferation of HUVEC cells // *Nano Lett.* 2004. V. 4. P. 2479–2481.

484. Mukherjee P., Bhattacharya R., Wang P., Wang L., Basu S., Nagy J.A., Atala A., Mukhopadhyay D., Soker S. Antiangiogenic properties of gold nanoparticles // *Clin. Cancer Res.* 2005. V. 11. P. 3530–3534.

485. Arvizo R.R., Rana S., Miranda O.R., Bhattacharya R., Rotello V.M., Mukherjee P. Mechanism of anti-angiogenic property of gold nanoparticles: Role of nanoparticle size and surface charge // *Nanomedicine.* 2011. V. 7. P. 580–587.

486. Bhattacharya R., Mukherjee P. Biological properties of "naked" metal nanoparticles // *Adv. Drug Deliv. Rev.* 2008. V. 60. P. 1289–1306.

487. Kalishwaralal K., Sheikpranbabu S., Barathmanikanth S., Haribalaganesh R., Ramkumarpandian S., Gurunathan S. Gold nanoparticles inhibit vascular endothelial growth factor-induced angiogenesis and vascular permeability via Src dependent pathway in retinal endothelial cells // *Angiogenesis.* 2011. V. 14. P. 29–45.

488. Roma-Rodrigues C., Heuer-Jungemann A., Fernandes A.R., Kanaras A.G., Baptista P.V. Peptide-coated gold nanoparticles for modulation of angiogenesis *in vivo* // *Int. J. Nanomedicine.* 2016. V. 11. P. 2633–2639.

489. Mukherjee P., Bhattacharya R., Bone N., Lee Y.K., Patra C.R., Wang S., Lu L., Secreto C., Banerjee P.C., Yaszemski M.J., Kay N.E., Mukhopadhyay D. Potential therapeutic application of gold nanoparticles in B-chronic lymphocytic leukemia (BCLL): Enhancing apoptosis // *J. Nanobiotechnol.* 2007. V. 5. 4.

490. Bhattacharya R., Patra C.R., Verma R., Kumar S., Greipp P.R., Mukherjee P. Gold nanoparticles inhibit the proliferation of multiple myeloma cells // *Adv. Mater.* 2007. V. 19. P. 711–716.

491. Wang L., Liu Y., Li W., Jiang X., Ji Y., Wu X., Xu L., Qiu Y., Zhao K., Wei T., Li Y., Zhao Y., Chen C. Selective targeting of gold nanorods at the mitochondria of cancer cells: Implications for cancer therapy // *Nano Lett.* 2011. V. 11. P. 772–780.

492. Alanazi F.K., Radwan A.A., Alsarra I.A. Biopharmaceutical applications of nanogold // *Saudi Pharm. J.* 2010. V. 18. P. 179–193.

493. Duncan B., Kim C., Rotello V.M. Gold nanoparticle platforms as drug and biomacromolecule delivery systems // *J. Control. Release.* 2010. V. 148. P. 122–127.

494. Ahmad M.Z., Akhter S., Jain G.K., Rahman M., Pathan S.A., Ahmad F.J., Khar R.K. Metallic nanoparticles: Technology overview & drug delivery applications in oncology // *Exp. Opin. Drug Deliv.* 2010. V. 7. P. 927–942.

495. Pissuwan D., Niidome T., Cortie M.B. The forthcoming applications of gold nanoparticles in drug and gene delivery systems // *J. Control. Release.* 2011. V. 149. P. 65–71.

496. Rana S., Bajaj A., Mout R., Rotello V.M. Monolayer coated gold nanoparticles for delivery applications // *Adv. Drug Deliv. Rev.* 2012. V. 64. P. 200–216.

497. Kumar A., Zhang X., Liang X.-J. Gold nanoparticles: Emerging paradigm for targeted drug delivery system // *Biotechnol. Adv.* 2013. V. 31. P. 593–606.

498. Vigderman L., Zubarev E.R. Therapeutic platforms based on gold nanoparticles and their covalent conjugates with drug molecules // *Adv. Drug Deliv. Rev.* 2013. V. 65. P. 663–676.

499. Mocan L. Drug delivery applications of gold nanoparticles // *Biotechnol. Mol. Biol. Nanomed.* 2013. V. 1. P. 1–6.

500. Jeong E.H., Jung G., Hong C.A., Lee H. Gold nanoparticle (AuNP)-based drug delivery and molecular imaging for biomedical applications // *Arch. Pharm. Res.* 2014. V. 37. P. 53–59.

501. Fratoddi I., Venditti I., Cametti C., Russo M.V. Gold nanoparticles and gold nanoparticle-conjugates for delivery of therapeutic molecules. Progress and challenges // *J. Mater. Chem. B.* 2014. V. 2. P. 4204–4220.

502. Llevot A., Astruc D. Applications of vectorized gold nanoparticles to the diagnosis and therapy of cancer// *Chem. Soc. Rev.* 2012. V. 41. P. 242–257.

503. Zhao Y., Jiang X. Multiple strategies to activate gold nanoparticles as antibiotics // *Nanoscale.* 2013. V. 5. P. 8340–8350.

504. Chen Y.H., Tsai C.Y., Huang P.Y., Chang M.Y., Cheng P.C., Chou C.H., Chen D.H., Wang C.R., Shiau A.L., Wu C.L. Methotrexate conjugated to gold nanoparticles inhibits tumor growth in a syngeneic lung tumor model // *Mol. Pharm.* 2007. V. 4. P. 713–722.

505. Li J., Wang X., Wang C., Chen B., Dai Y., Zhang R., Song M., Lv G., Fu D. The enhancement effect of gold nanoparticles in drug delivery and as biomarkers of drug-resistant cancer cells // *ChemMedChem.* 2007. V. 2. P. 374–378.

506. Patra C.R., Bhattacharya R., Wang E., Katarya A., Lau J.S., Dutta S., Muders M., Wang S., Buhrow S.A., Safgren S.L., Yaszemski M.J., Reid J.M., Ames M.M., Mukherjee P., Mukhopadhyay D. Targeted delivery of gemcitabine to pancreatic adenocarcinoma using cetuximab as a targeting agent // *Cancer Res.* 2008. V. 68. P. 1970–1978.

507. Podsiadlo P., Sinani V.A., Bahng J.H., Kam N.W., Lee J., Kotov N.A. Gold nanoparticles enhance the anti-leukemia action of a 6-mercaptopurine chemotherapeutic agent // *Langmuir.* 2008. V. 24. P. 568–574.

508. Azzam E.M.S., Morsy S.M.I. Enhancement of the antitumour activity for the synthesised dodecylcysteine surfactant using gold nanoparticles // *J. Surf. Deterg.* 2008. V. 11. P. 195–199.

509. Stiti M., Cecchi A., Rami M., Abdaoui M., Barragan-Montero V., Scozzafava A., Guari Y., Winum J.Y., Supuran C.T. Carbonic anhydrase inhibitor coated gold nanoparticles selectively inhibit the tumor-associated isoform IX over the cytosolic isozymes I and II // *J. Am. Chem. Soc.* 2008. V. 130. P. 16130–16131.

510. Agasti S.S., Chompoosor A., You C.C., Ghosh P., Kim C.K., Rotello V.M. Photoregulated release of caged anticancer drugs from gold nanoparticles // *J. Am. Chem. Soc.* 2009. V. 131. P. 5728–5729.

511. Dhar S., Daniel W.L., Giljohann D.A., Mirkin C.A., Lippard S.J. Polyvalent oligonucleotide gold nanoparticle conjugates as delivery vehicles for platinum(IV) warheads // *J. Am. Chem. Soc.* 2009. V. 131. P. 14652–14653.

512. Patra C.R., Bhattacharya R., Mukherjee P. Fabrication and functional characterization of goldnanoconjugates for potential application in ovarian cancer // *J. Mater. Chem.* 2010. V. 20. P. 547–554.

513. Brown S.D., Nativo P., Smith J.-A., Stirling D., Edwards P.R., Venugopal B., Flint D.J., Plumb J.A., Graham D., Wheate N.J. Gold nanoparticles for the improved anticancer drug delivery of the active component of oxaliplatin // *J. Am. Chem. Soc.* 2010. V. 132. P. 4678–4684.

514. Hosta L., Pla-Roca M., Arbiol J., López-Iglesias C., Samitier J., Cruz L.J., Kogan M.J., Albericio F. Conjugation of Kahalalide F with gold nanoparticles to enhance *in vitro* antitumoral activity // *Bioconjug. Chem.* 2009. V. 20. P. 138–146.

515. Dreaden E.C., Mwakwari S.C., Sodji Q.H., Oyelere A.K., El-Sayed M.A. Tamoxifen-poly(ethylene glycol)-thiol gold nanoparticle conjugates: Enhanced potency and selective delivery for breast cancer treatment // *Bioconjug. Chem.* 2009. V. 20. P. 2247–2253.

516. Eghtedari M., Liopo A.V., Copland J.A., Oraevsky A.A., Motamedi M. Engineering of hetero-functional gold nanorods for the *in vivo* molecular targeting of breast cancer cells // *Nano Lett.* 2009. V. 9. P. 287–291.

517. Park C., Youn H., Kim H., Noh T., Kook Y.H., Oh E.T., Park H.J., Kim C. Cyclodextrin-covered gold nanoparticles for targeted delivery of an anti-cancer drug // *J. Mater. Chem.* 2009. V. 19. P. 2310–2315.

518. Asadishad B., Vossoughi M., Alemzadeh I. Folate-receptor-targeted delivery of doxorubicin using polyethylene glycol-functionalized gold nanoparticles // *Ind. Eng. Chem. Res.* 2010. V. 49. P. 1958–1963.

519. Staroverov S.A., Gasina O.A., Kladiev A.A., Bogatyrev V.A. The impact of complex prospidin–colloidal gold on tumor cells // *Russian J. Biother.* 2010. V. 9. P. 22–23 (in Russian).

520. François A., Laroche A., Pinaud N., Salmon L., Ruiz J., Robert J., Astruc D. Encapsulation of docetaxel into PEGylated gold nanoparticles for vectorization to cancer cells // *ChemMedChem.* 2011. V. 6. P. 2003–2008.

521. Joshi P., Chakraborti S., Ramirez-Vick J.E., Ansari Z.A., Shanker V., Chakrabarti P., Singh S.P. The anticancer activity of chloroquine-gold nanoparticles against MCF-7 breast cancer cells // *Colloids Surf. B.* 2012. V. 95. P. 195–200.

522. Kao H.-W., Lin Y.-Y., Chen C.-C., Chi K.-H., Tien D.-C., Hsia C.-C., Lin W.-J., Chen F.-D., Lin M.-H., Wang H.-E. Biological characterization of cetuximab conjugated gold nanoparticles in a tumor animal model // *Nanotechnology.* 2014. V. 25. 295102.

523. Rizk N., Christoforou N., Lee S. Optimization of anti-cancer drugs and a targeting molecule on multifunctional gold nanoparticles // *Nanotechnology.* 2016. V. 27. 185704.

524. Pradhan A., Bepari M., Maity P., Roy S.S., Roy S., Choudhury S.M. Gold nanoparticles from indole-3-carbinol exhibit cytotoxic, genotoxic and antineoplastic effects through the induction of apoptosis // *RSC Adv.* 2016. V. 6. P. 56435–56449.

525. http://mayoresearch.mayo.edu/mayo/research/dev_lab/nanogold.cfm.
526. Cheng Y., Samia A.C., Meyers J.D., Panagopoulos I., Fei B., Burda C. Highly efficient drug delivery with gold nanoparticle vectors for in vivo photodynamic therapy of cancer // *J. Am. Chem. Soc.* 2008. V. 130. P. 10643–10647.
527. Kim C.-k., Ghosh P., Rotello V.M. Multimodal drug delivery using gold nanoparticles // *Nanoscale.* 2009. V. 1. P. 61–67.
528. Qiu L., Chen T., Öçsoy I., Yasun E., Wu C., Zhu G., You M., Han D., Jiang J., Yu R., Tan W. A cell-targeted, size-photocontrollable, nuclear-uptake nanodrug delivery system for drug-resistant cancer therapy // *Nano Lett.* 2015. V. 15. P. 457–463.
529. Wang C.-C., Wu S.-M., Li H.-W., Chang H.-T. Biomedical applications of DNA-conjugated gold nanoparticles // *ChemBioChem.* 2016. V. 17. P. 1052–1062.
530. Dreaden E.C., Mackey M.A., Huang X., Kangy B., El-Sayed M.A. Beating cancer in multiple ways using nanogold // *Chem. Soc. Rev.* 2011. V. 40. P. 3391–3404.
531. Gu H., Ho P.L., Tong E., Wang L., Xu B. Presenting vancomycin on nanoparticles to enhance antimicrobial activities // *Nano Lett.* 2003. V. 3. P. 1261–1263.
532. Rosemary M.J., MacLaren I., Pradeep T. Investigations of the antibacterial properties of ciprofloxacin@SiO2 // *Langmuir.* 2006. V. 22. P. 10125–10129.
533. Selvaraj V., Alagar M. Analytical detection and biological assay of antileukemic drug 5-fluorouracil using gold nanoparticles as probe // *Int. J. Pharm.* 2007. V. 337. P. 275–281.
534. Saha B., Bhattacharya J., Mukherjee A., Ghosh A.K., Santra C.R., Dasgupta A.K., Karmakar P. *In vitro* structural and functional evaluation of gold nanoparticles conjugated antibiotics // *Nanoscale Res. Lett.* 2007. V. 2. P. 614–622.
535. Grace A.N., Pandian K. Antibacterial efficacy of aminoglycosidic antibiotics protected gold nanoparticles—A brief study // *Colloids Surf. A Physicochem. Eng. Aspects.* 2007. V. 297. P. 63–70.
536. Grace A.N., Pandian K. Quinolone antibiotic–capped gold nanoparticles and their antibacterial efficacy against gram positive and gram negative organisms // *J. Bionanosci.* 2007. V. 1. P. 96–105.
537. Burygin G.L., Khlebtsov B.N., Shantrokha A.N., Dykman L.A., Bogatyrev V.A., Khlebtsov N.G. On the enhanced antibacterial activity of antibiotics mixed with gold nanoparticles // *Nanoscale Res. Lett.* 2009. V. 4. P. 794–801.
538. Rai A., Prabhune A., Perry C.C. Antibiotic mediated synthesis of gold nanoparticles with potent antimicrobial activity and their application in antimicrobial coatings // *J. Mater. Chem.* 2010. V. 20. P. 6789–6798.
539. Rastogi L., Kora A.J., Arunachalam J. Highly stable, protein capped gold nanoparticles as effective drug delivery vehicles for amino-glycosidic antibiotics // *Mat. Sci. Eng. C.* 2012. V. 32. P. 1571–1577.
540. Ahangari A., Salouti M., Heidari Z., Kazemizadeh A.R., Safari A.A. Development of gentamicin-gold nanospheres for antimicrobial drug delivery to *Staphylococcal* infected foci // *Drug Deliv.* 2013. V. 20. P. 34–39.
541. Payne J.N., Waghwani H.K., Connor M.G., Hamilton W., Tockstein S., Moolani H., Chavda F., Badwaik V., Lawrenz M.B., Dakshinamurthy R. Novel Synthesis of kanamycin conjugated gold nanoparticles with potent antibacterial activity // *Front. Microbiol.* 2016. V. 7. 607.

542. Brown A., Smith K., Samuels T.A., Lu J., Obare S., Scott M.E. Nanoparticles functionalized with ampicillin destroy multiple antibiotic resistant isolates of *Pseudomonas aeruginosa, Enterobacter aerogenes* and methicillin resistant *Staphylococcus aureus* // *Appl. Environ. Microbiol.* 2012. V. 78. P. 2768–2774.

543. Kalita S., Kandimalla R., Sharma K.K., Kataki A.C., Deka M., Kotoky J. Amoxicillin functionalized gold nanoparticles reverts MRSA resistance // *Mater. Sci. Eng. C.* 2016. V. 61. P. 720–727.

544. Ali H.R., Ali M.R., Wu Y., Selim S.A., Abdelaal H.F., Nasr E.A., El-Sayed M.A. Gold nanorods as drug delivery vehicles for rifampicin greatly improve the efficacy of combating *Mycobacterium tuberculosis* with good biocompatibility with the host cells // *Bioconjug. Chem.* 2016. V. 27. P. 2486–2492.

545. Du L., Suo S., Zhang H., Jia H., Liu K.J., Zhang X.J., Liu Y. The alternative strategy for designing covalent drugs through kinetic effects of pi-stacking on the self-assembled nanoparticles: A model study with antibiotics // *Nanotechnology.* 2016. V. 27. 445101.

546. Nie Z., Liu K.J., Zhong C.J., Wang L.F., Yang Y., Tian Q., Liu Y. Enhanced radical scavenging activity by antioxidant-functionalized gold nanoparticles: A novel inspiration for development of new artificial antioxidants // *Free Radic. Biol. Med.* 2007. V. 43. P. 1243–1254.

547. Bowman M.C., Ballard T.E., Ackerson C.J., Feldheim D.L., Margolis D.M., Melander C. Inhibition of HIV fusion with multivalent gold nanoparticles // *J. Am. Chem. Soc.* 2008. V. 130. P. 6896–6897.

548. Joshi H.M., Bhumkar D.R., Joshi K., Pokharkar V., Sastry M. Gold nanoparticles as carriers for efficient transmucosal insulin delivery // *Langmuir.* 2006. V. 22. P. 300–305.

549. Cho H.-J., Oh J., Choo M.-K., Ha J.-I., Park Y., Maeng H.-J. Chondroitin sulfate-capped gold nanoparticles for the oral delivery of insulin // *Int. J. Biol. Macromol.* 2014. V. 63. P. 15–20.

550. Ehsan O., Qadir M.I., Malik S.A., Abbasi W.S., Ahmad B. Efficacy of nanogold-insulin as a hypoglycemic agent // *J. Chem. Soc. Pak.* 2012. V. 34. P. 365–370.

551. Shilo M., Berenstein P., Dreifuss T., Nash Y., Goldsmith G., Kazimirsky G., Motiei M., Frenkel D., Brodie C., Popovtzer R. Insulin-coated gold nanoparticles as a new concept for personalized and adjustable glucose regulation // *Nanoscale.* 2015. V. 7. P. 20489–20496.

552. Chamberland D.L., Agarwal A., Kotov N., Fowlkes J.B., Carson P.L., Wang X. Photoacoustic tomography of joints aided by an Etanercept-conjugated gold nanoparticle contrast agent—An *ex vivo* preliminary rat study // *Nanotechnology.* 2008. V. 19. 095101.

553. Huang Y.-J., Shiau A.-L., Chen S.-Y., Chen Y.-L., Wang C.-R., Tsai C.-Y., Chang M.-Y., Li Y.-T., Leu C.-H., Wu C.-L. Multivalent structure of galectin-1-nanogold complex serves as potential therapeutics for rheumatoid arthritis by enhancing receptor clustering // *Eur. Cell Mater.* 2012. V. 23. P. 170–181.

554. Gomes A., Datta P., Sengupta J., Biswas A., Gomes A. Evaluation of anti-arthritic property of methotrexate conjugated gold nanoparticle on experimental animal models // *J. Nanopharm. Drug Deliv.* 2013. V. 1. P. 1–6.

555. Kabir N., Ali H., Ateeq M., Bertino M.F., Shah M.R., Franzel L. Silymarin coated gold nanoparticles ameliorates CCl4-induced hepatic injury and cirrhosis through down regulation of Hepatic stellate cells and attenuation of Kupffer cells // *RSC Adv.* 2014. V. 4. P. 9012–9020.

556. Abdellatif A.A.H., Rasoul S.A.E., Osman S. Gold nanoparticles decorated with octreotide for somatostatin receptors targeting // *J. Pharm. Sci. Res.* 2015. V. 7. P. 14–20.

557. Gao N., Sun H., Dong K., Ren J., Qu X. Gold-nanoparticle-based multifunctional amyloid-β inhibitor against Alzheimer's disease // *Chemistry.* 2015. V. 21. P. 829–835.

558. Miller A.D. Human gene-therapy comes of age // *Nature.* 1992. V. 357. P. 455–460.

559. Zelenin A.V. Gene therapy: On the threshold of the third millennium // *Herald of the RAS.* 2001. V. 71. P. 215–222.

560. Oishi M., Nakaogami J., Ishii T., Nagasaki Y. Smart PEGylated gold nanoparticles for the cytoplasmic delivery of siRNA to induce enhanced gene silencing // *Chem. Lett.* 2006. V. 35. P. 1046–1047.

561. Noh S.M., Kim W.K., Kim S.J., Kim J.M., Baek K.H., Oh Y.K. Enhanced cellular delivery and transfection efficiency of plasmid DNA using positively charged biocompatible colloidal gold nanoparticles // *Biochim. Biophys. Acta.* 2007. V. 1770. P. 747–752.

562. Lee J.S., Green J.J., Love K.T., Sunshine J., Langer R., Anderson D.G. Gold, poly(β-amino ester) nanoparticles for small interfering RNA delivery // *Nano Lett.* 2009. V. 9. P. 2402–2406.

563. Patel P.C., Giljohann D.A., Daniel W.L., Zheng D., Prigodich A.E., Mirkin C.A. Scavenger receptors mediate cellular uptake of polyvalent oligonucleotide-functionalized gold nanoparticles // *Bioconjug. Chem.* 2010. V. 21. P. 2250–2256.

564. Ding Y., Jiang Z., Saha K., Kim C.S., Kim S.T., Landis R.F., Rotello V.M. Gold nanoparticles for nucleic acid delivery // *Mol. Ther.* 2014. V. 22. P. 1075–1083.

565. Barnaby S.N., Lee A., Mirkin C.A. Probing the inherent stability of siRNA immobilized on nanoparticle constructs // *Proc. Natl. Acad. Sci. USA.* 2014. V. 111. P. 9739–9744.

566. Jiang Y., Huo S., Hardie J., Liang X.-J., Rotello V.M. Progress and perspective of inorganic nanoparticles based siRNA delivery system // *Expert Opin. Drug Deliv.* 2016. V. 13. P. 547–559.

567. Nam J.-M., Thaxton C.S., Mirkin C.A. Nanoparticle-based bio-bar codes for the ultrasensitive detection of proteins // *Science.* 2003. V. 301. P. 1884–1886.

568. Abalde-Cela S., Aldeanueva-Potel P., Mateo-Mateo C., Rodríguez-Lorenzo L., Alvarez-Puebla R.A., Liz-Marzán L.M. Surface-enhanced Raman scattering biomedical applications of plasmonic colloidal particles // *J. R. Soc. Interface.* 2010. V. 7. P. S435-S450.

569. Rodríguez-Lorenzo L., Krpetic Z., Barbosa S., Alvarez-Puebla R.A., Liz-Marzán L.M., Prior I.A., Brust M. Intracellular mapping with SERS-encoded gold nanostars // *Integr. Biol.* 2011. V. 9. P. 922–926.

570. Alvarez-Puebla R.A., Agarwal A., Manna P., Khanal B.P., Aldeanueva-Potel P., Carbó-Argibay E., Pazos-Pérez N., Vigderman L., Zubarev E.R., Kotov N.A., Liz-Marzán L.M. Gold nanorods 3D-supercrystals as surface enhanced Raman scattering spectroscopy substrates for the rapid detection of scrambled prions // *Proc. Natl. Acad. Sci. USA.* 2011. V. 108. P. 8157–8161.

571. Shiohara A., Wang Y., Liz-Marzán L.M. Recent approaches toward creation of hot spots for SERS detection // *J. Photochem. Photobiol. C.* 2014. V. 21. P. 2–25.

572. Jiang W., Kim B.Y.S., Rutka J.T., Chan W.C.W. Nanoparticle-mediated cellular response is size-dependent // *Nat. Nanotechnol.* 2008. V. 3. P. 145–150.

3

Biodistribution and Toxicity of Gold Nanoparticles

3.1 An Integrated Approach to Biodistribution and Nanotoxicology Experiments

Recent advances in making, measuring, and modeling plasmonic nanoparticles have led to a storm of publications dealing with the discovery and potential applications of plasmon-resonant nanoparticles and their bioconjugates to various fields. Alongside these applications, which have already been tested, gold nanoparticles (GNPs) have, in recent years, begun to be actively used in various fields of nanomedicine for diagnostic and therapeutic purposes. Noteworthy is the fact that GNPs are being increasingly administered to animals and humans parenterally. In particular, they serve as carriers for the delivery of drugs [1], genetic materials [2], and antigens [3], and they also are used as a medicinal or diagnostic agent *per se* for the treatment of tumors [4] or other diseases [5]. Two intravenous (IV) preparations—AurImmune (CYT-6091) and AuroLase—have already been clinically tried [6,7].

Almost synchronously with the beginning of the use of GNPs in medicine, acute questions were raised about the biodistribution and circulation of GNPs in the blood stream, their pharmacokinetics and elimination from the organism, and their possible toxicity to the organism as a whole or at the level of cytotoxicity and genotoxicity. The concerns about the possible consequences of nanoparticle application are by no means unfounded. Indeed, whereas the safety evaluation of engineered nanostructures is a new and emerging discipline called nanotoxicology, some important lessons can be learned from previous studies of other pathogenic particles and fibers. For example, in the 1930s to 1950s, 3-to-10-nm particles of thorium dioxide were extensively used as a contrasting agent ("thorotrast") in radiography. Later, however, it was found [8,9] that these particles can accumulate and persist in the organism for decades, causing unwanted radiation effects. There have been also reports showing the toxicity of ultrasmall GNPs (on the order of 1.5 nm) and the considerable accumulation of GNPs in the animal liver and spleen, with the kinetics of their elimination being very slow. For gold nanorods (GNRs), the initial stabilizer is cetyltrimethylammonium bromide (CTAB) [10], which in the free state is a known toxic surfactant.

"Colloidal metallic gold is not bio-inert"—such is the name Brown *et al.* [11] gave to their article so as to stress the importance of nanometer size in biological effects, even for such a seemingly inert material as gold.

Very little is known about the positive or negative effect of GNPs on plant cells and on the plant organism in general; the data of studies are controversial [12,13]. Recently, we proposed a new diagnostic test system [14] to analyze the toxicity of nanomaterials to the saltwater microalga *Dunaliella salina* through evaluation of cell death and changes in the culture growth rate at various toxicant concentrations.

Because many teams began their projects independently, there is a great scatter in experimental design, including particle size and shape, functionalization methods, animal types, particle administration doses and methods, and so on. Correspondently, there is a large scatter of data and conclusions on the levels and kinetics of biodistribution and on toxicity estimates. Besides these evident reasons, which are due simply to different experimental conditions, there also are more fundamental (and understudied still) questions related to the effects of the shape, structure, charge, and surface chemical modification of particles on their pharmacokinetics in the blood stream, their kinetics of accumulation and elimination from the organism, and their short- and long-term toxicity. That is why, despite the availability of several reviews on nanomaterial toxicity [15–45], there is a strong need in summarizing the constantly renewed data published to date, in order to help investigators evaluate the results already at hand and plan new studies. This task is addressed in this chapter, which, in addition to summarizing the published data, also presents the authors' original findings concerning the size dependence of GNP toxicity.

Although nanomedicine is generally considered a promising field of research with exciting prospects for the diagnostics and clinical treatment of human disease, the biodistribution behavior and toxicological effects of new nanomaterials must be carefully assessed before actual clinical use [46,47]. Figure 3.1 shows a summary of the main steps in current research on the biodistribution and toxicity of GNPs in experiments *in vitro* and *in vivo*. In short, the main steps include (1) the synthesis and characterization of nanoparticles with predetermined geometrical and structural parameters; (2) the functionalization of the nanoparticle surface with biocompatible ligands that ensure the required properties of the conjugate; (3) the design of an experiment involving animal or cell models, sampling, and the characterization of organ distribution, including analysis of the kinetics of nanoparticle accumulation and excretion; and (4) the identification and localization of particles at the cellular level, as well as elemental and structural analysis.

The scheme also shows the basic parameters for each step and the methods used for determining these parameters. The ultimate goal is to characterize the biological action of GNPs and to evaluate the risks involved by integrating all cellular-level information (cellular recognition and penetration, cytotoxicity, genotoxicity, and apoptosis/necrosis) and at the whole organism level (organ distribution, accumulation and clearance/excretion, degradation and metabolism, immunogenicity, and inflammation).

Consider now some basic concepts behind methodologies used for biodistribution and nanotoxicology experiments. Because this topic is out of the general scope of this chapter, here, we restrict ourselves to a short discussion.

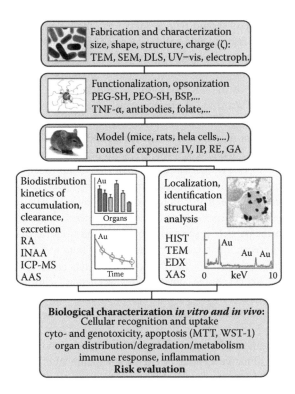

FIGURE 3.1 Scheme of biodistribution and nanotoxicology experiments. The first step is the fabrication of desired particles and the characterization of their size, shape, structure, charge (zeta potential) by transmission or scanning electron microscopy (TEM, SEM), DLS, UV–vis spectroscopy at the ensemble (suspension) and single-particle levels, electrophoresis, and other methods. The second step includes the functionalization of the particle surface with appropriate ligands, including thiolated PEG or poly(ethylene oxide) (PEO) molecules, TNF-α, antibodies, and folates, as well as opsonization with blood serum proteins (BSP; like albumin, *etc.*). Conjugates are administered to models in accordance with the experimental design, *i.e.*, by using selected doses and routes of exposure, including IV, intraperitoneal (IP), respiratory (RE) [intratracheal (IT)], or gastrointestinal (GA). The biodistribution into organs and the kinetics of accumulation/clearance are determined according to a selected time-dependent scheme of tissue sampling. Samples are analyzed by radioactive analysis (RA), instrumental neutron activation analysis (INAA), ICP–MS, and AAS. Particles and related structures also can be identified at the tissue (histological, HIST) and cellular levels by electron microscopy (SEM, TEM), EDX, and XAS. The final step is the integration of data for the biological characterization of GNP effects and the evaluation of possible risks through the use of cellular-level information (cellular recognition and penetration, cytotoxicity, genotoxicity, and apoptosis/necrosis; MTT and WST-1 assays) and at the whole organism level (organ distribution, accumulation and clearance/excretion, degradation and metabolism, immunogenicity, and inflammation).

Wet chemical synthesis should be considered the main approach to fabrication of GNPs through the reduction of gold halides such as $HAuCl_4$ by various chemical reducers. A detailed discussion of fabrication protocols can be found in a recent review [48] and a book [49].

For characterization of the GNP size, shape, and structure, transmission electron microscope (TEM) and scanning electron microscope (SEM) are the most reliable and direct methods. However, two factors can distort the output data: (1) the preparative transfer of particles from suspensions to TEM grids and (2) insufficient statistics of non-representative samples. Fortunately, for GNPs, both problems can be resolved easily by the use of properly washed samples and by measuring about 1000 particle images. The dynamic light scattering (DLS) technique provides information about the particle-size distribution and the particle charge in terms of zeta potential [50]. However, the basic model of DLS is a dilute suspension of Mie spheres. That is why the DLS data can be greatly affected by the particle shape and structure. In addition, a small portion of large contaminant or aggregated particles can greatly contribute to the recorded scattering intensity from fabricated particles. In practice, both effects restrict the accuracy of DLS data. Nevertheless, particle charge control is very important for such applications as the preparation of conjugates and the particle uptake by cells. Applications of ultraviolet-visible (UV–Vis) single-particle or ensemble spectroscopy can be found in the review in Ref. [51].

GNP functionalization with appropriate surface ligands can be achieved by a direct physical adsorption of native ligands, by chemical attachment of their thiolated derivatives, or by the use of intermediate linkers such as bifunctional poly(ethylene glycol) (PEG) with thiol (SH) and carboxyl (COOH) terminal chemical groups. Similarly, Kumar *et al.* [52] suggested an efficient method for conjugating antibodies onto GNP surface through a bifunctional linker, hydrazide–PEG–dithiol. Deyev *et al.* [53] developed an antibody engineering approach based on the use of the barnase-barstar module as a molecular constructor. This universal strategy can be successfully applied to functionalization of GNPs with m-antibodies or their specific fragments.

Animal and cellular models are discussed in Section 3.2 and throughout the text. Evidently, a particular method of GNP administration can influence the biodistribution and possible toxic effects of GNPs. At present, the majority of available data are for IV administration, whereas intraperitoneal, respiratory, or gastrointestinal methods were used in a few publications. To some extent, this can be explained by a search for ways for efficient delivery of GNP conjugates to target cells and tissues through IV injection [54]. For instance, the accumulation of GNPs in tumors depends on whether passive or active targeting is employed. Passive targeting relies exclusively on the EPR effect of tumors, whereas in active targeting, the GNP surface is functionalized by probing bio-molecules to produce an enhanced accumulation at the tumor sites [55]. Unfortunately, both mechanisms suffer from an intrinsic host immune system, which eliminates GNPs from blood circulation.

For biodistribution and toxicology experiments, one of the main tasks is an accurate determination of gold content and localization of GNPs in tissues. Currently, instrumental neutron activation analysis (INAA) [56] and inductively coupled plasma–mass spectrometry (ICP–MS) [55,57] are "gold standards" for quantifying gold concentrations in tissues. Whereas generally providing good accuracy with an about 1 ppb detection limit, INAA and ICP–MS are labor-intensive, take significant time to acquire results, and cannot be applied to *in vivo* studies. A bit lower sensitivity is provided by atomic absorption spectroscopy (AAS), which also needs sacrificed animals and labor-intensive sample pretreatment. Perhaps, the optoacoustic techniques could be a sensitive tool for *in vivo* biodistribution and toxicology applications [58]. However, no related reports have been published till now.

For identification and localization of GNPs in tissues, the histology, autometallography, SEM, and TEM techniques are in common use [56]. SEM or TEM analysis can be combined with energy dispersive x-ray spectroscopy (EDX) to obtain elemental data. Additionally, x-ray absorption spectroscopy (XAS and its variants X-ray absorption near edge structure (XANES) and extended x-ray absorption fine structure (EXAFS)) [59] can be used to obtain structural information on sulfur and gold atoms in samples under examination.

In some analogy with INAA and ICP–MS as applied to biodistribution studies, MTT assay is the current "gold standard" for *in vitro* nanotoxicology studies. The assay is based on the ability of cellular enzymes to reduce the tetrazolium dye MTT (3-(4,5-di$\underline{M}$ethyl-$\underline{T}$hiazol-2-yl)-2,5-diphenyl $\underline{T}$etrazolium bromide) to formazan depending on cellular respiratory activity and viability. A method similar to the MTT assay, WST-1, uses $\underline{W}$ater-$\underline{S}$oluble $\underline{T}$etrazolium salt (WST; 4-[3-(4-iodophenyl)-2-(4-nitrophenyl)-2H-5-tetrazolio]-1,3-benzene disulfonate)). In the case of *in vivo* toxicology, the published studies have reported some changes in animal behavior and histological *ex vivo* examinations.

3.2 Biodistribution of GNPs

3.2.1 Summary of the Published Data for the Period 1995 to 2016

Table 3.1 summarizes, in concise form, the GNP biodistribution information published between 1995 and 2016. Before proceeding to discuss these data, we wish to make several important general remarks. First, we have, naturally, not aimed at giving an exhaustive coverage of all published work. In particular, Table 3.1 includes only GNPs, with other metals being left out. Second, we have sought to cover mostly those biodistribution data that were obtained for IV administration, although some articles report on other methods as well. Furthermore, some published reports on biodistribution also contain toxicity assessments; therefore, the corresponding references may include both aspects. Finally, Table 3.1 presents only the information that we considered important for comparative estimations.

As the main parameters, we chose particle type and size; methods of surface functionalization; animals used with special mention of cases with implanted tumors; doses and method of administration; tissues, organs, or cells analyzed; time of collection of pharmacokinetics or biodistribution data; and methods used for analyzing samples for gold content. Although our literature data table is slightly different from that of Balasubramanian *et al.* [60], both tables contain all vital information about the main experimental conditions determining the biodistribution and toxicity effects. For instance, the particle type, size, and surface functionalization are known to be crucial

TABLE 3.1 Published Data on the Biodistribution of GNPs

Nature and Size of Particles[a] (nm)	Surface Coating	Model	Administration Methods[b] and Dosage	Cells, Organs, and Tissues[c]	Exposure Duration	Method[d]	References
CG[e]-16	Albumin	Swine	IV; 10–20 µg/g	Bl, Li, Lu	5–6 h	INAA, TEM	Darien [56]
CG-6–8, Aurasol	Maltodextrin	Human	Oral; 30 mg	Bl	5 min–8 h	ICP-MS	Abraham; Abraham [61, 62]
CG-13	–	Mice	IP; 20 µg/g	Bl, Br, Lu, He, Ki, Sp, Li, St, Si	3 h–4 d	INAA, TEM, EDX	Hillyer [63]
CG-4, 10, 28, 58	–	Mice	Water *ad libitum*; 200 µg/ml, 7 d	Bl, Br, He, Lu, Li, Ki	12 h after last dose (7 d)	INAA, TEM	Hillyer [64]
Gold clusters-1.9	–	Mice + Tu	IV; 1.35, 2.7 µg/g	Tu, Li, Ki, Bl, Mu	5–60 min	AAS	Hainfeld [65]
CG-33	Tumor necrosis factor (TNF); PEG-SH + TNF	Mice + Tu	IV; 0.2–1 µg TNF–CG vector	Bl, Br, Li, Lu, Tu	5, 180, and 360 min	EIA	Paciotti [66]
CG-50, 80, 100, 150	PEG-SH, galactose-PEG-SH	Mice	IV; 0.46–0.57 µg/g	Bl, Hep, Ku, Se	2 h	INAA	Bergen [67]
Gold clusters-1.9	–	Mice + Tu	IV; 2.7 mg/g	Bl, Ki, Li, Mu, Tu	24 h	AAS	Hainfeld [68]
GNR-65 × 11	PEG-SH, CTAB	Mice	IV; 2.5–5 µg/g	Bl, Li, Lu, Ki, Sp	0.5–72 h	ICP-MS	Niidome [69]
^{198}Au CG-15–20	Gum arabic	Mice	IV; 10–20 µCi	Bl, Br, Li, Ki, Si, Sp	1, 4, and 24 h	NaI well counter	Katti [70]
Au-dendrimer PAMAM composites 5, 11, 22	Poly(amidoamine) (PAMAM)	Mice + Tu	IV; 0.8–1.6 µg/g (16 µg/g of composites)	Bl, Br, He, Ki, Li, Lu, Mu, Pa, Sp, Tu	5 min, 1 h, 1, 4, 7 d	INAA	Balogh [71]
CG-2, 40	–	Mice	IV; 0.6–3.2 µg/g	Ad, Br, Ki, Li, Sp, Ov, Si, Ln	1, 4, 24 h	TEM	Sadauskas [72]

(Continued)

TABLE 3.1 (CONTINUED) Published Data on the Biodistribution of GNPs

Nature and Size of Particles[a] (nm)	Surface Coating	Model	Administration Methods[b] and Dosage	Cells, Organs, and Tissues[c]	Exposure Duration	Method[d]	References
GNS SiO₂/Au 110/10	PEG	Mice	IV; 10 µg/g	Bl, Bo, Br, Ki, Mu, Li, Lu, Sp	4 h, 1, 4, 7, 14, 21, 28 d	INAA	James [73]
CG-10, 50, 100, 250	–	Rats	IV; 0.3–0.5 µg/g	Br, He, Ki, Li, Lu, Ln, Sp, Te, Thy, Ad	24 h	ICP-MS	De Jong [74]
GNS SiO₂/Au 120/ (12–15)	PEG–SH	Mice + Tu	IV; 0.5 µg/g	Tu	24 h	SEM	Diagaradjane [75]
GNR-60 × 15 GNS SiO₂/Au 80/25	PEG–SH	Mice + Tu	IV; 2 µg/g	Bl, Li, Mu, Tu	5 min–24 h	AAS	Kogan [76]
CG-10,15, 30	PEG–SH	Human placenta (*in vitro*)	Human placental perfusions, 0.1–0.5 µg/ml in the maternal perfusate, (1–2) × 10⁹ particles/ml	Human placenta	6 h	ICP-MS	Myllynen [77]
GNS Au/Au₂S 20–50	Mercaptoundecanoic acid + cisplatin	Mice + Tu	IV; 10 µg/g	Bl, Bo, Br, He, Ki, Mu, Li, Lu, Sp, Tu	1 and 7 d	ICP-MS	Huang [78]
GNR-65 × 11	PEG–SH Phosphatidylcholine	Mice	IV; 1.3–6 µg/g	Whole body	6 min	Near-IR images	Niidome [79]
Gold clusters-1.4, CG-18	Na salt of monosulfonated triphenylphosphine	Rats	IV; IT 0.01–0.1 µg/g	Bl, Br, He, Ki, Li, Lu, Sk, Sp, Ut, Ur, Fe, Ca	24 h	INAA TEM	Semmler-Behnke [80]
CG-15, 102, 198	–	Rats	Diffusion; 4.72 × 10¹⁸ particles/ml	Sk, Si	24 h	ICP-MS EDX	Sonavane [81]
CG-15, 50, 100, 200	Na alginate	Mice	IV; 1000 µg/g	Bl, Br, He, Ki, Li, Lu, Sp, St, Pa	24 h	ICP MS	Sonavane [82]

(*Continued*)

TABLE 3.1 (CONTINUED) Published Data on the Biodistribution of GNPs

Nature and Size of Particles[a] (nm)	Surface Coating	Model	Administration Methods[b] and Dosage	Cells, Organs, and Tissues[c]	Exposure Duration	Method[d]	References
Hollow GNS-30	IgG	Mice + Tu	IV; 7.3×10^{10} particles/ml, 0.13 ml	Bl, Bo, He, Ki, Li, Lu, Mu, Si, Sp, St, Tu	24 h	[111]In RA	Melancon [83]
GNR-55 × 9	PEG-SH	Mice + Tu	IV; 0.2, 1, 10 μg/g	Bl, Ki, Li, Lu, Sk, Sp, Tu	0.5, 3, 6, 24, 72 h	ICP-MS	Akiyama [84]
CG-13	PEG-SH	Mice	IV; 0.85, 4.26 μg/g	Bl, Br, Ki, Li, Lu, Sp, Te	5, 30 min, 4, 24 h, 7 d	ICP-MS	Cho [85]
CG-15-20 CG-6-10	Gum arabic Maltose	Swine	IV; 2 μg/g	Bl, Ki, Li, Lu, Sp, Ur	0.5-170 h	AAS	Fent [86]
CG-33	TNF PEG-SH	Mice + Tu	IV; 6.25 μg/g	Ki, Li, Lu, Sp, Tu	5 min, 1, 4, 12, 24 h, 3, 60, 80, 120 d	AAS TEM	Goel [87]
CG-20, 100	–	Mice		Retina			Kim [88]
CG-2, 40, 100	–	Mice	IT-1; (0.6-3 μg) IT-5; (5 × 3 weeks)	Li, Lu	24 h	ICP-MS INAA	Sadauskas [89]
CG-40	–	Mice	1.5 μg/g	Li	24 h, 1, 3, 6 months	ICP-MS	Sadauskas [90]
CG-15, 50, GNS SiO$_2$/Au 120/20	PEG-SH	Rats	IV; 1.3 μg/g	Bl, Br, Ki, Li, Lu, Sp	24 h	AAS	Terentyuk [91]
CG-20, 40, 80	PEG	Mice	IV; 0.05-4 μg/g	Bl, Bo, He, Ki, Li, Lu, Mu, Si, Sp, St, Tu	10 min-60 h (Bl -48 h)	[111]In RA	Zhang [92]

(Continued)

TABLE 3.1 (CONTINUED) Published Data on the Biodistribution of GNPs

Nature and Size of Particles[a] (nm)	Surface Coating	Model	Administration Methods[b] and Dosage	Cells, Organs, and Tissues[c]	Exposure Duration	Method[d]	References
CG-15 GNR-50 × 15	HS-poly(ethylene oxide)–(OH, NH2, COOH), CTAB, serum, DNase, cytochalasin D	Human immune cells		Gr, Mn, Mph, Dc, Ly	15, 60 min	TEM	Bartneck [93]
CG-20	–	Rats	IV; 0.01–0.015 µg/g	28 organs	1, 7 d, 1, 2 months	ICP–MS	Balasubramanian [60]
GNR-56 × 13	CTAB, BSA	Rats	IV; 0.560 µg/g	Bl, Br, Bo, He, Ki, Li, Lu, Mu, Sp, Ur, Fe	0.5, 1–24 h, 3–28 d	ICP–MS TEM, EDX, XAS	Wang [59]
^{198}Au CG-1.4, 2.8, 5, 18, 80, 200	Triphenylphosphine Mercaptoacetic acid Cysteamine	Rats	IV; 1.6–43.7 µg per rat	Sp, Lu, He, Br, Ut, Ki, Bl, Ur	24 h	γ-spectrometry	Hirn [94]
^{198}Au CG-1.4, 2.8, 5, 18, 80, 200	Triphenylphosphine Mercaptoacetic acid Cysteamine	Rats	Oral; 1–27 µg per rat	Sp, Lu, He, Br, Ut, Ki, Bl, Ur	24 h	γ-spectrometry	Schleh [95]
CG-40 GNR-80 × 30 GNR-50 × 50	CTAB CTAB	Mice	IV; 10⁷ particles/g	Li, Sp, Lu, Ki	14 d	ICP–MS	Sun [96]
CG-30	PEG TNF-α	Mice	IV; 11 µg/g	Bl, Li, Sp, Tm	1, 8, 24 h	ICP–MS ICP–AES	Shah [97]
CG-20	–	Mice	IP; 7.85 µg/g	Li, Sp, He	1, 24, 72 h	ICP–MS	Chen [98]
CG-25	Polyvinyl alcohol	Mice	IV; 0.3 µg/g	Li, He, Br, Lu, Ki	24 h	ICP–MS	Wojnicki [99]
GNR-40 × 10	PEG	Mice	IV; 4 µg/g	Sp, Li, Br, Lu, Ur	1 h; 1, 7, 30 d	TXRF	Fernández-Ruiz [100]

(Continued)

TABLE 3.1 (CONTINUED)   Published Data on the Biodistribution of GNPs

Nature and Size of Particles[a] (nm)	Surface Coating	Model	Administration Methods[b] and Dosage	Cells, Organs, and Tissues[c]	Exposure Duration	Method[d]	References
CG-20	CALNN	Rats	IV; 0.7 µg/g	Li, He, Lu, Ki, Sp, Te, Si, Bo, Mu, Thy, Br	30 min, 28 d	AAS	Fraga [101]
CG-15	PEG	Mice	IV; 1 µg/g	Ln, Ki, Br, Te, Li, Sp, Lu, He, Bl	30 min, 4, 24 h, 7, 14 d, 1, 3, 6 months	ICP-MS	Lee [102]
CG-15	–	Mice	IV; IT; 50 µl per mice	Li, He, Lu, Ki, Sp, Te, Si, Bo, Mu, Thy, Br	1, 3, 6 h	ICP-MS	Koyama [103]
198Au CG-5	Poly(isobutylene-alt-maleic anhydride)	Rats	IV; 0.04 µg/g	Li, He, Lu, Ki, Sp, Br, Ut, Bl, Ur	24 h	111In RA	Kreyling [104]
198Au CG-15	–	Rats	Oral; IV; 9–90 µg per rat	Li, Sp, Lu, Bl	24 h	γ-spectrometry	Rambanapasi [105]
CG-5-15	–	Rats	Oral; 1.3 µg/g	Br, Ki, Si, Li, Lu, Te, Sp, St	14 d	ICP-MS	Jo [106]
CG-2, 5, 10	–	Mice	IV; 1.25 µg/g	Lu, Li, Sp, He, Ki, Br, Bl, Ur	1, 15, 30, 90 d	ICP-MS	Naz [107]
CG-14	–	Rats	IV; 0.9, 9 and 90 µg per rat	He, Ki, Li, Lu, Sp, Ca, Bo	14 d	INAA	Rambanapasi [108]

[a] Numbers are sizes in nm.

[b] Administration methods: IV, intravenous; IP, intraperitoneal; IT, intratracheal.

[c] Cells, organs, and tissues: Ad, adrenal; Bl, blood; Bo, bones; Br, brain; He, heart; Li, liver; Ln, lymph nodes; Lu, lung; Ki, kidney; Ov, ovary; Si, small intestine; Se, sinusoidal endothelial cells; Sk, skin; Sp, spleen; St, stomach; Te, testes; Thy, thymus; Ut, uterus; Tm, tumor; Hep, hepatocytes; Ku, Kupffer cells; Mu, muscle; Gr, granulocytes; Mn, monocytes; Mph, macrophages; Dc, dendritic cells; Ly, lymphocytes; Ur, urine; Fe, feces; Ca, carcass.

[d] Methods: RA, radioactive analysis; INAA, instrumental neutron activation analysis; ICP-MS, inductively coupled plasma–mass spectrometry; AAS, atomic absorption spectroscopy; TEM, SEM, transmission and scanning electron microscopy; EDX, energy-dispersive x-ray; XAS, x-ray absorption spectroscopy; EIA, enzyme immunoassay; TXRF, total-reflection X-ray fluorescence spectrometry.

[e] Nanoparticles: CG, colloidal gold (gold nanospheres); GNR, gold nanorods; GNS, gold nanoshells.

factors for biodistribution over animal organs and related toxicity [16,29]. Note that the dosage data are given in terms of the quantity of gold (μg) administered per unit average animal weight (g) and often have been obtained by the article authors by recalculation using the volume of suspension administered, particle concentration and size, and so on. For this reason, with account taken of the scatter in actual animal weight, the figures in Table 3.1 should be considered only as speculative, giving an idea of the order of administered doses, with an accuracy of approximately 30%.

3.2.2 Particles, Models, and Methods

The main object, for which the most data are available, is by far colloidal gold (CG) particles. The range of sizes studied spans more than two orders of magnitude—from the minimal 1.4 nm (the atomic cluster Au_{55}) [80] to the maximal 250 nm [74]. Apart from data for these particles, there is information on the biodistribution of GNRs [69,75,79,93], nanoshells of the SiO_2/Au [73,75,91] or Au/Au_2S [78] type, hollow nanoshells [82], nanostars [109], and composites (GNPs with diameters of 5 to 15 nm in various nanometer-sized templates [71,86]).

For surface functionalization, the most popular reagent (approximately 30% of all publications) is a thiol-terminated PEG–SH, which forms a strong donor–acceptor link to the surface atoms of gold particles. Several authors have used PEG molecules as linkers for subsequent labeling with a radioactive chelator [92] or with probing molecules for specific targets (*e.g.*, tumor necrosis factor [TNF] [87]). Sporadic publications also reported the use of albumin, maltodextrin, poly(ethylene oxide), and other substances given in Table 3.1. Initially, GNRs are obtained with a surface already functionalized with CTAB molecules. Biocompatibility is ensured by replacing CTAB with PEG [69,76] or with other nontoxic ligands [110].

The main models studied include mice and rats, whereas other models (pigs or humans) and also the human eye and placenta (an experiment *in vitro* [77]) have been examined sporadically.

Unfortunately, one of the most variable parameters in different studies is the administered dose of GNPs. As seen from Table 3.1, the possible values range from a hundredth fractions of a microgram per gram of animal [80,60] to 2700 μg/g [68]. Of course, this latter case is an exception, because Hainfeld *et al.* [68] merely investigated the possibilities of x-ray contrasting by using 1.9-nm-diameter nanoparticles, without attention to problems of biodistribution within the limits of reasonable physiological doses. Yet, despite the great scatter in doses (more than five orders), it is evident from Figure 3.2 that most studies have been performed in the range of 0.1 to 10 μg/g, which can possibly be taken as the main range for comparing data from various studies.

The content of gold in tissue or blood samples has been analyzed by using, to an approximately equal degree, neutron activation methods (INAA or NAA), ICP–MS, and AAS. The sensitivity of these three methods declines just in this order. Melancon *et al.* [83], Zhang *et al.* [92], and Kreyling *et al.* [104] have used the radioactive tracer ^{111}In to ensure, as claimed by those authors, the greatest sensitivity of those of all known methods. For the most part, ICP–MS ensured a detection limit of approximately 0.001 μg/kg [59], which is quite sufficient for an accurate estimation of biodistribution.

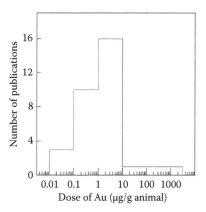

FIGURE 3.2   Histogram of the publication distribution over the doses used.

The final remark in this section pertains to the correlation of the biodistribution data. Unfortunately, only qualitative comparison for the distribution character itself is often possible, for two reasons. First, with rare exception, no balance studies (*i.e.*, accounting for all particle distribution and excretion pathways) have been made. Second, the data have been presented either as percentage of the administered dose or as µg per gram of organ sample.

3.2.3 Dependence of Biodistribution on the Size of GNPs and on the Method of their Administration

3.2.3.1 Early Studies: The Role of Kupffer Cells

A very important question is the dependence of biodistribution kinetics on the size, dose, and method of administration of nanoparticles. The first studies on the biodistribution of CG were made with mice [111] and rats [112,113] in the 1970s to 1980s. It was found that after being introduced parenterally, CG particles are taken up by hepatocytes, secreted through the bile, and eliminated through feces. The work of Hardonk *et al.* [112] can probably be regarded as one of the first demonstrations of the size effect *in vivo* with the help of functionalized GNPs (polyvinylpyrrolidone and bovine serum albumin (BSA) were used). More specifically, the maximum quantity of 17-nm gold particles in hepatocytes was observed at 2 h after injection, whereas 79-nm particles were not found in the hepatocytes. In feces, GNPs were found within 4 to 12 days after injection, and at 6 days after injection, gold was identified in Kupffer cells.

Along with Hardonk *et al.* [112], an important role of Kupffer cells in the elimination of GNPs was found by Sadauskas *et al.* [72], who injected mice intravenously with 2- and 40-nm GNPs. TEM data showed that after injection, the nanoparticles accumulated in the macrophages of the liver (90%), whereas in spleen macrophages, their quantity was much smaller (10%). In other organs (kidneys, brain, lungs, adrenals, ovaries, and placenta), no gold particles were found by TEM. These data are in harmony with those of Katti *et al.* [70], who analyzed the biodistribution of the radioactive colloidal gold [198]Au.

Sadauskas *et al.* [72] concluded that nanoparticles penetrate only phagocytes, first and foremost the Kupffer cells of the liver, without passing the placenta and blood–brain barriers (BBBs), and 2-nm particles can be excreted *via* urine. Subsequently [90], those authors reported that 40-nm GNPs are localized in lysosomes (endosomes) of Kupffer cells and can persist there for as long as 6 months.

3.2.3.2 Gastrointestinal and Peritoneal Administrations

We now turn to consider the data of Table 3.1. In 2001, Hillyer and Albrecht [64] did a pioneering study on the size dependence of biodistribution of 4- to 58-nm GNPs administered to mice through the alimentary tract in the form of a supplement (concentration, 200 µg/ml) to drinking water (Figure 3.3). They found that the smallest, 4-nm, particles can penetrate through the alimentary tract (by means of persorption through enterocytes followed by their degradation in the villi) and can be redistributed throughout nine organs studied. Penetration from the alimentary tract into other organs decreased strongly with increasing particle size (10, 28, and 58 nm), and the gold level for 4-nm particles was maximal (approximately 0.075 µg/g) in the kidneys and approximately 0.035 to 0.02 µg/g in other organs (small intestine, lungs, stomach, spleen, and liver). Four-nanometer particles were present even in the brain at a statistically significant (relative to the other sizes) quantity (4.7 ng/g). This finding seems to be the first experimental evidence that nanoparticles can overcome the BBB.

In 1999, Hillyer and Albrecht presented the first data [63] on the biodistribution of 13-nm gold particles injected into mice intraperitoneally every day for 4 days at high enough doses (20 µg/g) (our estimate of the dose for the average mouse weight of 28 g and for 1 ml of a 10-fold concentrated injection suspension obtained by reducing 0.01% HAuCl$_4$ as described in the citrate protocol of Frens). The differences concerning the method and doses used for administration led to a completely different picture of distribution, as compared with the authors' 2001 data [64] (Figure 3.3, last bar).

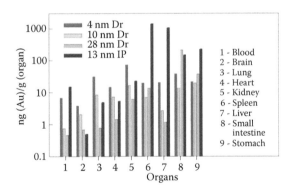

FIGURE 3.3 Biodistribution of 4-, 10-, and 28-nm GNPs in mouse organs at 12 h after administration (drinking *ad libitum*, solution concentration of 200 µg/ml, 7 days) and of 13-nm GNPs at 72 h after IP administration (4 × 20 µg/g per day). (Adapted from Hillyer, J.F., Albrecht, R.M., *Microsc. Microanal.*, 4, 481–490, 1999. With permission.)

The majority of 13-nm particles accumulated in the liver and spleen at 1060 to 1400 µg/g, which is 50,000 times greater than the amounts observed after administration through the alimentary tract [64]. Concentrations an order of magnitude lower were observed in the stomach and small intestine (100 to 200 µg/g), and those still an order of magnitude lower (5 to 20 µg/g) were found in the kidneys, blood, heart, and lungs. Noteworthy is the fact that the particle content in the brain (50 ng/g) was an order of magnitude greater than the gold content for 4-nm particles and twofold greater than that for 10-nm particles. Thus, both these studies unequivocally confirm the possibility that particles with a diameter of approximately 10 nm or smaller can penetrate the BBB.

3.2.3.3 IV Administration of GNPs

We now shift to discuss the data on the biodistribution of GNPs administered intravenously. In those experiments, another important point is the kinetics of GNP circulation in the blood stream.

In one of the very first experiments, Darien *et al.* [56] measured the distribution of 16-nm albumin-stabilized particles that were infused into pigs (weight, 20+ kg) intravenously at an approximate dose of 10 to 20 µg/g. The measurements were made at 5 to 6 h after injection. With the help of INAA, most particles (270 µg/g) were found to be present in the lungs, followed by 88 µg/g in the liver and 1 µg/g in blood plasma.

De Jong *et al.* [74] were the first to study, by ICP–MS, the size effect of GNP biodistribution at 24 h after IV injection into rats, with a broad range of GNP sizes being used (10, 50, 100, and 250 nm) (Table 3.2).

For all GNP sizes, maximum accumulation was shown to occur in the liver and spleen, both in absolute figures and in terms of percentage content (of the dose), exceeding 30% on the average. Such a high percentage points clearly to a redistribution effect and an accumulation mechanism operating in precisely these two organs when particles circulate in the blood stream. A clear difference was observed between accumulation of

TABLE 3.2   Biodistribution of GNPs Expressed as µg (Au)/g (Sample) or as % of the Given Dose at 24 h after IV Administration to rats at a Dose of 0.3 to 0.5 µg (Au)/g (Animal)

Tissue/Organs	Au Concentration, µg (Au)/g (Sample) or % of the Given Dose			
	10 nm	50 nm	100 nm	250 nm
Blood	1.5/36	1.4/31	2.2/45	0.97/16
Liver	2.7/46	1.2/21	3.3/44	3.0/31
Spleen	2.2/2.2	1.4/1.3	1.8/1.4	2.0/1.2
Lungs	0.19/0.3	1.7/2.3	0.044/0.15	0.035
Kidneys	0.34/1.0	0.058/0.2	0.030/0.1	0.026/0.1
Testes	0.055/0.2	ND	ND	0.006
Thymus	0.21/0.2	ND	0.006	0.035
Heart	0.16/0.2	0.049/0.1	0.009	ND
Brain	0.13/0.2	ND	ND	ND

Source: Adapted from De Jong, W.H., Hagens, W.I., Krystek, P., Burger, M.C., Sips, A.J., Geertsma, R.E., *Biomaterials*, 29, 1912–1919, 2008. With permission.

10-nm particles and that of larger particles. Only 10-nm particles were found not only in the liver and spleen but also in the kidneys, testes, thymus, heart, lungs, and brain.

A similar experiment, using GNPs with diameters of 15, 50, 100, and 200 nm, was conducted by Makino's team [82]. However, those authors chose a different model. Biodistribution in mouse organs was determined by ICP–MS at 24 h after IV administration of GNPs suspended in a solution of sodium alginate (Table 3.3).

The major difference in experimental design between the works of De Jong *et al.* [74] and Sonavane *et al.* [82] consists in the doses used (0.3 to 0.5 µg/g versus 1000 µg/g). Note that from our point of view, the dose reported by Sonavane *et al.* [82] (1 g/kg = 1 mg/g, with the volume of injected suspension being approximately 0.1 ml [3 ml/100 g] [82]) seems nonphysiological and difficult to implement technically. The mass concentration of the suspension should be on the order of 30 mg/0.1 ml = 300 mg/ml, which is approximately 5000 times higher than the standard concentration of a sol obtained from 0.01% $HAuCl_4$ by the method of Frens and having an optical density of approximately 1.0 in a 1-cm-thick cuvette at a resonance wavelength of 520 nm.

Comparison of the data of De Jong *et al.* [74] and Sonavane *et al.* [82] shows some similarity in that on the average, considerable accumulation of particles of all sizes has been identified in the liver. There are, however, serious differences. One is that accumulation was not as dominant in the spleen of mice [82] as it was in the spleen of rats [74]. There was considerable accumulation of 15-nm particles in the kidneys and lungs [82], and only the largest, 200-nm GNPs dominated in the spleen as well as in the liver. Yet another noticeable difference between the data of De Jong *et al.* [74] and Sonavane *et al.* [82] is in the blood level of GNPs. For rats, the blood content of GNPs at 24 h was quite comparable with the content in the main accumulating organs [74], whereas for mice, blood levels an order of magnitude lower than the contents in the accumulating organs were obtained [82]. Finally, both studies demonstrated the penetration of small (10 to 15 nm) particles through the BBB; however, 50-nm GNPs were identified in the mouse brain [82] but not in the rat brain [74].

TABLE 3.3   Biodistribution of GNPs Expressed as µg (Au)/g (Sample) or as % of the Given Dose at 24 h after IV Administration to Mice at a Dose of 1 mg (Au)/g (Animal)

Tissue/Organs	Au Concentration, µg (Au)/g (Sample) or % of the Given Dose			
	15 nm	50 nm	100 nm	200 nm
Blood	0.56/0.004	0.59/0.002	ND	0.11/0.0002
Liver	52.3/0.40	21.3/0.08	27.1/0.11	58.8/0.14
Spleen	5.5/0.041	11.5/0.045	12.9/0.050	28.9/0.070
Lungs	32.3/0.24	18.7/0.073	15.2/0.059	19.4/0.047
Kidneys	25.5/0.193	3.75/0.014	1.29/0.005	9.35/0.022
Stomach	3.6/0.027	0.25/0.0009	0.80/0.003	0.15/0.0003
Pancreas	ND	1.7/0.006	7.52/0.029	6.77/0.016
Heart	1.05/0.007	0.97/0.003	3.24/0.012	2.86/0.006
Brain	10.0/0.075	9.1/0.036	6.0/0.023	0.15/0.0003

Source: Adapted from Sonavane, G., Tomoda, K., Makino, K., *Colloids Surf. B.*, 66, 274–280, 2008. With permission.

These differences are possibly due primarily to the sharp distinction in the doses, animal models, and particle preparation method used. Note that according to the data of De Jong *et al.* [74], the colloids, when diluted with phosphate-buffered saline, changed their color, which clearly points to particle aggregation and to a change in the character of circulation. It is obvious that the biodistribution data obtained by those authors were somewhat distorted by aggregation.

In the works considered previously, the lower size limit was sufficiently large (approximately 10 nm), not counting the article [65] on x-ray contrasting with high doses of 1.9-nm particles. Semmler-Behnke *et al.* [80] examined, by INAA, the biodistribution of the smallest, 1.4-nm GNPs, as compared with that of the usual colloidal 18-nm GNPs (Table 3.4). The 1.4-nm GNPs were used as a complex of Au_{55} and sulfonated triphenylphosphine ligand molecules. This object was chosen because it can irreversibly bind to the B form of DNA [114] and can cause a toxic effect in culture cell experiments.

In this case, the size effect of distribution is evidenced by accumulation in the liver and kidneys. If for 1.4-nm GNPs, the main accumulation target is the liver, 18-nm GNPs circulate in blood for a longer time, accumulating mostly in the liver, skin, and kidneys. A considerable amount of these particles is distributed throughout the animal body (or "carcass," in the authors' terminology [80]).

It is of interest to compare the data of De Jong *et al.* [74] and Semmler-Behnke *et al.* [80], obtained for similarly sized particles (15 and 18 nm) and animals of the same type. There is some similarity in that most accumulation was found in the liver, but there is a substantial difference in the accumulation levels in the organs. More specifically, the data of De Jong *et al.* [74], showing comparable levels in the liver, kidneys, and lungs, are not corroborated by Semmler-Behnke *et al.* [80]. Furthermore, Semmler-Behnke *et al.*

TABLE 3.4 Biodistribution of 1.4- and 18-nm GNPs at 24 h after IV Administration to Rats, Expressed as % of the Given Dose

Tissue/Organs	Concentration of Au (% of the given dose)	
	1.4 nm	18 nm
Blood	<0.1	3.7
Liver	94	48
Spleen	2.2	1.3
Lungs	0.1	0.7
Kidneys	<0.1	5.5
Uterus	<0.1	0.2
Heart	<0.1	0.2
Brain	<0.1	<0.1
Urine	<0.1	8.6
Feces	0.5	5.0
Skin	0.2	7.9
Carcass	2.1	19

Source: Adapted from Semmler-Behnke, M., Kreyling, W.G., Lipka, J., Fertsch, S., Wenk, A., Takenaka, S., Schmid, G., Brandau, W., *Small*, 4, 2108–2111, 2008. With permission.

[80] observed no penetration of the BBB, either for the smallest (1.4 nm) or for the 18-nm GNPs.

One of the most detailed (in terms of the number of organs tested and the length of the time interval) studies on the biodistribution of unmodified 20-nm GNPs was made recently by Balasubramanian *et al.* [60]. They determined, during 2 months, the concentration of gold in the organs (a total of 28), feces, and urine of rats subjected to a single IV administration of GNPs at a very low dose (0.01 to 0.015 μg/g; ICP–MS was used). In agreement with most published reports, Balasubramanian *et al.* [60] found that GNPs accumulated the most in the liver (50 to 70 ng/g) and spleen (8 to 10 ng/g). Smaller amounts of gold were present in the kidneys (5 ng/g) and testes (0.6 ng/g). Contrary to the findings of the studies considered earlier, very little gold was found in the lungs, and in the brain, even trace amounts were not detected.

3.2.3.4 Penetration of GNPs through the BBB

Comparing data from Balasubramanian *et al.* [60] with the reports of smaller GNP amounts found in the brain, one can speculate that the penetration of nanoparticles through the BBB is critically size dependent, with the upper penetration limit being approximately 20 nm. Sonavano *et al.* [82] proposed a mechanism for the size dependence of overcoming of the BBB. Almost 100% of the surface area of the capillary basement membrane is covered by end-feet of astrocytes. These astrocytic end-feet are separated from the capillary endothelium by approximately 20 nm. It follows that smaller-sized GNPs can in principle penetrate through this gap. Another mechanism is linked to the nanoparticle-assisted delivery of doxorubicin to the brain [115]. It was hypothesized that transport may be due to the interaction of apolipoprotein A-I, anchored to the nanoparticle surface, with the scavenger receptor class B type I, located at the BBB. The adsorption of apolipoprotein on GNPs may facilitate transport through the BBB. Li *et al.* [116] provided an enhanced understanding of the mechanism of action of GNPs: the nanoparticles inactivated protein kinase C zeta type, suppressed threonine phosphorylation on occludin and zonula occludens-1, and perturbed occludin/zonula occludens-1 complex formation, and thus caused tight junction proteins disassembly and tight junction protein degradation. Although it is evident that more research is needed, we want to support the hypothesis for the 20-nm critical size for passing the BBB by comparison with the data of the recent work [88], in which the possibility of GNPs penetrating through the blood–retinal barrier of the mouse was examined. Particles of 100 nm size were not found in retinal tissues, whereas 20-nm particles were present in almost all retinal layers, including neurons (70%), endothelial cells (17%), and glial cells (8%). In addition to these observations, one should note an interesting report [117] in which ultrafine nanoparticles were shown to be able to enter the body by translocation of inhaled nanoparticles to the brain via the olfactory neuronal pathway.

The penetration of GNPs through the hematoplacental barrier and the potential fetotoxicity of GNPs at their exposure in pregnancy are important research fields of nanotoxicity. Surprisingly enough, there were no published reports until recently [118,119]. Yang *et al.* [118] investigated the effect of gestational age and nanoparticle composition on fetal accumulation of maternally administered bare and PEG-coated 13-nm GNPs in mice. We have studied the placental barrier permeability of white rats for PEGylated

5- and 30-nm GNPs, which was injected to pregnant female rats intravenously on a ten day of gestation at a dose about 0.8 mg(Au)/kg(animal). GNPs in tissues were visualized by the silver nitrate autometallography and the total Au content in fetuses was evaluated by AAS. In particular, GNPs were observed in the fetus, liver, and spleen, whereas AAS revealed enhanced total gold content of 30-nm particles in fetuses. To the best of our knowledge, it is the first *in vivo* demonstration of the GNP penetration through the rat placental barrier. Despite of the presence of GNPs in fetuses, no morphological changes were observed in the organs of fetuses examined.

We end this section with one general conclusion following from the entire array of data: from the standpoint of the accumulation level and slow kinetics of excretion, the most vulnerable organ is undoubtedly the liver (and, to a lesser extent, the spleen). This peculiarity of biodistribution may lead to acute inflammations of the liver [85]. On the other hand, the particle-organ distribution strongly depends on the administration method, as evident from comparison of the data for peritoneal (Figure 3.3) and IV injections [74].

3.2.4 Dependence of Biodistribution on the Functionalization of the GNP Surface

Along with size, the surface functionalization of GNPs is undeniably a key factor that governs the fate of particles in the animal organism after IV administration. First of all, nonstabilized GNPs will aggregate already at the time of administration under conditions of elevated ionic concentration, and this inevitably will lead to differences in the kinetics of particle circulation in the blood stream. Moreover, the use of special stabilizers with probing molecules may enhance the accumulation of GNPs in the target organ, *e.g.*, in a tumor.

As the first example, we compare the data from an early experiment with pigs [56] and those from a recent work [86] using the same model and particles in approximately the same size range (15 to 20 nm). The experiment described in [85] differed mainly in the dose applied (2 µg/g) and in the period of kinetics monitoring (up to 170 h), and it also used different surface stabilizers (gum arabic and maltose). Gum arabic- and maltose-coated particles showed differing patterns of accumulation in blood, tissues (liver and lung), and urine, with the difference being up to 50% or even greater. Specifically, gum-arabic-coated GNPs accumulated mostly in the liver, whereas those coated with maltose predominated in the lung.

Bergen *et al.* [66] compared the biodistribution kinetics of nonstabilized particles with diameters of 50, 80, 100, and 150 nm with that of particles conjugated to different stabilizers (PEG and PEG + galactose), which ensured the net positive (only for 50 nm) or negative charge of GNPs (Figure 3.4). Unfortunately, no experiment to assess the influence of charge on the biodistribution of 50-nm GNPs was presented. At 2 h after mice had been given an injection of approximately 0.5 µg/ml GNPs, the blood level was maximal for the PEGylated particles. In terms of the ratio of gold amount (ng) to protein amount (mg), the gold level was on the order of 5 to 15 ng/mg for 50-, 80-, and 100-nm PEGylated particles and was noticeably lower (approximately 0.25 ng/mg) for 150-nm particles. The adsorption of albumin and other plasma proteins on particles stabilizes

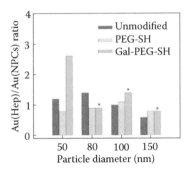

FIGURE 3.4 Liver cell biodistribution analysis of GNPs at 2 h after IV injection into mice at a dose of about 0.5 μg/g. Values are the ratio of Au in hepatocytes (HEP) to Au in nonparenchymal cells (NPCs). No significant statistical difference as compared to the PEG–SH control ($p > 0.05$). (Adapted from Khlebtsov, N.G., Dykman, L.A., *Chem. Soc. Rev.*, 40, 1647–1671, 2011. With permission.)

them partly, which explains the fact that the particle level in blood is detectable, even if low (on the average, it is approximately two orders of magnitude lower than that for the PEG-protected GNPs). However, for particles conjugated to the galactose + PEG complex, the gold level in blood was minimal (three and more orders of magnitude lower than that in the case of PEG without galactose).

Bergen *et al.* [67] explained this unexpected result by the fact that the galactose molecules serve as a target for liver hepatocytes. Incidentally, this conclusion can be applied in full measure only to conjugates of 50-nm particles (Figure 3.4), because for them, the ratio of the gold content in hepatocytes to the gold content in nonparenchymal cells (mostly Kupffer cells and sinusoidal endothelial cells) was approximately 2.5. Moreover, only for these particles was the gold level in hepatocytes 16 times higher for the conjugate GNPs-50 + galactose + PEG–SH, as compared with the control (GNPs-50 + PEG–SH). Thus, these results unambiguously demonstrated the specific delivery *in vivo* of the complex of 50-nm GNPs with the surface modifier galactose + PEG–SH to hepatocytes (but not to Kupffer cells).

Elci *et al.* [120] demonstrated that surface charge dictates the suborgan distributions of nanoparticles in the kidney, liver, and spleen of mice intravenously injected with functionalized GNPs. Images of the kidney show that positively charged GNPs accumulate extensively in the glomeruli, the initial stage in filtering for the nephron, suggesting that these GNPs may be filtered by the kidney at a different rate than the neutral or negatively charged GNPs. They find that positively and negatively charged GNPs accumulate extensively in the red pulp of the spleen. In contrast, uncharged GNPs accumulate in the white pulp and marginal zone of the spleen to a greater extent than the positively or negatively charged nanoparticles. Moreover, these uncharged GNPs are also more likely to be found associated with Kupffer cells in the liver. Positively charged GNPs accumulate extensively in liver hepatocytes, whereas negatively charged GNPs show a broader distribution in the liver. Together, these observations suggest that neutral GNPs having 2-nm cores may interact with the immune system to a greater extent than charged GNPs do, highlighting the value of determining the suborgan distributions of nanomaterials for delivery and imaging applications.

The role of surface PEG modifiers in biodistribution and in obtainment of enhanced accumulation in an implanted tumor was explored by Zhang *et al.* [92], by using 20-, 40-, and 80-nm GNPs. The modifier also included a surface radioactive chelator with the tracer [111]In. The authors showed that the characteristic time of the half-life clearance of particles from mouse blood, with the dose ranging from 0.04 to 4 µg/g, is less than 1 min for 80-nm GNPs, approximately 10 min for 40-nm GNPs, and approximately 30 to 40 min for 20-nm GNPs. The 80-nm particles were present in the liver and spleen as early as 10 min after administration and were not present in blood, kidneys, bladder, or intestines. On the contrary, the 20-nm particles circulated in blood for a longer time, accumulated somewhat less in the liver and spleen, and was present in the heart, kidneys, and intestines. In addition, only 20-nm particles accumulated in tumor tissue, which can naturally be explained by the longer time of circulation and by the retention effect in a tumor with increased blood supply.

The studies considered previously have used either nonmodified GNPs or GNPs conjugated to BSA, PEG, and PEG derivatives. Goel *et al.* [87] investigated the biodistribution of 33-nm GNPs coated with TNF-α and PEG–SH. The studies were made with healthy mice and with mice bearing implanted solid tumors (Figure 3.5). The authors found that in healthy mice, PEG-coated GNPs accumulated mostly in the liver and spleen. The high concentrations in the liver and spleen were recorded at 24 h and subsequently changed little for nearly 3 to 4 months. Slow accumulation kinetics was observed in the kidneys as well, even if the total average level was much lower (approximately 5- to 10-fold). In the lungs, trace amounts of gold were found. Particles coated with TNF-α accumulated in the tumor as well. Despite the biospecific functionalization, however, the maximal level of accumulation in the tumor (at 12 h) was almost an order of magnitude lower than it was in the liver and spleen, and it was comparable only to accumulation in the kidneys and lung.

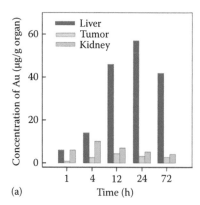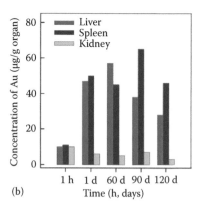

(a)   Time (h)   (b)   Time (h, days)

FIGURE 3.5 Time-dependent biodistribution of the conjugates CG-33 + TNF-α + PEG–SH during 1 to 72 h (a) and 1 h; and during 1 day to 120 days (b) after IV injection in mice at a dose of about 6.25 µg/g. (Adapted from Khlebtsov, N.G., Dykman, L.A., *Chem. Soc. Rev.*, 40, 1647–1671, 2011. With permission.)

An important observation made in that study is that a considerable amount of GNPs in the liver was recorded even at 120 days (!) after injection of 6.25 µg/g GNPs (Figure 3.5b). Thus, both the rapid elimination of particles from blood and their long-time retention in the organism are associated with the functioning of the hepatobiliary system.

3.2.5 Biodistribution of Other Types of Nanoparticles

Along with colloidal GNPs of approximately spherical shape, active use has recently been made in various biomedical nanotechnologies of nanoparticles that have different shapes and structures, viz., nanorods [121–123], nanoshells [7,124–126], nanocages [127], nanostars [128], and other types [51,129]. Novel particle types are needed, first of all, for tuning the plasmon resonance to the desired spectral range and also for tuning the ratio between the efficiencies of particle absorption and scattering [51]. The first requirement is associated with the agreement between the optical properties of particles and the window of biotissue transparency (750 to 1000 nm), whereas the second requirement is dictated by a concrete application task. For example, the maximum efficiency of absorption is important for photothermal therapy [130], and the maximum efficiency of scattering is essential for bioimaging [126].

The first study devoted to the biodistribution of GNRs (65 × 11 nm) was made by Niidome *et al.* [69]. Using ICP–MS, they observed that at 30 min after IV injection into mice, 54% of the administered PEGylated GNRs was present in blood, whereas upon administration of CTAB-stabilized GNRs, most of the gold was found in the liver. In the lungs, spleen, and kidneys, trace amounts of PEGylated GNRs were identified. At 24 h, the PEGylated GNRs were found only in the liver. Thus, replacement of CTAB with PEG led in this case to a dramatic change in the character of biodistribution at the earliest stages. The PEG-coated particles showed a circulation time that was an order of magnitude longer than that for the CTAB-coated particles, and correspondingly, they also showed much slower accumulation in the liver.

Kogan *et al.* [76] explored the possibility of achieving enhanced accumulation, through ensuring longer particle circulation, of PEG-coated GNRs (50 × 15 nm) in mice bearing implanted tumors. They used BDF_1 mice with implanted Lewis lung carcinoma. As early at 5 min after administration, blood was found to contain gold at 14 µg/g. This content remained almost the same for an hour, but GNRs gradually accumulated in the liver (approximately 2.5 to 3.5 µg/g) and tumor (3 to 9 µg/g). Thus, already within an hour, GNR accumulation in the tumor was quite significant relative to that in other organs.

In 2009, Niidome *et al.* [79] published an in-depth study of the role of PEG and the mechanism of GNR "permeability–retention" in a tumor. They showed that GNRs grafted with various amounts of PEG differ in their retention and distribution throughout organs. In particular, as the grafting degree was increased, the amount of gold (in percentage of the injected dose) decreased in the spleen but increased in the liver and tumor. At a high PEG:Au molar ratio (more than 1.5), the amount of gold per unit weight was larger in the tumor than in the liver. Niidome *et al.* [79] also investigated the effect of 4 to 200 µg of gold in mice. With an increase in the dose, accumulation was approximately constant, except in the liver, where it decreased. In absolute figures, the amount of gold per gram of organ was approximately proportional to the dose. In the spleen and tumor, however, the amount of gold per gram of sample divided by the dose mass decreased slightly. Overall, the authors concluded that PEG grafting at a PEG:Au molar ratio of more than 1.5 and an increase in the dose to approximately 40 µg per mouse ensure high accumulation of GNRs in the tumor.

An all-round analysis of all biodistribution stages was made recently by Wang *et al.* [59] for CTAB-capped GNRs with sizes of 56 × 13 nm, which were injected intravenously into rats at approximately 0.6 µg/g (Table 3.5).

Preliminarily, those authors demonstrated that BSA incubation (imitating the conditions *in vivo*) changes the sign of the zeta potential of GNRs yet preserves nanorod stability. The kinetic data for particle accumulation (0.5 h to 29 days) showed a rapid decline in the blood concentration of GNRs at the very initial moment (from 9.3 µg/ml to 262 ng/ml for 30 min) followed by a gradual decrease during 1 to 3 days and by a gradual but very significant accumulation in the liver (up to 60% of the dose) and spleen (up to 1.3%). Of note is the fact that up until day 28 of observation, the amount of GNRs was almost unchanged. The strong initial accumulation in the lungs (on the order of 2.5%) was accompanied by a gradual decrease to 0.6% of the dose by day 28. Accumulation in the kidneys and other organs was moderate, with the kinetics of elimination being slow. It is significant that GNRs were found with certainty in the brain. Thus, the smallness of even one size (GNR thickness, 13 nm) was sufficient for overcoming the BBB with a critical size of 20 nm (discussed earlier). Analysis by TEM and EDX demonstrated the preferential localization of the nanorods in the lysosomes of spleen macrophages and liver Kupffer cells.

There are few data on the biodistribution of gold nanoshells (GNSs). Using neutron activation, James *et al.* [73] examined the biodistribution of PEG-coated GNSs with a

TABLE 3.5   Biodistribution of PEG-Coated GNRs among Rat Organs at 0.5 h to 28 Days after Intravenous Injection at a Dose of 0.56 µg/g Animal

Organ	0.5 h	1 h	4 h	16 h	1 d	3 d	7 d	14 d	28 d
Liver[a]	15	30	42	55	60	60	66	62	67
Spleen[a]	0.5	0.6	0.75	0.9	1.2	1.0	1.0	1.3	1.25
Lung[a]	2.5	2.0	1.7	1.3	0.8	0.9	0.8	0.7	0.6
Kidney[a]	0.18	0.12	0.115	0.11	0.12	0.1	0.08	0.07	0.1
Heart[a]	0.07	0.1	0.06	0.07	0.07	0.06	0.05	0.04	0.01
Brain[b]	0.08	0.08	0.05	0.06	0.05	0.02	0.01	0.02	0.02
Blood[b]	260	260	250	250	60	70	30	20	10
Feces[b]	–	–	85	28	29	28	16	7	4
Bone[b]	40	30	20	15	15	18	19	10	5
Muscle[b]	27	20	20	19	20	20	20	11	5

Source: Adapted from Wang, L., Li, Y.-F., Zhou, L., Liu, Y., Meng, L., Zhang, K., Wu, X., Zhang, L., Li, B., Chen, C., *Anal. Bioanal. Chem.*, 396, 1105–1114, 2010. With permission.

[a] % of the given dose.

[b] ng/g.

TABLE 3.6 Biodistribution of PEG-Coated 110/10 nm Silica/Gold Nanoshells among Mice Organs at 4 h to 28 days after Intravenous Injection at a Dose of 10 µg/g Animal

Time	Concentration of Au (µg/g Organ)							
	Blood	Liver	Kidney	Spleen	Lung	Muscle	Brain	Bone
Control	0.0009	0.0007	0.0011	0.0174	0.0021	0.0230	0.0011	0.0049
4 h	313.7	103.8	52.22	952.2	88.58	3.796	7.187	9.531
1 day	29.17	311.8	27.61	1890	12.71	1.060	0.547	5.912
7 days	0.0187	313.4	21.49	2863	6.066	1.916	0.0684	7.319
14 days	0.0290	324.5	19.30	2039	3.738	0.779	0.0310	5.365
21 days	0.0430	252.0	23.53	1738	4.748	1.593	0.1243	8.333
28 days	0.0567	227.2	24.70	1703	3.781	1.023	0.0293	6.875

Source: Adapted from James, W.D., Hirsch, L.R., West, J.L., O'Neal, P.D., Payne, J.D., *J. Radioanal. Nucl. Chem.*, 271, 455–459, 2007. With permission.

silica core diameter of 110 nm and a gold layer thickness of 10 nm, which were injected intravenously into healthy mice (Table 3.6) and into mice bearing implanted tumors.

Note that the difference of these data from those discussed previously is the anomalously high accumulation of GNSs in the spleen by day 7, with very slow clearance (up to the 50% level only after a month). Similar kinetics of accumulation and clearance was observed for the liver as well, but the average total level was approximately nine times lower than that in the spleen. Such slow clearance of GNSs from organs of the reticuloendothelial system closely resembles the slow clearance of nanorods [59]. The high level of GNSs in blood was retained for less than a day, and from the lung, most of the particles were cleared by the end of the day and then the particle content decreased slowly. In the case of tumor-bearing mice, the maximum accumulation was obtained at one day postinjection. It is important to emphasize that even without any specific functionalization, the tumor accumulation of GNSs exceeded the corresponding levels in blood, kidney, muscle, brain, and bone (by 7.8, 6.6, 19.7, 3.7, 188, and 17.4 times, respectively), of course with the exception of the spleen and liver. In our work [91], we studied the size effect of biodistribution at 24 h after IV injection into rats of PEGylated GNPs with diameters of 15 and 50 nm and GNSs with an outer diameter of 160 nm and a core diameter of 120 nm (Figure 3.6).

Contrary to the data of James *et al.* [73], most accumulation after a day was recorded in the liver, and the gold level in the spleen was comparable for 50-nm GNPs and 160-nm GNSs. At present, it is difficult to explain such difference in results, which, true, were achieved for different animals (mice [73] and rats [91]) and at substantially different doses (10 µg/g [73] versus 1.3 µg/g [91]).

Melancon *et al.* [83], using the radiotracer ^{111}In, studied the biodistribution in tumored mice of hollow (without a silica core) 30-nm GNSs coated with immunoglobulins. They demonstrated that at 24 h, the maximum buildup of GNSs was in the liver, spleen, and kidney—approximately 20% of the dose administered per organ. Noticeably less gold was present in the heart, lung, stomach, small intestine, muscle, and bone (approximately 5% per organ on the average). About 10% GNSs accumulated in the tumor. Thus, these data agree qualitatively with the findings for the biodistribution of usual 30-nm

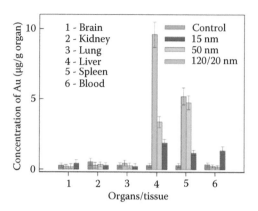

FIGURE 3.6 Biodistribution of PEG–SH-coated 15-nm and 50-nm CG nanoparticles and 120/20-nm SiO$_2$/Au nanoshells at 24 h after IV injection into rats at a dose of 0.25 µg/g animal. (Adapted from Terentyuk, G.S., Maslyakova, G.N., Suleymanova, L.V., Khlebtsov, B.N., Kogan, B.Ya., Akchurin, G.G., Shantrocha, A.V., Maksimova, I.L., Khlebtsov, N.G., Tuchin, V.V., *J. Biophoton.*, 2, 292–302, 2009. With permission.)

GNPs and do not show any peculiarities dictated by the lighter weight of the hollow particles.

We end this section with a short discussion of the results from a recent work by Bartneck *et al.* [93] on the biotoxicity and distribution of a broad assortment of particles. We have shown previously that the uptake of nanoparticles by Kupffer cells is a major mechanism of particle accumulation and clearance. Bartneck *et al.* [93] disclosed an alternative clearance mechanism, reporting that GNPs and GNRs can be eliminated from peripheral blood via an extracellular network ("trap") formed by neutrophil granulocytes. It is well known that nanoparticles are quickly internalized by immune cells if the particle surface has not been treated to minimize the interaction [131], *e.g.*, with poly(ethylene oxide). According to the data of Bartneck *et al.* [93], generated in a study of the interaction of variously shaped and sized particles with human immune cells, the mechanism of nanoparticle uptake can be classified as macropinocytosis rather than as phagocytosis. Particle shape was found to have a strong influence on uptake by cells of the immune system; in particular, nanorods may be taken up more rapidly than nanospheres and other particles. These data appear to be quite important in light of the previously revealed [3] adjuvant properties of GNPs.

3.3 Toxicology Studies

3.3.1 Preliminary Remarks

The tremendous upgrowth of nanobiotechnologies that is currently witnessed inevitably gives rise to questions related to the safety of nanomaterials. In connection with this, such terms have emerged as "nanotoxicology," "nanomaterial toxicity," and even "nanosafety" [132]. So far, however, the data on GNP toxicity have been contradictory. This fact can be explained, as in the case of biodistribution, by the large scatter in experimental design, using various particle sizes and shapes, functionalization methods, animal types, particle doses and administration methods, and so on.

It should be said that issues related to the toxicity of metallic gold have been raised earlier as well [133–135]. In those early articles, it was noted that injection of metallic gold into laboratory animals may lead to inflammatory responses, accumulation of gold in the reticular cells of lymphoid tissue, and activation of cellular and humoral immunity. In his brilliant review of pioneering works, Pacheco [133] presented many data on the effect of CG on nonspecific immune responses. Specifically, he noted that at 2 h after rabbits had been injected intravenously with 5 ml of GNPs, there was a considerable increase in the total number of leukocytes in 1 ml of blood (from 9,900 to 19,800) against the background of a slight decrease in mononuclear forms (from 5,200 to 4,900) and a large increase in polynuclear forms (from 4,700 to 14,900). No such phenomena were observed upon administration of other colloidal metals. As the mechanism of such action of the noble metal, the conversion of Au(0) to Au(I) in the cells of the immune system under the effect of certain sulfur-containing amino acids was discussed. The immunological properties of nanomaterials were reviewed by Dobrovolskaia and McNeil [136]. Because this section addresses the toxicity of only one type of nanomaterial, GNPs, we refer the readers to recent reviews [21,26,30] with wider coverage of nanoparticle types.

There are two experimental approaches to investigating the cytotoxicity of nanoparticles, *viz.* the *in vitro* one (using animal cell cultures) and the *in vivo* one. Here, we will consider each of these approaches separately. Table 3.7 presents, in concise form, the literature information on GNP toxicity.

3.3.2 Cytotoxicity *In Vitro*

3.3.2.1 Effects of Particle Size, Shape, Concentration, and Surface Modification

Cytotoxicity *in vitro* is studied with various animal cell cultures, most commonly the fibroblast of the human skin (HeLa), human leukemia (K562), human hepatocarcinoma (HepG2), human breast carcinoma (SK-BR-3), and others. The toxic effect is most often assessed by the MTT assay.

As in the case of biodistribution, surface properties may be crucial in the cytotoxicity of the same particles. For example, Goodman *et al.* [137] investigated the toxicity of anionic and cationic 2-nm GNPs to Cos-1 cells (CV-1 kidney fibroblasts from the African green monkey that were transformed by the SV40 virus) and to erythrocytes. Cationic particles functionalized with quaternary amines proved to be sevenfold more toxic (the lethal concentration LC_{50}, 1 µM) than anionic particles, prepared by substituting carboxylic groups for amine groups (LC_{50}, 7.37 µM).

Connor *et al.* [138] also used the MTT assay to study the toxicity of 18-nm GNPs to K562 cells. The nanoparticle surface was modified with several ligands (citrate, biotin, and CTAB). Citrate- and biotin-modified particles were not toxic when used up to a concentration of 250 µM, whereas CTAB-coated particles were already toxic at 0.05 µM. However, being washed free of CTAB, the particles lost their toxic properties. Clearly,

TABLE 3.7   Literature Data on the Toxicity of GNPs

Particles (size in nm)	Surface Coating	Model (cell lines, animals)	Dosage	Exposure Duration	Method	References
		Cytotoxicity *in vitro*				
GNR Au–Ni 200 × 100	DNA/transferrin	HEK293	44 µg/ml	4 h	WST-1	Salem [139]
CG-2	Quaternary ammonium	Cos-1	0.38, 0.75, 1.5, 3 µM	1, 2.5, 6, 24 h	MTT	Goodman [137]
CG-20	BSA + peptides	HeLa 3T3/NIH HepG2	150 pM	3 h	LDH	Tkachenko [140
CG-4, 12, 18	Biotin, CTAB	K562	0.002–250 µM	1 h–3 d	MTT	Connor [138]
CG-3.5	Lysine, poly-L-lysine	RAW264.7	10, 25, 50, 100 µM	24, 48, and 72 h	MTT	Shukla [141]
CG-3	Tiopronin TAT-peptide	HTERT-BJ1	0–15 µM	24 h	MTT	de la Fuente [142]
CG-14	–	CF-31	0–0.8 mg/ml	2, 4, 6 d	Confocal microscopy, scanning modulation force microscopy, TEM, migration assay	Pernodet [143]
CG-15	Coumarin–PEG–SH	MDA-MB-231	100 µg/ml	24 h	MTT	Shenoy [144]
CG-5, 10, 14	BSA PEG	NB2a	1.0×104–5.4×1012 particles/ml	24 h	Light microscopy	Ikah [145]
GNR-65 × 11	CTAB, PEG–SH	HeLa	0.01–0.5 mM	24 h	MTT	Niidome [69]

(Continued)

TABLE 3.7 (CONTINUED) Literature Data on the Toxicity of GNPs

Particles (size in nm)	Surface Coating	Model (cell lines, animals)	Dosage	Exposure Duration	Method	References
GNR-65 × 11	CTAB, phosphatidylcholine	HeLa	0.1–1.5 mM	24 h	MTT	Takahashi [146]
Gold clusters-0.8, 1.2, 1.4, 1.8, CG-15	Triphenylphosphine	HeLa Sk-Mel-28 L929 J774A1	1–10,000 µM	6, 12, 18, 24 h	MTT	Pan [147]
CG-18	–	HeLa	0.2–2 nM	3, 6 h	MTT	Khan [148]
CG-33	–	BHK21 HepG2 A549	10, 30, 60, 120 nM	36, 72 h	MTT	Patra [149]
CG-30	PEG-SH	HepG2	0.01, 0.1, 1, 10, 10, 1,000 µg/ml	24 h	MTT	Kim [150]
GNS Au/Cu	–	Vero	0.001, 0.01, 0.1, 1, 10, 50, 100, 200 µg/ml	6, 24 h	WST-1	Su et [151]
GNR-32 × 7	CTAB, phosphatidylcholine, animomercaptotriazole, mercaptoundecanoic acid	MCF 10A	0.25 mM	1, 6, 12, 24 h	Trypan blue staining	Yu [152]
Nanowires 0.58, 1.8, 4.5, 8.6 µm × 200 nm	Thiols with amino, alkyl, or carboxyl end groups, serum	NIH 3T3 HeLa	103–106 particles/ml	24 h	MTT	Kuo [153]
CG-5	Cyclodextrin	V79	up to 45 µM	0–20 h	Impedance spectroscopy	Male [154]
CG-20	Cysteine	HepG2		1, 2, 4, 6 h, 1, 2, 4 d	Multiplexed cytotoxicity assay	Jan [155]

(Continued)

TABLE 3.7 (CONTINUED)　Literature Data on the Toxicity of GNPs

Particles (size in nm)	Surface Coating	Model (cell lines, animals)	Dosage	Exposure Duration	Method	References
GNR-72 × 28	CTAB, polystyrenesulfonate	Hep G2 LLC-PK1 KB	0.5–100 µg/ml	6, 24, 48 h	MTT LDH	Leonov [156]
CG-5, 12, 20, 30, 50, 70 GNRs, aspect ratios from 2.1 to 3.5	CTAB, polystyrenesulfonate	HaCaT		24, 48 h	MTT	Wang [157]
GNR-40 × 18	CTAB, copolymer poly(diallyldimethylammonium chloride)–poly(4-styrenesulfonic acid)	HeLa	10, 20, 50, 150 µM	6 h	Trypan blue staining	Hauck [158]
CG-5	–	Hep3B Panc-1	1, 10, 67 µM	4 h	MTT	Gannon [159]
CG-40	–	NRK-52E Hep-G2	0–200 µg/ml	24 h	MTT	Simon-Deckers [160]
CG-13, 30, 60	CALNN peptide	HeLa	0.02, 0.04, 0.08, 0.16, 0.32 nM	24 h	Trypan blue staining	Sun [161]
CG-10-50	–	HDF-f	10, 50, 100, 200, 300 µM	72 h	MTT, histology	Qu [162]
CG-5, 15, 30	–	hBMSCs HuH-7	30, 40 µg/ml	5 d	MTT	Fan [163]
GNRs, aspect ratios from 2.1 to 4.1	CTAB, polyacrylic acid, poly(allylamine) hydrochloride	HT-29	0.01–5 nM	2, 3, 4, 5 d	MTT	Alkilany [164]

(Continued)

TABLE 3.7 (CONTINUED)   Literature Data on the Toxicity of GNPs

Particles (size in nm)	Surface Coating	Model (cell lines, animals)	Dosage	Exposure Duration	Method	References
CG-3, 5, 5, 8, 12, 17, 37, 50, 100	–	HeLa	100 nM–0.4 mM		MTT	Chen [165]
CG-18	Poly-N-isopropylacrylamide-co-acrylamide co-polymer	MCF7	0–0.74 nM	48 h	Sulforhodamine staining	Salmaso [166]
CG-18	Folic acid	HeLa	0.014, 0.14, 0.68 mM	24 h	MTT	Li [167]
CG-7	Polyvinylpyrrolidone	HEK293 MDA-MB-231	0.01–100 µM	24 h	MTT	Zhou [168]
CG-15	BSA	NIH3T3	$10^{-7}, 10^{-6}, 10^{-5}, 10^{-4}$ M	3 h	MTT	Murawala [169]
CG-4, 13	–	A549 Jurkat THP-1	2 nM	24, 48 h	CellTiterBlue, toxilight	Pfaller [170]
GNR-60 × 13	Cystamine, herceptin	SK-BR-3	0.05, 0.1, 0.5 nM	24 h	MTT	Chen [171]
CG-8	Polyvinylpyrrolidone	MDCK, HepG2	10, 100, 400, 600, 800, 1000 nM	24 h	Neutral red uptake, colony forming efficiency test	Ponti [172]
CG-10	Polyvinylpyrrolidone	DCs	0.5 mM	4, 24, 48 h	Flow cytometry	Villiers [173]
CG-10	Glycolipid	HepG2	51.4 µg/ml (100 µM)	3 h	MTT, comet assay	Singh [174]
GNR-30 × 10	CTAB, poly(sodium-4-styrenesulfonate)/polyethyleneimine	HeLa	25, 50, 100, 200 µM	48 h	MTT	Chen [175]
GNR-45 × 18, 40 × 12	CTAB, polystyrenesulfonate, PEG–SH	SKBR3, CHO, C2C12, HL60	20–174 pM	24 h	MTT	Rayavarapu [176]

(Continued)

TABLE 3.7 (CONTINUED)   Literature Data on the Toxicity of GNPs

Particles (size in nm)	Surface Coating	Model (cell lines, animals)	Dosage	Exposure Duration	Method	References
CG-13, 45	–	CF-31	0–189 µg/ml (13 nm); 0–26 µg/ml (45 nm)	2–17 d	Confocal microscopy, SEM, TEM	Mironava [177]
CG-43 GNR-38 × 17	CTAB	MDCK II	15 µg/ml	20 h	MTS, TEM, dark-field microscopy	Tarantola [178]
CG-17	Serum protein	A549	2.43 |g/ml	24 h	MTT, LDH	Choi [179]
CG-4	Poly(methacrylic acid)	Cl7.2, PC12	10–200 nM	2, 4, 8, 12, 24 h	Confocal microscopy, LDH, ROS level	Soenen [180]
CG-14, 20	–	BEAS-2B, CHO, HEK 293	5 nM	1 h	XTT, LDH	Vetten [181]
GNR, aspect ratios 3 and 32	Tannic acid, PEG-COOH	HaCaT	20 µg/ml	24 h	MTT, TEM	DeBrosse [182]
CG-20	–	MRC5	1 nmol/L	72 h	LDH, Trypan blue staining	Ng [183]
CG-5	Polyethylene imine	A549, MCF-7	0.4-1.6 µM	48 h	MTT	Mohan [184]
GNR-41 × 10	CTAB	AGS, A549, NIH3T3, PK-15, Vero	72–720 µg/L	72 h	Trypan blue staining, MTS,	Chuang [185]
CG-3, 5, 6, 8, 10, 17, 30, 25	–	MCF-7, PC-3	10–130 µg/L	24 h	MTT	Vijayakumar [186]
CG-5, 13	–	HK-2	1, 25, 50 nM	24 h	LDH	Ding [187]

(Continued)

TABLE 3.7 (CONTINUED) Literature Data on the Toxicity of GNPs

Particles (size in nm)	Surface Coating	Model (cell lines, animals)	Dosage	Exposure Duration	Method	References
GNR effective radii 5–20 nm	PEG	SKOV3, IGROV-1, A2780/S, HeLa	10–100 μM	24, 72, 168 h	MTT	Tatini [188]
CG-20	–	C2C12	100–1000 ng/ml	24 h	MTT	Wahab [189]
CG-5, 30	–	Caco-2	10, 50, 100, 200, 300 μM	24, 72 h	Colony forming efficiency (CFE), Trypan blue staining	Bajak [190]
CG-30, 50, 90	–	NHDF	1–25 μg/ml	24, 48, 72 h	MTT, LDH	Mateo [191]
GNR-52 × 25	CTAB	A549	2.5–15 μg/ml	4 h	MTT, LDH	Tang [192]
CG-1.5 CG-4 CG-14	Mercapto-alkyl acids Mercaptosuccinic acid BSA	hESCs	0.6 μg/ml 10 μg/ml 10 μg/ml	1–14 d	Trypan blue staining, MTT	Senut [193]
Cytotoxicity *in vivo*						
CG-13	PEG–SH	Mice	IV; 0.17, 0.85, 4.26 mg/kg	5, 30 min, 4, 24 h, and 7 d	ICP MS, histology, TEM	Cho [85]
CG-4, 100	PEG–SH	Mice	IV; 4.26 mg/kg	30 min	Histology, TEM, TUNEL assay	Cho [194]
CG-3, 5, 8, 12, 17, 37, 50, 100	BSA, lysozyme, peptides	Mice	IP; 8 mg/kg/week	50 d	Histology, CARS microscopy	Chen [165]

(Continued)

TABLE 3.7 (CONTINUED) Literature Data on the Toxicity of GNPs

Particles (size in nm)	Surface Coating	Model (cell lines, animals)	Dosage	Exposure Duration	Method	References
CG-11	–	Zebrafish embryos	0.025, 0.05, 0.1, 0.2, 0.4, 0.6, 0.8, 1.0, 1.2 nM	24, 48, 72, 96, 120 h	Histology, TEM	Browning [195]
CG-5	–	Mice Rats	IP; 57 µg IP; 285 µg	2, 4, 24 h 3, 7, 10 d	Histology	Pocheptsov [196]
CG-20	–	Rats	IV; 0.01-0.015 mg/kg	1, 7 d, 1, and 2 month	RNA microarray analyses	Balasubramanian [60]
CG-13	–	Mice	IP; 40, 200, 400 mg/kg/day	8 d	Histology, serum biochemical analysis, hematological analysis	Lasagna-Reeves [197]
CG-10, 20, 50	–	Rats	IP; 50, 100 µl	3, 7 d	Histology	Abdelhalim [198]
CG-5, 10, 30, 60	PEG	Mice	IP; 4 mg/kg	28 d	Histology, serum biochemical analysis, hematological analysis	Zhang [199]
GNS-155	PEG	Mice Dogs	IV; 2 mg/kg IV; 35 mg/kg	1, 7, 28, 56, 182, 404 d	Histology	Gad [200]
CG-14	–	Rats	IV; 0.9, 9 and 90 µg per rat	14 d	Histology, INAA	Rambanapasi [108]

in this case, the toxicity of the conjugate results entirely from the presence of CTAB on the particle surface.

Particle size is an important physical parameter that controls endocytosis effectiveness [201] or binding to B-form DNA [80]; therefore, it is natural to expect the size dependence of cytotoxicity. Pan *et al.* [147] examined in detail the size dependence of toxicity by using four cell lines (HeLa, Sk-Mel-28, L929, and J774A1); gold atomic clusters of 0.8 nm (eight gold atoms), 1.2 nm (35), 1.4 nm (55) and 1.8 nm (150); and 15-nm nanoparticles stabilized with triphenylphosphine derivatives. According to the MTT assay data, the clusters of 1.4 nm proved the most cytotoxic, with the inhibitory concentration IC_{50} ranging from 30 to 46 µM. The IC_{50} values for the clusters of 0.8, 1.2, and 1.8 nm were much higher (250, 140, and 230 µM, respectively). The 15-nm GNPs were not cytotoxic even at concentrations 100-fold higher than the IC_{50} for the small clusters. In addition, the authors demonstrated that the action of the 1.4-nm clusters led to cell necrosis after 12 h of incubation and that the use of the 1.2-nm clusters led to apoptosis. The toxicity of 1.4-nm gold clusters to healthy and tumorous human cells has also been reported by Tsoli *et al.* [202]. In the work cited earlier [80], it was noted that the high toxicity of the 1.4-nm Au_{55} cluster may be due to the size similarity to B-form DNA. Thus, those studies have shown that the transition to the sizes of classic colloidal particles (15 nm) sharply decreases cytotoxicity, as compared to atomic clusters of approximately 1 to 2 nm, which can irreversibly bind to DNA and, possibly, to other key molecules as well.

Naturally, the toxicity of colloidal GNPs also may depend on the GNP concentration, which can control the number of particles penetrating the cell. Pernodet *et al.* [143] explored the interaction of a fibroblast cell culture with 14-nm citrate-coated GNPs added at various concentrations (0.2, 0.4, 0.6, and 0.8 mg/ml). Analysis of the cells by electron microscopy revealed that at high concentrations, the GNPs penetrate the membranes and accumulate in vacuoles. This is accompanied by damage to actin filaments, affecting the motility of the cells and their proliferative and adhesive abilities. In a further study [177], it was shown that 45-nm GNPs were more markedly cytotoxic than 13-nm GNPs. The suggestion was made that this might be due to the greater damaging effect of the 45-nm GNPs on vacuoles and, correspondingly, to the greater release of these GNPs into the cytoplasm, with the normal cell function being disrupted. In addition, toxicity was found to depend on the concentration of GNPs and on the duration of their effect on the cells.

No toxicity of 10- to 50-nm citrate-coated GNPs to embryonal fibroblasts was found by the MTT assay up to the maximum particle concentration of 300 µM [162]; however, such high concentrations induced changes in cell morphology. No toxicity of 5-nm fluorescent GNPs (concentration, up to 45 µM) to fibroblasts was reported by Male *et al.* [154], who used impedance microscopy (as compared with the cytotoxicity of quantum dots). Similar data have been recorded by Jan *et al.* [155]; however, the authors note that although not being cytotoxic, GNPs nevertheless inhibit the proliferation and intracellular calcium release of HepG2 cells.

No cytotoxicity was found by the LDH assay (determination of Lactate DeHydrogenase activity) for 20-nm BSA-coated GNPs in HeLa (80% of living cells) and 3T3/NIH (95%) cultures [140]. Likewise, as found by the MTT assay, 15-nm BSA-coated GNPs were not toxic to 3T3/NIH [169]. The MTT assay revealed no toxicity of 15-nm PEGylated nanoparticles (up to 100 µg/ml) interacting with the cell line MDA-MB-231 [144] or of 30-nm particles interacting with the line HepG2 [150]. Khan *et al.* [148] not only revealed no toxic effect of 18-nm GNPs (maximal concentration, 2 nM) on HeLa cells in an MTT assay but also showed the absence of any GNP effect on cellular transcriptional processes. Using the same model, method, and particle size, Li *et al.* [167] found no cytotoxicity of GNPs coated with folic acid.

Salmaso *et al.* [166] reported that neither nonfunctionalized nor polyacrylamide-coated 18-nm GNPs were toxic to human breast adenocarcinoma cells. Likewise, polyvinylpyrrolidone-coated colloidal 8-nm particles showed no toxicity to human embryo kidney and human breast cancer cells [168] or to canine kidney and human hepatocarcinoma cells [172].

Analysis of the viability of dendritic cells incubated with 10-nm GNPs [173] demonstrated that the nanoparticles were not cytotoxic even at high concentrations. Moreover, the dendritic cell phenotype remained the same after cellular interaction with the GNPs, even though the nanoparticles were present in the endosomes of the dendritic cells. Increased cytokine secretion was observed, attesting to the activation of immune response.

Shukla *et al.* [141] focused on the cytotoxic and immunogenic action of CG with a mean particle diameter of 3.5 nm. Their study was made on RAW264.7 macrophage cells. They found that the GNPs, while penetrating into the intracellular space, are not cytotoxic or immunogenic (up to a concentration of 100 µM). Atomic force microscopy and TEM revealed the GNPs in the lysosomes and perinuclear space of the cells. The authors concluded that GNPs are fully biocompatible. Similar results were obtained by de la Fuente and Berry [142] for 3-nm GNPs; Gannon *et al.* [159] for 5-nm GNPs; Wang *et al.* [157] for 5-, 12-, 20-, 30-, 50-, and 70-nm GNPs; Chen *et al.* [165] for 3-, 5-, 8-, 12-, 17-, 37-, 50-, and 100-nm GNPs; Pfaller *et al.* [170] for 4- and 13-nm GNPs; and Simon-Deckers *et al.* [160] for 40-nm GNPs.

Thus, the works reviewed earlier did not reveal, on the whole, any noticeable toxicity of typical citrate-coated colloidal particles with sizes of 5 to 20 nm up to concentrations on the order of 100 µM, at which the penetration of GNPs into cells and their localization in intracellular structures are observed. With regard to the data cited earlier, one could noticeably extend the boundaries of absence of toxic GNP effect both in the direction of the lower limit (down to 3 nm) and in the direction of large particles (up to 100 nm), if one considers the limiting dose to be approximately 10^{12} particles/ml.

There are, however, intriguing data [149] to indicate that with all other factors being equal, cytotoxicity may depend on the type of cell. With the MTT assay, Patra *et al.* [149] established that 33-nm GNPs were not toxic to hamster kidney cells (BHK21) or to human hepatocellular carcinoma (HepG2) but were toxic to human carcinoma lung cell line A549. Interestingly, toxicity was directly proportional to the injected nanoparticle dose. Ikah *et al.* [145] showed that GNPs are toxic to neuroblastoma cells.

Contrary to the previous conclusion as to the low cytotoxicity of GNPs, which was made on the basis of an analysis of representative publications, Fan *et al.* [163], studying the toxicity of 5-, 15-, and 30-nm GNPs on human bone marrow stem cells (hBMSCs line) and on human hepatocarcinoma (HuH-7), concluded otherwise. Cytotoxicity was investigated by the MTT assay for 5 days, by using various nanoparticle concentrations.

In both cultures, after administration of 15- and 30-nm particles at up to 70 μg/ml, the authors found more than 80% of living cells. However, when 5-nm particles were used—even at up to 30 μg/ml—the number of viable cells did not exceed 60%. Thus, in this model, the lower size limit with respect to toxicity proved to be approximately 15 nm. From our point of view, the concentrations of 30 μg/ml for 5-nm particles (150 μM Au, 2.5×10^{13} particles/ml, corresponding to an optical density of about 0.5 in a 1-cm-thick cuvette at a plasmon resonance wavelength of 515 nm) is sufficiently high that one could speak of a truly toxic action of low doses. Therefore, we believe that the results of Fan *et al.* [163] are in reasonable conformity with our conclusion if one takes an approximate value of 10^{12} particles/ml as the upper limit of the particle number concentration. To support this conclusion, we mention, additionally to the works cited above, the paper of Singh *et al.* [174] in which the cytotoxicities and genotoxicities of glycolipid-coated GNPs were probed. Using HepG2 cells, the authors demonstrated that at up to 100 μM, 10-nm GNPs (concentration, 2×10^{12} particles/ml) were not cytotoxic (survivability, 80% of cells during 3 h [MTT assay]) or genotoxic (comet assay assessment of damage to DNA).

Undoubtedly, surface functionalization of GNPs also affects cytotoxicity. Ryan *et al.* [203] analyzed the cytotoxicity of 5-nm nanoparticles depending on the number of peptide molecules adsorbed on them. They showed that when the number of adsorbed molecules was increased from 30 to 150 per particle, cell survivability declined from 98% to 66%. Sun *et al.* [161] determined cytotoxicity by using 13-, 30-, and 60-nm GNPs conjugated to CALNN peptides (which facilitate intracellular penetration). They found that at 0.02 to 0.08 nM GNPs, the survivability of HeLa cells in a Trypan blue assay ranged from 80% to 95%. However, at 0.16 nM GNPs, survivability decreased to 60%, and at 0.32 nM, it became less than 10%.

Schaeublin *et al.* [204] demonstrated that three different GNPs (positively charged, neutral, and negatively charged) showed that cell morphology was disrupted by all three GNPs and that they demonstrated a dose-dependent toxicity; the charged GNPs displayed toxicity as low as 10 μg/ml and the neutral at 25 μg/ml.

Xia *et al.* [205] reported that (2-[4-(2-hydroxyethyl)-1-piperazinyl]ethanesulfonic acid)-prepared GNPs had better biosafety and biocompatibility than citric-prepared GNPs did. This study not only demonstrated the key fact of reducing agent but also accumulated more evidence for the study of biosafety and biocompatibility of nanoparticles.

Favi *et al.* [206, 207] investigated the effect of various shapes on the cytotoxicity of GNPs. Results showed that the gold nanospheres with diameters of approximately 60 nm showed greater toxicity with fibroblast and endothelial cells compared with the gold nanostars with diameters of approximately 35 nm and long GNRs 534 × 65 nm (fabricated by electron beam physical vapor deposition; without CTAB).

In summary, although on the face of it, the picture of *possible* cytotoxicity of gold nanospheres is quite contradictory, we have come to recognize that the entire array of experimental data permits a conclusion that is in line with most studies—namely, that if the upper particle concentration limit does not exceed 10^{12} particles/ml, GNPs are not cytotoxic down to the small diameters of 3 to 5 nm. The lower toxicity boundary is associated with GNP diameters of 1 to 2 nm and is determined by the ability of GNPs to irreversibly bind to key biomolecules (DNA, *etc.*) and to change the functioning of cellular molecular processes.

3.3.2.2 Biological Mechanisms of Toxicity

To date, there have been only fragmentary data on the mechanisms governing cytotoxicity. For example, Jia *et al.* [208] speak of a GNP-induced release of nitrogen oxide into serum, whereas Li *et al.* [209] point to oxidative stress, which damages cells *in vitro*. Uboldi *et al.* [210] consider the residues of sodium citrate adsorbed to the GNP surface to be responsible for causing toxicity. This hypothesis does not seem to be fully justified in light of the known phenomenon of opsonization—the adsorption of high-molecular-weight substances to particles when they get into the blood stream or a culture medium [211–213]. Furthermore, it was shown recently [214] that citrate-coated GNPs stimulate the respiratory activity of macrophages and the activity of macrophage mitochondrial enzymes. Yan *et al.* [215] reported that nanoparticle-induced reactive oxygen species (ROS) play a key role in cellular and tissue toxicity.

In contrast to the available body of literature relating to gold nanospheres, there is much less toxicological information on variously structured nanorods, including GNRs. Salem *et al.* [139] reported data on the cytotoxicity of gold/nickel nanorods prepared on hard templates. Such particles do not have the toxic molecules of CTAB on their surfaces, which is a key factor in this aspect. The studies were made on human embryonic kidney cells (HEK293). Nanorods were added to a concentration of 44 µg/ml, and the WST-1 assay was used (an analog of the MTT assay that uses 4-[3-(4-iodophenyl)-2-(4-nitrophenyl)-2H-5-tetrazolio]-1,3-benzene disulfonate). It was found that under such conditions, the LD_{50} equaled 750 µg/ml; this served to indicate that the nanomaterials used in the study were nontoxic. Compared with the previously discussed typical limit for GNPs (on the order of tens of µg/ml), the value of 750 µg/ml is, undoubtedly, very high.

Niidome *et al.* [69] used GNRs obtained with a protocol of synthesis induced by CTAB-coated seed in a CTAB-containing growth solution. Using the MTT assay on HeLa cells, they demonstrated that CTAB-coated rods were highly cytotoxic at as low as 0.05 mM (80% of dead cells). However, when CTAB was replaced with PEG–SH, the authors observed 95% of viable cells, with the nanorod concentration being 0.5 mM.

The CTAB-coated GNRs exhibited a stronger cytotoxicity in terms of the metabolic activity and apoptosis induction than the PEG-coated GNRs [216].

Huff *et al.* [217] explained this fact by the decreased ability of the PEGylated nanorods to penetrate into cells. However, this assumption is only partly legitimate. Besides the decrease in GNP penetration of the cells, the main factor is possibly the lowered cytotoxicity of the particles themselves resulting from the replacement of CTAB molecules by the thiol-modified molecules of PEG. The experiments described by Leonov *et al.* [156], Wang *et al.* [157], and Parab *et al.* [218] are illustrative in this respect. Those authors observed a considerable decline in nanorod cytotoxicity not only when CTAB was replaced with polystyrene sulfonate (PSS; Figure 3.7) but also when the nanorods were washed three times.

By contrast, unwashed nanorods were highly toxic. Comparison of free CTAB (without any particles) with the additives revealed that in this case, the triply washed GNRs

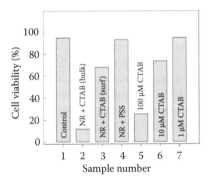

FIGURE 3.7 Viability of cells without any additives (control 1), cells incubated with GNRs (NR, aspect ratio of 3) containing bulk (2) or surface (3) CTAB, and cells incubated with GNRs coated with PSS (4). Samples 5, 6, and 7 are 100, 10, and 1 µM CTAB solutions, respectively. (Adapted from Wang, S., Lu, W., Tovmachenko, O., Rai, U.S., Yu, H., Ray, P.C., *Chem. Phys. Lett.*, 463, 145–149, 2008. With permission.)

were approximately similar in their action to 10 µM free CTAB, and PSS-coated particles were as toxic as 1 µM CTAB or as the control. The major difficulty in assessing the toxicity of CTAB-coated GNRs is their low stability and their tendency toward aggregation, which almost certainly is accompanied by an additional release of CTAB into the surrounding medium.

Pissuwan *et al.* [219] presented data showing that the cytotoxicity of CTAB-coated GNRs declined after the nanorods had been washed free of CTAB with distilled water and also after they had been conjugated to immunoglobulins. The authors concluded that the toxicity of such systems is determined by free CTAB and that CTAB-coated GNRs are noncytotoxic. However, they also pointed out that washed GNRs are poorly stable.

In a recent detailed study [176], the cellular responses of mammary adenocarcinoma (SKBR3), Chinese hamster ovary (CHO), mouse myoblast (C2C12), and human leukemia (HL60) cell lines to polymer-treated GNRs with sizes of 44.8 × 28.5 nm and 41.8 × 11.7 nm were evaluated by the MTS assay. The initially CTAB-coated GNRs were modified with either PSS or PEG. The PEGylated particles were shown to have superior biocompatibility compared with the PSS-coated nanorods, which demonstrated substantial cytotoxicity. According to TEM analysis, there was no cellular uptake of the PEGylated particles compared with their PSS counterparts. The PEGylated GNRs also exhibited better colloidal stability in the presence of the cell growth medium. By contrast, the PSS-coated GNRs were poorly stable, and their toxicity correlated with the surface area of the nanorods studied.

The cytotoxicity of GNRs also can be reduced by using, apart from PEG and PSS, the following substances (which replace CTAB or create an additional surface layer): phosphatidylcholine [146], aminomercaptotriazole and mercaptoundecanoic acid [152], the copolymer poly(diallyldimethylammonium chloride)/polystyrene sulfonic acid [158,220], Pluronic F-127 [221], mercaptohexadecanoic acid [222], silica shell [223], and combination coatings (PSS/polyethyleneimine [175], cystamine/herceptine [171], and PEG/herceptine [224]).

Alkilany *et al.* [164] published a thorough study of the cellular penetration and cyto-toxicity of GNRs. They used the MTT assay to examine the cytotoxicity of CTAB-capped GNRs interacting with human colon carcinoma cells (HT-29). The authors showed that coating the nanorods with the polymers polyacrylic acid (PAA) and poly(allylamine) hydrochloride (PAH) substantially decreased cytotoxicity (Figure 3.8).

In summary, the available data permit us to draw the justified conclusion that the replacement of CTAB with nontoxic stabilizers (PEG, PSS, PAH, *etc.*) is effective at reducing the cytotoxicity of CTAB-stabilized GNRs [225].

Regarding the toxicity of nanoshells, the literature data are fragmentary. Hirsch *et al.* [226] and Loo *et al.* [227] reported the absence of any cytotoxic action by PEG- or immunoglobulin-coated nanoshells on human breast epithelial carcinoma and adenocarci-noma cells. Stern *et al.* [228] showed the absence of cytotoxicity when human prostate cancer cells were treated with GNSs for 5 to 7 days. Liu *et al.* [229] presented data for the absence of any cytotoxic effect of GNSs on human hepatocellular carcinoma cells.

Su *et al.* [151] described a dose-dependent cytotoxicity of Au–Cu nanoshells. As determined by the WST assay, African green monkey (Vero) cells had 100% viability at low nanoshell concentrations (50 µg/ml) and 67% viability at high concentrations (200 µg/ml).

Kuo *et al.* [153] explored the cytotoxicity of gold nanowires with lengths of 0.58, 1.8, 4.5, and 8.6 µm and a thickness of 200 nm. The study was made with normal fibroblasts (NIH3T3) and HeLa cells. Before incubation with the cells, the nanowire surface was modified with thiols bearing amino, alkyl, or carboxyl end groups or was conjugated to serum. The authors found that depending on functionalization, cytotoxicity decreased in the ligand sequence serum > amino thiol > alkyl thiol > carboxyl thiol. HeLa cells, on the whole, proved more stable than normal fibroblasts. Nanowire size had no effect on cytotoxicity; however, smaller-diameter nanoparticles showed greater penetration ability.

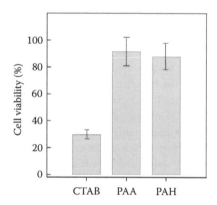

FIGURE 3.8 Four-day viability of HT-29 cells exposed to 0.4 nM of CTAB-, PAA-, and PAH-coated GNRs with an aspect ratio of 4.1. Error bars represent one standard deviation. (Adapted from Khlebtsov, N.G., Dykman, L.A., *Chem. Soc. Rev.*, 40, 1647–1671, 2011. With permission.)

3.3.3 Cytotoxicity *In Vivo*

An *in vivo* toxicity study using 13-nm PEG-coated GNPs was made by Cho *et al.* [85]. GNPs were intravenously administered to mice onefold at doses of 0.17, 0.85, and 4.26 mg/kg. For up to 7 days after injection, the nanoparticles were found to accumulate in the liver and spleen and to have long blood circulation times. In addition, TEM analysis revealed PEG-coated GNPs in numerous cytoplasmic vesicles and lysosomes of liver Kupffer cells and spleen macrophages. One of the main findings was an observation of acute inflammation and apoptosis in the liver.

In the work [196], 5-nm GNPs were intraperitoneally administered to mice (1 ml) and rats (5 ml) onefold at doses of about 2 µg/kg. Samples of the liver, kidneys, spleen, brain, and testes were analyzed at 2, 4, and 24 h and at 3, 7, and 10 days after injection. Histological examination of mouse samples revealed a maximal amount of GNPs in the spleen (between white and red pulp), whereas in rat samples, the maximal amount of GNPs was observed in liver Kupffer cells. For brain samples of both animal models, GNPs were detected in brain shells only. Four hours after GNP injection, multiple mitoses were observed in the mouse liver. In addition, some foci of extramedullary hematopoiesis appeared at 3 days after injection and disappeared after 7 to 19 days.

Cho *et al.* [194] studied the genetic effects of PEG-coated 4- and 100-nm GNPs after IV administration to mice at a dose of 4.26 mg/kg. Thirty minutes after a single injection, the mice were killed and their liver samples were taken for histological and genetic examination. Whereas histological analysis of the liver tissues did not indicate any pathological changes in all treatment groups, the 4- and 100-nm AuNP treatment groups shared 67.1% and 50.9% of the significantly changed genes, respectively. Commonly expressed genes by a single IV injection of 4- or 100-nm AuNPs were categorized as being involved in apoptosis, the cell cycle, inflammation, and metabolism. Besides, the authors found some changes in the specifically expressed genes of 4- or 100-nm AuNPs. Although the genes were different from each other, 4- and 100-nm particles showed similar gene categories, such as the cell cycle, response to stress, signal transduction, and the metabolic process. The similarity in genetic effects of 4- and 100-nm particles looks somewhat unexpected, as the biodistribution of such particles is different (see above). Recent microarray analysis of rat liver and spleen samples [60] also points to significant effects on genes related to detoxification, lipid metabolism, the cell cycle, defense response, and circadian rhythm.

The published data on size-dependent toxicity effects *in vivo* are quite restricted. Recently, Chen *et al.* [165] reported a detailed study using Balb/C mice and 3-, 5-, 8-, 12-, 17-, 37-, 50-, and 100-nm GNPs, administered intraperitoneally at 8 mg/kg/week (Figure 3.9). It was found that the 3-, 5-, 50- and 100-nm particles did not show harmful effects.

However, 8- to 37-nm GNPs induced fatigue, loss of appetite, change of fur color, and weight loss in mice. Starting on day 14, the mice in this group exhibited a camel-like back and crooked spine; the majority of the mice in these groups died within 21 days. Pathological examination of the major organs of the mice in the diseased groups indicated an increase in the number of Kupffer cells in the liver, loss of structural integrity in the lungs, and diffusion of white pulp in the spleen. The authors also reported a

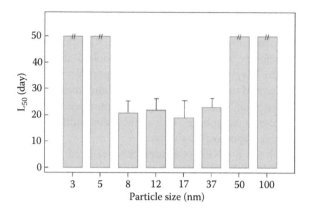

FIGURE 3.9 The average lifespan (L_{50}) of mice receiving GNPs with diameters from 3 to 100 nm. The L_{50} is the time beyond which half of the mice died. The break marks on top of the bars indicate no death observed during the experimental period for mice injected with 3-, 5-, 50-, and 100-nm particles. (Adapted from Chen, Y.-S., Hung, Y.-C., Liau, I., Huang, G.S., *Nanoscale Res. Lett.*, 4, 858–864, 2009.)

significant decrease in the toxicity of 17-nm GNPs after conjugation with proteins and peptides and low toxicity of 3 to 5 nm GNPs (no sickness or lethality).

Keeping in mind the previous data on size-dependent biodistributions, it is difficult to explain the main findings of Chen *et al.* [165]. Indeed, there are no reasons in principle for the observed difference in the toxicity of 5- and 8-nm GNPs. Moreover, according to the published data, both 37- and 50-nm particles have similar biodistribution over animal organs; therefore, one should expect similar toxicity effects. Perhaps, the possible reasons for such unexpected results are related to the rather high introduced dose of GNPs. For instance, one can compare the data of Chen *et al.* [165] with the observations of Lasagna-Reeves *et al.* [197], who studied the biodistribution and toxicity of 13-nm GNPs at everyday intraperitoneal administration to mice at doses of 0.04, 0.2, and 0.4 mg/kg/day for 8 days. It should be emphasized that these doses were 3 to 30 times lower than the daily average dose used by Chen *et al.* [165].

According to Lasagna-Reeves *et al.* [197], the gold concentration in blood did not increase with an increase in the applied dose, whereas the gold content in all organs increased proportionally to the injected dose. Although the gold content was lowest in the brain, the authors speculated that GNPs can penetrate the BBB and accumulate in the neural tissue. For our discussion, the most important conclusion of the work [197] is that no evidence of toxicity was observed in any of the diverse studies performed, including survival, behavior, animal weight, organ morphology, blood biochemistry, and tissue histology. Thus, accumulation of 13-nm GNPs was shown to be dose dependent and did not cause any acute or subacute physiological damage, in contrast to the data of Chen *et al.* [165] for 12-nm GNPs. Clearly, this difference should be attributed to a 20-fold increase in the injected dose used in the study [165].

In recent years, *Danio rerio* embryos have become a popular model for toxicity experiments. In particular, Browning *et al.* [195] showed that 11-nm GNPs can be passively transferred by diffusion into the chorionic space of the embryos and can retain their random-walk motion through chorionic space and into the inner mass of the embryos without detectable toxic effects on embryo development. Bar-Ilan *et al.* [230] arrived at similar conclusions, based on experiments with 3-, 10-, 50-, and 100-nm silver nanoparticles and GNPs. Whereas all sizes of Ag caused toxicity in zebrafish embryos, albeit in a size-dependent manner for some concentrations and time points, GNPs clearly did not.

After addition of GNPs to blood [231] and urine [232], they were able to penetrate into white blood and renal cells, respectively. Furthermore, in a donor spermatic fluid mixed with a GNP solution, 25% of spermatozoa were immotile, and the appearance of GNPs in the sperm head and tails was observed [233].

Velikorodnaya *et al.* [234] investigated the effect of 5-nm GNPs on the proliferation and apoptosis during spermatogenesis in rats. It was found that the *per os* administering of GNPs to white rats for 8 weeks at a dose of 0.57 mg/kg/day did not change the expression of proliferation proteins (KI-67 and D1 cyclin), and did not alter the activity of apoptosis (TUNEL-method) in the spermatogenic epithelium.

Kim *et al.* [88] demonstrated that intravenously administered 20-nm GNPs could pass through the blood–retinal barrier and were distributed in all retinal layers without cytotoxicity. Specifically, 20-nm nanoparticles were observed in neurons (75% ± 5%), endothelial cells (17% ± 6%), and periendothelial glial cells (8% ± 3%). By contrast, 100-nm nanoparticles were not detected in the retina.

In our recent study [235], toxicity effects were studied with PEG-coated silica/GNSs administered intravenously to rats at 75, 150, 225, and 300 μg/kg. Fifteen days after injection, the rats were killed and various organs were taken for pathomorphological examination. For the minimal dose of 75 μg/kg, the relative mass of the thymus was reliably increased, whereas the relative kidney mass was lower than that in the control group. Similarly, the relative mass of the kidney and liver was decreased and the relative mass of the spleen was increased for animals examined 15 days after a single injection of 300 μg/kg nanoshells. Some macroscopic changes in the liver and spleen, as well as multiple macrofocal effusion of blood, were observed for rats after injection of nanoshells at 225 and 300 μg/kg. Note that such morphological changes were detected only in some of the treated groups (from 1/6 to 3/6 animals). For the majority of the rats, we observed necrosis of hepatocytes, cells with pyknosis of the nucleus, and other histological modifications, in comparison with normal samples. In liver samples with maximal doses of 225 and 300 μg/kg, we observed convoluted tubules with epithelial cells containing lysed nuclei and vacuolized cytoplasm, as well as distenation of vessels in follicles and hyperemia of red pulp. For doses of 75 and 150 μg/kg, we observed a reliable increase in the B lymphocyte population, a notable decrease in the subpopulation of T helpers, and an increase in the amount of T suppressor cells. For higher doses (225 and 300 μg/kg), B- and T-cell amounts were not affected, but a reliable decrease in circulating immune complexes was detected in the blood serum of both animal groups.

3.4 Concluding Remarks

Of course, gold particles are not the only nanomaterial for the ever-increasing applications in such fields as the development of carriers for active pharmaceutical drugs, bioimaging, phototermal therapy, and so on. Along with gold, such materials as silver, palladium, metal oxides, silica, fullerenes, carbon nanotubes, and structured semiconductor quantum dots are widely tested. For targeted delivery of drugs, special carriers can be avoided in many cases, as some drug formulations can be prepared in nanoparticle form. The advantages of nanoparticle formulations for drug delivery are well known. Nanoparticles combine biocompatibility and bioavailability, long-term circulation, enhanced stability, and high capability in respect to the target drug, which may be lyophobic or lyophilic. All these properties often result in lower toxicity.

It is evident from Figure 3.1 that intense work on the biodistribution and toxicity of GNPs has actually been performed only during the past five years. This period is too short for a careful and comprehensive study of all aspects of nanobiosafety related to the fabrication and application of gold nanomaterials.

With inspiration from the review by De Jong and Borm [236], one of the main questions of GNP toxicology can be formulated as follows: do GNPs cause nanospecific toxic effects that are qualitatively different from the toxic effects of larger particles or bulk chemicals? Keeping in mind all the data discussed here, our answer is undoubtedly yes, as GNPs can represent a potential hazard to human health. Of course, this does not mean that any bare GNP or conjugate should be dangerous *a priori*. However, in this regard, one can agree with Fadeel and Garsia-Bennet [29] that it is better to be "safe than sorry"; therefore, each novel Au-nanoparticle formulation should be tested on a case-by-case basis.

Unfortunately, the absence of coordination in research programs at the very early stages of work has led to undesirable scatter in experimental conditions and, as a result, in the ultimate conclusions. Nevertheless, even the available data are sufficient to make some important albeit preliminary conclusions.

The available instrumental methods for the determination of gold content in organs (RA, INAA, ICP–MS, and AAS), the localization and identification of GNPs at the cellular (TEM and EDX) and structural (XAS) levels, and the evaluation of cytotoxicity *in vitro* (MTT and WST-1) have been time-tested, and they are certainly adequate to all tasks. It is important, however, to have in mind that common tests for studies of drug and material toxicity may not be suitable for the evaluation of all potential risks related to nanosizes. Perhaps, some additional, specific nanoassays and methods should be developed.

It is particle size that seems to be a key, albeit not single, parameter defining the biological action of GNPs. The following arguments may support this statement. First, all organs of the reticuloendothelial system are the basic primary target for accumulation of GNPs within the size range of 10 to 100 nm, and the uniformity of distribution increases with a decrease in particle size. A rapid decrease in the concentration of nanoparticles in blood and their long-term retaining in animal body is related to the functioning of the hepatobiliar system. As the excretion of accumulated particles from the liver and spleen can take up to 3 to 4 months, the question as to the injected doses and possible inflammation processes is still of great importance.

Second, the published data allow one to assume that the penetration of GNPs across the blood–brain or blood–retinal barriers is critically size dependent, with an upper penetration limit of about 15 to 20 nm.

Third, small GNPs with diameters of 1 to 2 nm have potentially high toxicity because of the possibility of irreversible binding to key biopolymers (an example is the strong binding of 1.4-nm Au_{55} clusters with B-form DNA). At the same time, many experiments with cell cultures *in vitro* did not reveal notable toxicity of 3- to 100-nm CG particles, provided that the upper limit of the applied dose did not exceed a value of the order of 10^{12} particles/ml.

Keeping in mind the many reports on *in vitro* toxicity, it would be important to elucidate if *in vitro* observations are representative of *in vivo* observations. Such comparisons between *in vitro* and *in vivo* studies would help to identify target sites that may be at risk from nanoparticle exposure. Unfortunately, the published data on experiments *in vivo* are rare and controversial. At present, one can only assume that appreciable toxicity does not occur during short-period (approximately weeklong) administration of GNPs at a daily dose lower than 0.5 mg/kg. Both our and literature data on GNP toxicity studies with rats demonstrate a rather complicated picture with some features of toxicity or negative effects.

As a rule, the biodistribution over organs and tissues can be strongly affected by particle surface modifiers. Note that the functionalization effect is not a simple consequence of prolonged systemic circulation. A well-designed surface modifier can ensure an appreciable accumulation contrast in a target organ. An instructive example is the enhanced accumulation in liver hepatocytes of nanoparticles covered with galactose-containing PEG or maltose. Another well-known example is the targeted accumulation of TNF-modified GNPs in solid tumors. Recently, Eck *et al.* [237] reported on PEGylated GNPs conjugated to anti-CD4 monoclonal antibodies that provides molecularly selective x-ray contrast enhancement of peripheral lymph nodes in living mice. Although this study demonstrates the general feasibility of biologically specific x-ray imaging in living animals, a multitude of unresolved questions still remain. For example, it is not clear how contrast enhancement by targeted GNPs could be achieved in tissues with lower antigen concentrations or different uptake mechanisms.

Thus, there is an evident need to continue and further develop the current studies in several directions. First of all, research groups should coordinate their projects in order to establish important correlations between the particle parameters (size, shape, structure, charge, surface functionalization, *etc.*), experimental design (animal model, doses, method and time schedule of administration, observation time, and organs, cells, and subcellular structures examined) and observed biological effects. To this end, one should also introduce standards for particles and methods that can be used for nanoparticle toxicity assessment. For example, keeping in mind the data of Figure 3.2, one can recommend an administration dose range from 0.1 to 10 (μg/g animal).

Further *in vivo* studies are needed to determine the biodistribution–accumulation–retention–clearance effects and their role in toxicity. Moreover, these studies are of great importance in view of the application of novel nanomaterials to targeted drug delivery. Biological processes and such factors as cell type, function, and age should be included in the interpretation of viability data following GNP exposure [238]. Conflicting results and findings are abundant in the literature calling for more careful experimental design, result interpretation, and detailed reporting [239,240]. Finally, special attention should be paid to the interaction of nanoparticles and conjugates with the immune system, as this is likely to determine some nonspecific immune responses and the delivery of engineered particles and drugs to target organs, tissues, or cells.

References

1. Ghosh P., Han G., De M., Kim C.K., Rotello V.M. Gold nanoparticles in delivery applications // *Adv. Drug Deliv. Rev.* 2008. V. 60. P. 1307–1315.
2. Donnelly J.J., Wahren B., Liu M.A. DNA vaccines: Progress and challenges // *J. Immunol.* 2005. V. 175. P. 633–639.
3. Dykman L.A., Sumaroka M.V., Staroverov S.A., Zaitseva I.S., Bogatyrev V.A. Immunogenic properties of colloidal gold // *Biol. Bull.* 2004. V. 31. P. 75–79.
4. Akhter S., Ahmad M.Z., Ahmad F.J., Storm G., Kok R.J. Gold nanoparticles in theranostic oncology: Current state-of-the-art // *Expert Opin. Drug Deliv.* 2012. V. 9. P. 1225–1243.
5. Mieszawska A.J., Mulder W.J.M., Fayad Z.A., Cormode D.P. Multifunctional gold nanoparticles for diagnosis and therapy of disease // *Mol. Pharm.* 2013. V. 10. P. 831–847.
6. Libutti S.K., Paciotti G.F., Byrnes A.A., Alexander H.R. Jr., Gannon W.E., Walker M., Seidel G.D., Yuldasheva N., Tamarkin L. Phase I and pharmacokinetic studies of CYT-6091, a novel PEGylated colloidal gold-rhTNF nanomedicine // *Clin. Cancer Res.* 2010. V. 16. P. 6139–6149.
7. Lal S., Clare S.E., Halas N.J. Nanoshell-enabled photothermal cancer therapy: Impending clinical impact // *Acc. Chem. Res.* 2008. V. 41. P. 1842–1851.
8. Krasinskas A.M., Minda J., Saul S.H., Shaked A., Furth E.E. Redistribution of thorotrast into a liver allograft several years following transplantation: A case report // *Mod. Pathol.* 2004. V. 17. P. 117–120.
9. Moghimi S.M., Hunter A.C., Murray J.C. Long-circulating and target-specific nanoparticles: Theory to practice // *Pharmacol. Rev.* 2001. V. 53. P. 283–318.
10. Yu Y.-Y., Chang S.-S., Lee C.-L., Wang C.R.C. Gold nanorods: Electrochemical synthesis and optical properties // *J. Phys. Chem. B.* 1997. V. 101. P. 6661–6664.
11. Brown C.L., Whitehouse M.W., Tiekink E.R.T., Bushell G.R. Colloidal metallic gold is not bio-inert // *Inflammopharmacology.* 2008. V. 16. P. 133–137.
12. Selivanov N.Y., Selivanova O.G., Sokolov O.I., Sokolova M.K., Sokolov A.O., Bogatyrev V.A., Dykman L.A. Effect of gold and silver nanoparticles on the growth of the *Arabidopsis thaliana* cell suspension culture // *Nanotechnol. Russia.* 2017. V. 12. P. 116–124.
13. Dykman L.A., Shchyogolev S.Y. Interactions of plants with noble metal nanoparticles // *Agric. Biol.* 2017. V. 52. P. 13–24.
14. Golubev A.A., Prilepskii A.Y., Dykman L.A., Khlebtsov N.G., Bogatyrev V.A. Colorimetric evaluation of the viability of the microalga *Dunaliella salina* as a test tool for nanomaterial toxicity // *Toxicol. Sci.* 2016. V. 151, No. 1. P. 115–125.
15. Hoet P.M., Brüske-Hohlfeld I., Salata O.V Nanoparticles—Known and unknown health risks // *J. Nanobiotechnology.* 2004. V. 2. 12.
16. Oberdörster G., Oberdörster E., Oberdörster J. Nanotoxicology: An emerging discipline evolving from studies of ultrafine particles // *Environ. Health Perspect.* 2005. V. 113. P. 823–839.

17. Oberdörster G., Stone V., Donaldson K. Toxicology of nanoparticles: A historical perspective // *Nanotoxicology.* 2007. V. 1. P. 2–25.
18. Fischer H.C., Chan W.C.W. Nanotoxicity: The growing need for *in vivo* study // *Curr. Opin. Biotechnol.* 2007. V. 18. P. 565–571.
19. Medina C., Santos-Martinez M.J., Radomski A., Corrigan O.I., Radomski M.W. Nanoparticles: Pharmacological and toxicological significance // *Br. J. Pharm.* 2007. V. 150. P. 552–558.
20. Stern S.T., McNeil S.E. Nanotechnology safety concerns revisited // *Toxicol. Sci.* 2008. V. 101. P. 4–21.
21. Lewinski N., Colvin V., Drezek R. Cytotoxicity of nanoparticles // *Small.* 2008. V. 4. P. 26–49.
22. Murphy C.J., Gole A.M., Stone J.W., Sisco P.N., Alkilany A.M., Goldsmith E.C., Baxter S.C. Gold nanoparticles in biology: Beyond toxicity to cellular imaging // *Acc. Chem. Res.* 2008. V. 41. P. 1721–1730.
23. Casals E., Vázquez-Campos S., Bastús N.G., Puntes V. Distribution and potential toxicity of engineered inorganic nanoparticles and carbon nanostructures in biological systems // *Trends Anal. Chem.* 2008. V. 27. P. 672–683.
24. Aillon K.L., Xie Y., El-Gendy N., Berkland C.J., Forrest M.L. Effects of nanomaterial physicochemical properties on *in vivo* toxicity // *Adv. Drug Deliv. Rev.* 2009. V.139.881 pt61. P. 457–466.
25. Panyala N.R., Peña-Méndez E.M., Havel J. Gold and nano-gold in medicine: Overview, toxicology and perspectives // *J. Appl. Biomed.* 2009. V. 7. P. 75–91.
26. Marquis B.J., Love S.A., Braun K.L., Haynes C.L. Analytical methods to assess nanoparticle toxicity // *Analyst.* 2009. V. 134. P. 425–439.
27. Yen H.-J., Hsu S.-h., Tsai C.-L. Cytotoxicity and immunological response of gold and silver nanoparticles of different sizes // *Small.* 2009. V. 5. P. 1553–1561.
28. Yu L., Andriola A. Quantitative gold nanoparticle analysis methods: A review // *Talanta.* 2010. V. 82. P. 869–875.
29. Fadeel B., Garcia-Bennett A.E. Better safe than sorry: Understanding the toxicological properties of inorganic nanoparticles manufactured for biomedical applications // *Adv. Drug Deliv. Rev.* 2010. V. 62. P. 362–374.
30. Johnston H.J., Hutchison G., Christensen F.M., Peters S., Hankin S., V. Stone A review of the *in vivo* and *in vitro* toxicity of silver and gold particulates: Particle attributes and biological mechanisms responsible for the observed toxicity // *Crit. Rev. Toxicol.* 2010. V. 40. P. 328–346.
31. Alkilany A.M., Murphy C. Toxicity and cellular uptake of gold nanoparticles: What we have learned so far? // *J. Nanopart. Res.* 2010. V. 13. P. 2313–2333.
32. Khlebtsov N.G., Dykman L.A. Biodistribution and toxicity of engineered gold nanoparticles: A review of *in vitro* and *in vivo* studies // *Chem. Soc. Rev.* 2011. V. 40. P. 1647–1671.
33. Khlebtsov N.G., Dykman L.A. Biodistribution and toxicity of gold nanoparticles // *Nanotechnol. Russia.* 2011. V. 6. P. 17–42.
34. Zhang Q., Hitchins V.M., Schrand A.M., Hussain S.M., Goering P.L. Uptake of gold nanoparticles in murine macrophage cells without cytotoxicity or production of pro-inflammatory mediators // *Nanotoxicology.* 2011. V. 5. P. 284–295.

35. Li Y.-F., Chen C. Fate and toxicity of metallic and metal-containing nanoparticles for biomedical applications // *Small*. 2011. V. 7. P. 2965–2980.

36. Sabella S., Galeone A., Vecchio G., Cingolani R., Pompa P.P. AuNPs are toxic *in vitro* and *in vivo*: A review // *J. Nanosci. Lett.* 2011. V. 1. P. 145–165.

37. Kong B., Seog J.H., Graham L.M., Lee S.B. Experimental considerations on the cytotoxicity of nanoparticles // *Nanomedicine*. 2011. V. 6. P. 929–941.

38 Sharifi S., Behzadi S., Laurent S., Forrest M.L., Stroeve P., Mahmoudi M. Toxicity of nanomaterials // *Chem. Soc. Rev.* 2012. V. 41. P. 2323–2343.

39. Arora S., Rajwade J.M., Paknikar K.M. Nanotoxicology and i*n vitro* studies: The need of the hour // *Toxicol. Appl. Pharmacol.* 2012. V. 258. P. 151–165.

40. Nel A., Xia T., Meng H., Wang X., Lin S., Ji Z., Zhang H. Nanomaterial toxicity testing in the 21st century: Use of a predictive toxicological approach and high-throughput screening // *Acc. Chem. Res.* 2013. V. 46. P. 607–621.

41. Ivask A., Kurvet I., Kasemets K., Blinova I., Aruoja V., Suppi S., Vija H., Käkinen A., Titma T., Heinlaan M., Visnapuu M., Koller D., Kisand V., Kahru A. Size-dependent toxicity of silver nanoparticles to bacteria, yeast, algae, crustaceans and mammalian cells *in vitro* // *PLoS One*. 2014. V. 9. e102108.

42. Fratoddi I., Venditti I., Cametti C., Russo M.V. How toxic are gold nanoparticles? the state-of-the-art // *Nano Res*. 2015. V.8. P. 1771–1799.

43. Ali A., Suhail M., Mathew S., Shah M.A., Harakeh S.M., Ahmad S., Kazmi Z., Alhamdan M.A.R., Chaudhary A., Damanhouri G.A., Qadri I. Nanomaterial induced immune responses and cytotoxicity // *J. Nanosci. Nanotechnol.* 2016. V. 16. P. 40–57.

44. Carnovale C., Bryant G., Shukla R., Bansal V. Size, shape and surface chemistry of nano-gold dictate its cellular interactions, uptake and toxicity // *Prog. Mater. Sci.* 2016. V. 83. P. 152–190.

45. Jia Y.-P., Ma B.-Y., Wei X.-W., Qian Z.-Y. The *in vitro* and *in vivo* toxicity of gold nanoparticles // *Chin. Chem. Lett.* 2017. V. 28. P. 691–702.

46. Almeida J.P.M., Chen A.L., Foster A., Drezek R. *In vivo* biodistribution of nanoparticles // *Nanomedicine*. 2011. V. 6. P. 815–835.

47. Taylor U., Rehbock C., Streich C., Rath D., Barcikowski S. Rational design of gold nanoparticle toxicology assays: A question of exposure scenario, dose and experimental setup // *Nanomedicine (Lond.)* 2014. V. 9. P. 1971–1989.

48. Boisselier E., Astruc D. Gold nanoparticles in nanomedicine: Preparations, imaging, diagnostics, therapies and toxicity // *Chem. Rev.* 2009. V. 38. P. 1759–1782.

49. *Gold Nanoparticles: Properties, Characterization and Fabrication* / Ed. Chow P.E. New York: Nova Science Publisher. 2010.

50. Berne B.J., Pecora R. Dynamic light scattering with applications to chemistry, biology, and physics. New York: Dover Publ. 2000.

51. Khlebtsov N.G., Dykman L.A. Optical properties and biomedical applications of plasmonic nanoparticles // *J. Quant. Spectrosc. Radiat. Transfer*. 2010. V. 111. P. 1–35.

52. Kumar S., Aaron J., Sokolov K. Directional conjugation of antibodies to nanoparticles for synthesis of multiplexed optical contrast agents with both delivery and targeting moieties // *Nat. Protocols*. 2008. V. 3. P. 314–320.

53. Deyev S.M., Waibel R., Lebedenko E.N., Schubiger A.P., Pluckthunet A. Design of multivalent complexes using the barnase-barstar module // *Nat. Biotechnol.* 2003. V. 21. P. 1486–1492.

54. *Cancer Nanotechnology: Methods and Protocols* / Eds. Grobmyer S.R., Moudgil B.M. New York: Springer. 2010.

55. Erickson T., Tunnell J.W. Gold nanoshells in biomedical applications // In: *Nanomaterials for the Life Sciences: Mixed Metal Nanomaterials* / Ed. Kumar C. Weinheim: Wiley-VCH. 2009. P. 1–44.

56. Darien B.J., Sims P.A., Kruse-Elliott K.T., Homan T.S., Cashwell R.J., Cooley A.J., Albrecht R.M. Use of colloidal gold and neutron activation in correlative microscopic labeling and label quantitation // *Scanning Microsc.* 1995. V. 9. P. 773–780.

57. Krystek P. A review on approaches to bio distribution studies about gold and silver engineered nanoparticles by inductively coupled plasma mass spectrometry // *Microchem. J.* 2012. V. 105. P. 39–43.

58. Wang L. Prospects of photoacoustic tomography // *Med. Phys.* 2008. V. 35. P. 5758–5767.

59. Wang L., Li Y.-F., Zhou L., Liu Y., Meng L., Zhang K., Wu X., Zhang L., Li B., Chen C. Characterization of gold nanorods *in vivo* by integrated analytical techniques: Their uptake, retention, and chemical forms // *Anal. Bioanal. Chem.* 2010. V. 396. P. 1105–1114.

60. Balasubramanian S.K., Jittiwat J., Manikandan J., Ong C.-N., Yu L.E., Ong W.-Y. Biodistribution of gold nanoparticles and gene expression changes in the liver and spleen after intravenous administration in rats // *Biomaterials.* 2010. V. 8. P. 2034–2042.

61. Abraham G.E., Himmel P.B. Management of rheumatoid arthritis: Rationale for the use of colloidal metallic gold // *J. Nutr. Environ. Med.* 1997. V. 7. P. 295–305.

62. Abraham G.E. Clinical applications of gold and silver nanocolloids // *Original Internist.* 2008. V. 15. P. 132–158.

63. Hillyer J.F., Albrecht R.M. Correlative instrumental neutron activation analysis, light microscopy, transmission electron microscopy, and X-ray microanalysis for qualitative and quantitative detection of colloidal gold spheres in biological specimens // *Microsc. Microanal.* 1999. V. 4. P. 481–490.

64. Hillyer J.F., Albrecht R.M. Gastrointestinal persorption and tissue distribution of differently sized colloidal gold nanoparticles // *J. Pharm. Sci.* 2001. V. 90. P. 1927–1936.

65. Hainfeld J.F., Slatkin D.N., Smilowitz H.M. The use of gold nanoparticles to enhance radiotherapy in mice // *Phys. Med. Biol.* 2004. V. 49. P. N309–N315.

66. Paciotti G.F., Myer L., Weinreich D., Goia D., Pavel N., McLaughlin R.E., Tamarkin L. Colloidal gold: A novel nanoparticle vector for tumor directed drug delivery // *Drug Deliv.* 2004. V. 11. P. 169–183.

67. Bergen J.M., von Recum H.A., Goodman T.T., Massey A.P., Pun S.H. Gold nanoparticles as a versatile platform for optimizing physicochemical parameters for targeted drug delivery // *Macromol. Biosci.* 2006. V. 6. P. 506–516.

68. Hainfeld J.F., Slatkin D.N., Focella T.M, Smilowitz H.M. Gold nanoparticles: A new X-ray contrast agent // *Br. J. Radiol.* 2006. V. 79. P. 248–253.

69. Niidome T., Yamagata M., Okamoto Y., Akiyama Y., Takahishi H., Kawano T., Katayama Y., Niidome Y. PEG-modified gold nanorods with a stealth character for *in vivo* application // *J. Control. Release*. 2006. V. 114. P. 343–347.

70. Katti K.V., Kannan R., Katti K., Kattumori V., Pandrapraganda R., Rahing V., Cutler C., Boote E.J., Casteel S.W., Smith C.J., Robertson J.D., Jurrison S.S. Hybrid gold nanoparticles in molecular imaging and radiotherapy // *Czech. J. Phys. Suppl. D*. 2006. V. 56. P. D23–D34.

71. Balogh L., Nigavekar S.S., Nair B.M., Lesniak W., Zhang C., Sung L.Y., Kariapper M.S.T., El-Jawahri A., Llanes M., Bolton B., Mamou F., Tan W., Hutson A., Minc L., Khan M.K. Significant effect of size on the *in vivo* biodistribution of gold composite nanodevices in mouse tumor models // *Nanomedicine*. 2007. V. 3. P. 281–296.

72. Sadauskas E., Wallin H., Stoltenberg M., Vogel U., Doering P., Larsen A., Danscher G. Kupffer cells are central in the removal of nanoparticles from the organism // *Part. Fibre Toxicol*. 2007. V. 4. 10.

73. James W.D., Hirsch L.R., West J.L., O'Neal P.D., Payne J.D. Application of INAA to the build-up and clearance of gold nanoshells in clinical studies in mice // *J. Radioanal. Nucl. Chem*. 2007. V. 271. P. 455–459.

74. De Jong W.H., Hagens W.I., Krystek P., Burger M.C., Sips A.J., Geertsma R.E. Particle size-dependent organ distribution of gold nanoparticles after intravenous administration // *Biomaterials*. 2008. V. 29. P. 1912–1919.

75. Diagaradjane P., Shetty A., Wang J.C., Elliott A.M., Schwartz J., Shentu S., Park H.C., Deorukhkar A., Stafford R.J., Cho S.H., Tunnell J.W., Hazle J.D., Krishnan S. Modulation of *in vivo* tumor radiation response via gold nanoshell-mediated vascular-focused hyperthermia: Characterizing an integrated antihypoxic and localized vascular disrupting targeting strategy // *Nano Lett*. 2008. V. 8. 1492–1500.

76. Kogan B.Y., Andronova N.V., Khlebtsov N.G., Khlebtsov B.N., Rudoy V.M., Dement'eva O.V., Sedykh E.V., Bannykh L.N. Pharmacokinetic study of PEGylated plasmon resonant gold nanoparticles in tumor-bearing mice // *NSTI-Nanotech*. 2008. V. 2. P. 65–68.

77. Myllynen P.K., Loughran M.J., Howard C.V., Sormunen R., Walshe A.A., Vähäkangas K.H. Kinetics of gold nanoparticles in the human placenta // *Reprod. Toxicol*. 2008. V. 26. P. 130–137.

78. Huang X.-L., Zhang B., Ren L., Ye S.-F., Sun L.-P., Zhang Q.-Q., Tan M.-C., Chow G.-M. *In vivo* toxic studies and biodistribution of near infrared sensitive Au–Au2S nanoparticles as potential drug delivery carriers // *J. Mater. Sci*. 2008. V. 19. P. 2581–2588.

79. Niidome T., Akiyama Y., Shimoda K., Kawano T., Mori T., Katayama Y., Niidome Y. *In vivo* monitoring of intravenously injected gold nanorods using near-infrared light // *Small*. 2008. V. 4. P. 1001–1007.

80. Semmler-Behnke M., Kreyling W.G., Lipka J., Fertsch S., Wenk A., Takenaka S., Schmid G., Brandau W. Biodistribution of 1.4- and 18-nm gold particles in rats // *Small*. 2008. V. 4. P. 2108–2111.

81. Sonavane G., Tomoda K., Sano A., Ohshima H., Terada H., Makino K. *In vitro* permeation of gold nanoparticles through rat skin and rat intestine: Effect of particle size // *Colloids Surf. B*. 2008. V. 65. P. 1–10.

82. Sonavane G., Tomoda K., Makino K. Biodistribution of colloidal gold nanoparticles after intravenous administration: Effect of particle size // *Colloids Surf. B.* 2008. V. 66. P. 274–280.

83. Melancon M.P., Lu W., Yang Z., Zhang R., Cheng Z., Elliot A.M., Stafford J., Olson T., Zhang J.Z., Li C. *In vitro* and *in vivo* targeting of hollow gold nanoshells directed at epidermal growth factor receptors for photothermal ablation therapy // *Mol. Cancer Ther.* 2008. V. 7. P. 1730–1739.

84. Akiyama Y., Mori T., Katayama Y., Niidome T. The effects of PEG grafting level and injection dose on gold nanorod biodistribution in the tumor-bearing mice // *J. Control. Release.* 2009. V. 139. P. 81–84.

85. Cho W.-S., Cho M., Jeong J., Choi M., Cho H.-Y., Han B.S., Kim S.H., Kim H.O., Lim Y.T., Chung B.H., Jeong J. Acute toxicity and pharmacokinetics of 13 nm-sized PEG-coated gold nanoparticles // *Toxicol. Appl. Pharmacol.* 2009. V. 236. P. 16–24.

86. Fent G.M., Casteel S.W., Kim D.Y., Kannan R., Katti K., Chanda N., Katti K. Biodistribution of maltose and gum Arabic hybrid gold nanoparticles after intravenous injection in juvenile swine // *Nanomedicine.* 2009. V. 5. P. 128–135.

87. Goel R., Shah N., Visaria R., Paciotti G.F., Bischof J.C. Biodistribution of TNF-α-coated gold nanoparticles in an *in vivo* model system // *Nanomedicine.* 2009. V. 4. P. 401–410.

88. Kim J.H., Kim J.H., Kim K.-W., Kim M.H., Yu Y.S. Intravenously administered gold nanoparticles pass through the blood–retinal barrier depending on the particle size, and induce no retinal toxicity // *Nanotechnology.* 2009. V. 20. 505101.

89. Sadauskas E., Jacobsen N.R., Danscher G., Stoltenberg M., Vogel U., Larsen A., Kreyling W., Wallin H. Biodistribution of gold nanoparticles in mouse lung following intratracheal instillation // *Chem. Cent. J.* 2009. V. 3. 16.

90. Sadauskas E., Danscher G., Stoltenberg M., Vogel U., Larsen A., Wallin H. Protracted elimination of gold nanoparticles from mouse liver // *Nanomedicine.* 2009. V. 5. P. 162–169.

91. Terentyuk G.S., Maslyakova G.N., Suleymanova L.V., Khlebtsov B.N., Kogan B.Ya., Akchurin G.G., Shantrocha A.V., Maksimova I.L., Khlebtsov N.G., Tuchin V.V. Circulation and distribution of gold nanoparticles and induced alterations of tissue morphology at intravenous particle delivery // *J. Biophoton.* 2009. V. 2. P. 292–302.

92. Zhang G., Yang Z., Lu W., Zhang R., Huang Q., Tian M., Li L., Liang D., Li C. Influence of anchoring ligands and particle size on the colloidal stability and *in vivo* biodistribution of polyethylene glycol-coated gold nanoparticles in tumor-xenografted mice // *Biomaterials.* 2009. V. 30. P. 1928–1936.

93. Bartneck M., Keul H.A., Zwadlo-Klarwasser G., Groll J. Phagocytosis independent extracellular nanoparticle clearance by human immune cells // *Nano Lett.* 2010. V. 10. P. 59–63.

94. Hirn S., Semmler-Behnke M., Schleh C., Wenk A., Lipka J., Schäffler M., Takenaka S., Möller W., Schmid G., Simon U., Kreyling W.G. Particle size-dependent and surface charge-dependent biodistribution of gold nanoparticles after intravenous administration // *Eur. J. Pharm. Biopharm.* 2011. V. 77. P. 407–416.

95. Schleh C., Semmler-Behnke M., Lipka J., Wenk A., Hirn S., Schäffler M., Schmid G., Simon U., Kreyling W.G. Size and surface charge of gold nanoparticles determine absorption across intestinal barriers and accumulation in secondary target organs after oral administration // *Nanotoxicology*. 2012. V. 6. P. 36–46.

96. Sun Y.-N., Wang C.-D., Zhang X.-M., Ren L., Tian X.-H. Shape dependence of gold nanoparticles on *in vivo* acute toxicological effects and biodistribution // *J. Nanosci. Nanotechnol.* 2011. V. 11. P. 1210–1216.

97. Shah N.B., Vercellotti G.M., White J.G., Fegan A., Wagner C.R., Bischof J.C. Blood-nanoparticle interactions and *in vivo* biodistribution: Impact of surface PEG and ligand properties // *Mol. Pharm.* 2012. V. 9. P. 2146–2155.

98. Chen H., Dorrigan A., Saad S., Hare D.J., Cortie M.B., Valenzuela S.M. *In vivo* study of spherical gold nanoparticles: Inflammatory effects and distribution in mice // *PLoS One*. 2013. V. 8. e58208.

99. Wojnicki M., Luty-Błocho M., Bednarski M., Dudek M., Knutelska J., Sapa J., Zygmunt M., Nowak G., Fitzner K. Tissue distribution of gold nanoparticles after single intravenous administration in mice // *Pharm. Rep.* 2013. V. 65. P. 1033–1038.

100. Fernández-Ruiz R., Redrejo M.J., Friedrich E.J., Ramos M., Fernández T. Evaluation of bioaccumulation kinetics of gold nanorods in vital mammalian organs by means of total reflection X-ray fluorescence spectrometry // *Anal. Chem.* 2014. V. 86. P. 7383–7390.

101. Fraga S., Brandão A., Soares M.E., Morais T., Duarte J.A., Pereira L., Soares L., Neves C., Pereira E., Bastos M. de L., Carmo H. Short- and long-term distribution and toxicity of gold nanoparticles in the rat after a single-dose intravenous administration // *Nanomedicine*. 2014. V. 10. P. 1757–1766.

102. Lee J.K., Kim T.S., Bae J.Y., Jung A.Y., Lee S.M., Seok J.H., Roh H.S., Song C.W., Choi M.J., Jeong J., Chung B.H., Lee Y.-G., Jeong J., Cho W.-S. Organ-specific distribution of gold nanoparticles by their surface functionalization // *J. Appl. Toxicol.* 2015. V. 35. P. 573–580.

103. Koyama Y., Matsui Y., Shimada Y., Yoneda M. Biodistribution of gold nanoparticles in mice and investigation of their possible translocation by nerve uptake around the alveolus // *J. Toxicol. Sci.* 2015. V. 40. P. 243–249.

104. Kreyling W.G., Abdelmonem A.M., Ali Z., Alves F., Geiser M., Haberl N., Hartmann R., Hirn S., de Aberasturi D.J., Kantner K., Khadem-Saba G., Montenegro J.-M., Rejman J., Rojo T., de Larramendi I.R., Ufartes R., Wenk A., Parak W.J. *In vivo* integrity of polymer-coated gold nanoparticles // *Nat. Nanotechnol.* 2015. V. 10. P. 619–623.

105. Rambanapasi C., Barnard N., Grobler A., Buntting H., Sonopo M., Jansen D., Jordaan A., Steyn H., Zeevaart J.R. Dual radiolabeling as a technique to track nanocarriers: The case of gold nanoparticles // *Molecules*. 2015. V. 20. P. 12863–12879.

106. Jo M.-R., Bae S.-H., Go M.-R., Kim H.-J., Hwang Y.-G., Choi S.-J. Toxicity and biokinetics of colloidal gold nanoparticles // *Nanomaterials*. 2015. V. 5. P. 835–850.

107. Naz F., Koul V., Srivastava A., Gupta Y.K., Dinda A.K. Biokinetics of ultrafine gold nanoparticles (AuNPs) relating to redistribution and urinary excretion: A long-term *in vivo* study // *J. Drug Target.* 2016. V. 26. P. 1–10.

108. Rambanapasi C., Zeevaart J.R., Buntting H., Bester C., Kotze D., Hayeshi R., Grobler A. Bioaccumulation and subchronic toxicity of 14 nm gold nanoparticles in rats // *Molecules.* 2016. V. 21. 763.

109. Dam D.H., Culver K.S., Kandela I., Lee R.C., Chandra K., Lee H., Mantis C., Ugolkov A., Mazar A.P., Odom T.W. Biodistribution and *in vivo* toxicity of aptamer-loaded gold nanostars // *Nanomedicine.* 2015. V. 11. P. 671–679.

110. Tong L., Wei Q., Wei A., Cheng J.-X. Gold nanorods as contrast agents for biological imaging: Optical properties, surface conjugation and photothermal effects // *Photochem. Photobiol.* 2009. V. 85. P. 21–32.

111. Singer J.M., Adlersberg L., Sadek M. Long-term observation of intravenously injected colloidal gold in mice // *J. Reticuloendothel. Soc.* 1972. V. 12. P. 658–671.

112. Hardonk M.J., Harms G., Koudstaal J. Zonal heterogeneity of rat hepatocytes in the *in vivo* uptake of 17 nm colloidal gold granules // *Histochemistry.* 1985. V. 83. P. 473–477.

113. Renaud G., Hamilton R.L., Havel R. Hepatic metabolism of colloidal gold-low-density lipoprotein complexes in the rat: Evidence for bulk excretion of lysosomal contents into bile // *Hepatology.* 1989. V. 9. P. 380–392.

114. Liu Y., Meyer-Zaika W., Franzka S., Schmid G., Tsoli M., Kuhn H. Gold cluster degradation by the transition of B-DNA into A-DNA and the formation of nano-wires // *Angew. Chem. Int. Ed.* 2003. V. 42. P. 2853–2857.

115. Petri B., Bootz A., Khalansky A., Hekmatara T., Muller R., Uhl R., Kreuter J., Gelperina S. Chemotherapy of brain tumour using doxorubicin bound to surfactant-coated poly(butyl cyanoacrylate) nanoparticles: Revisiting the role of surfactants // *J. Control. Release.* 2007. V. 117. P. 51–58.

116. Li C.-H., Shyu M-K., Jhan C., Cheng Y.-W., Tsai C.-H., Liu C.-W., Lee C.-C., Chen R.-M., Kang J.-J. Gold nanoparticles increase endothelial paracellular permeability by altering components of endothelial tight junctions, and increase blood-brain barrier permeability in mice // *Toxicol. Sci.* 2015. V. 148. P. 192–203.

117. Oberdörster G., Sharp Z., Atudorei V., Elder A., Gelein R., Kreyling W., Cox C. Translocation of inhaled ultrafine particles to the brain // *Inhal. Toxicol.* 2004. V. 16. P. 437–445.

118. Yang H., Sun C., Fan Z., Tian X., Yan L., Du L., Liu Y., Chen C., Liang X.-j., Anderson G.J., Keelan J.A., Zhao Y., Nie G. Effects of gestational age and surface modification on materno-fetal transfer of nanoparticles in murine pregnancy // *Sci. Rep.* 2012. V. 2. 847.

119. Tsyganova N.A., Khairullin R.M., Terentyuk G.S., Khlebtsov B.N., Bogatyrev V.A., Dykman L.A., Erykov S.N., Khlebtsov N.G. Penetration of pegylated gold nanoparticles through rat placental barrier // *Bull. Exp. Biol. Med.* 2014. V. 157. P. 383–385.

120. Elci S.G., Jiang Y., Yan B., Kim S.T., Saha K., Moyano D.F., Tonga G.Y., Jackson L.C., Rotello V.M., Vachet R.W. Surface charge controls the sub-organ biodistributions of gold nanoparticles // *ACS Nano.* 2016. V.10. P. 5536–5542.

121. El-Sayed I.H., Huang X., El-Sayed M.A. Surface plasmon resonance scattering and absorption of anti-EGFR antibody conjugated gold nanoparticles in cancer diagnostics: Applications in oral cancer // *Nano Lett.* 2005. V. 5. P. 829–834.

122. Huang X., El-Sayed I.H., Qian W., El-Sayed M.A. Cancer cell imaging and photothermal therapy in the near-infrared region by using gold nanorods // *J. Am. Chem. Soc.* 2006. V. 128. P. 2115–2120.

123. Von Maltzahn G., Park J.-H., Agrawal A., Bandaru N.K., Das S.K., Sailor M.J., Bhatia S.N. Computationally guided photothermal tumor therapy using long-circulating gold nanorod antennas // *Cancer Res.* 2009. V. 69. P. 3892–3900.

124. Oldenburg S., Averitt R.D., Westcott S., Halas N.J. Nanoengineering of optical resonances // *Chem. Phys. Lett.* 1998. V. 288. P. 243–247.

125. Khlebtsov B.N., Khlebtsov N.G. Enhanced solid-phase immunoassay using gold nanoshells: Effect of nanoparticle optical properties // *Nanotechnology.* 2008. V. 19. 435703.

126. Khanadeev V.A., Khlebtsov B.N., Staroverov S.A., Vidyasheva I.V., Skaptsov A.A., Ilineva E.S., Bogatyrev V.A., Dykman L.A., Khlebtsov N.G. Quantitative cell bio-imaging using gold nanoshell conjugates and phage antibodies // *J. Biophotonics.* 2011. V. 4. P. 74–83.

127. Skrabalak S.E., Chen J., Sun Y., Lu X., Au L., Cobley C.M., Xia Y. Gold nanostructures: Engineering their plasmonic properties for biomedical applications // *Acc. Chem. Res.* 2008. V. 41. P. 1587–1595.

128. Nehl C.L., Liao H., Hafner J.H. Optical properties of star-shaped gold nanoparticles // *Nano Lett.* 2006. V. 6. P. 683–688.

129. Sharma V., Park K., Srinivasarao M. Colloidal dispersion of gold nanorods: Historical background, optical properties, seed-mediated synthesis, shape separation and self-assembly // *Mater. Sci. Eng.* 2009. V. 65. P. 1–38.

130. Cole J.R., Mirin N.A., Knight M.W., Goodrich G.P., Halas N.J. Photothermal efficiencies of nanoshells and nanorods for clinical therapeutic applications // *J. Phys. Chem. C.* 2009. V. 113. P. 12090–12094.

131. Gref R., Minamitake Y., Peracchia M.T., Trubetskoy V., Torchilin V., Langer R. Biodegradable long-circulating polymeric nanospheres // *Science.* 1994. V. 263. P. 1600–1603.

132. Maynard A.D., Aitken R.J., Butz T., Colvin V., Donaldson K., Oberdörster G., Philbert M.A., Ryan J., Seaton A., Stone V., Tinkle S.S., Tran L., Walker N.J., Warheit D.B. Safe handling of nanotechnology // *Nature.* 2006. V. 444. P. 267–269.

133. Pacheco G. Studies on the action of metallic colloids on immunization // *Mem. Inst. Oswaldo Cruz.* 1925. V. 18. P. 119–149.

134. Merchant B. Gold, the noble metal and the paradoxes of its toxicology // *Biologicals.* 1998. V. 26. P. 49–59.

135. Eisler R. Mammalian sensitivity to elemental gold (Au0) // *Biol. Trace Element Res.* 2004. V. 100. P. 1–18.

136. Dobrovolskaia M.A., McNeil S.E. Immunological properties of engineered nanomaterials // *Nat. Nanotechnol.* 2007. V. 2. P. 469–478.

137. Goodman C.M., McCusker C.D., Yilmaz T., Rotello V.M. Toxicity of gold nanoparticles functionalized with cationic and anionic side chains // *Bioconjugate Chem.* 2004. V. 15. P. 897–900.

138. Connor E.E., Mwamuka J., Gole A., Murphy C.J., Wyatt M.D. Gold Nanoparticles are taken up by human cells but do not cause acute cytotoxicity // *Small*. 2005. V. 1. P. 325–327.

139 Salem A.K., Searson P.C., Leong K.W. Multifunctional nanorods for gene delivery // *Nat. Mater*. 2003. V.2. P. 668–671.

140. Tkachenko A.G., Xie H., Liu Y., Coleman D., Ryan J., Glomm W.R., Shipton M.K., Franzen S., Feldheim D.L. Cellular trajectories of peptide-modified gold particle complexes: Comparison of nuclear localization signals and peptide transduction domains // *Bioconjugate Chem*. 2004. V. 15. P. 482–490.

141. Shukla R., Bansal V., Chaudhary M., Basu A., Bhonde R.R., Sastry M. Biocompatibility of gold nanoparticles and their endocytotic fate inside the cellular compartment: A microscopic overview // *Langmuir*. 2005. V. 21. P. 10644–10654.

142. de la Fuente J.M., Berry C.C. Tat peptide as an efficient molecule to translocate gold nanoparticles into the cell nucleus // *Bioconjugate Chem*. 2005. V. 16. P. 1176–1180.

143. Pernodet N., Fang X., Sun Y., Bakhtina A., Ramakrishnan A., Sokolov J., Ulman A., Rafailovich M. Adverse effects of citrate/gold nanoparticles on human dermal fibroblasts // *Small*. 2006. V. 2. P. 766–773.

144. Shenoy D., Fu W., Li J., Crasto C., Jones G., Dimarzio C., Sridhar S., Amiji M. Surface functionalization of gold nanoparticles using hetero-bifunctional poly(ethylene glycol) spacer for intracellular tracking and delivery // *Int. J. Nanomedicine*. 2006. V. 1. P. 51–58.

145. Ikah D.S.K., Howard C.V., McLean W.G., Brust M., Tshikhudo T.R. Neurotoxic effects of monodispersed colloidal gold nanoparticles // *Toxicology*. 2006. V. 219. P. 238.

146. Takahashi H., Niidome Y., Niidome T., Kaneko K., Kawasaki H., Yamada S. Modification of gold nanorods using phosphatidylcholine to reduce cytotoxicity // *Langmuir*. 2006. V. 22. P. 2–5.

147. Pan Y., Neuss S., Leifert A., Fischler M., Wen F., Simon U., Schmid G., Brandau W., Jahnen-Dechent W. Size-dependent cytotoxicity of gold nanoparticles // *Small*. 2007. V. 3. P. 1941–1949.

148. Khan J.A., Pillai B., Das T.K., Singh Y., Maiti S. Molecular effects of uptake of gold nanoparticles in HeLa cells // *ChemBioChem*. 2007. V. 8. P. 1237–1240.

149. Patra H.K., Banerjee S., Chaudhuri U., Lahiri P., Dasgupta A.K. Cell selective response to gold nanoparticles // *Nanomedicine*. 2007. V. 3. P. 111–119.

150. Kim D., Park S., Lee J.H., Jeong Y.Y., Jon S. Antibiofouling polymer-coated gold nanoparticles as a contrast agent for *in vivo* X-ray computed tomography imaging // *J. Am. Chem. Soc*. 2007. V. 129. P. 7661–7665.

151. Su C.-H., Sheu H.-S., Lin C.-Y., Huang C.-C., Lo Y.-W., Pu Y.-C., Weng J.-C., Shieh D.-B., Chen J.-H., Yeh C.-S. Nanoshell magnetic resonance imaging contrast agents // *J. Am. Chem. Soc*. 2007. V. 129. P. 2139–2146.

152. Yu C., Varghese L., Irudayaraj J. Surface modification of cetyltrimethylammonium bromide-capped gold nanorods to make molecular probes // *Langmuir*. 2007. V. 23. P. 9114–9119.

153. Kuo C.-W., Lai J.-J., Wei K.H., Chen P. Studies of surface-modified gold nanowires inside living cells // *Adv. Funct. Mater*. 2007. V. 17. P. 3707–3714.

154. Male K.B., Lachance B., Hrapovic S., Sunahara G., Luong J.H.T. Assessment of cytotoxicity of quantum dots and gold nanoparticles using cell-based impedance spectroscopy // *Anal. Chem.* 2008. V. 80. P. 5487–5493.

155. Jan E., Byrne S.J., Cuddihy M., Davies A.M., Volkov Y., Gun'ko Y.K., Kotov N.A. High-content screening as a universal tool for fingerprinting of cytotoxicity of nanoparticles // *ACS Nano.* 2008. V. 2. P. 928–938.

156. Leonov A.P., Zheng J., Clogston J.D., Stern S.T., Patri A.K., Wei A. Detoxification of gold nanorods by treatment with polystyrenesulfonate // *ACS Nano.* 2008. V. 2. P. 2481–2488.

157. Wang S., Lu W., Tovmachenko O., Rai U.S., Yu H., Ray P.C. Challenge in understanding size and shape dependent toxicity of gold nanomaterials in human skin keratinocytes // *Chem. Phys. Lett.* 2008. V. 463. P. 145–149.

158. Hauck T.S., Ghazani A.A., Chan W.C.W. Assessing the effect of surface chemistry on gold nanorod uptake, toxicity, and gene expression in mammalian cells // *Small.* 2008. V. 4. P. 153–159.

159. Gannon C.J., Patra C.R., Bhattacharya R., Mukherjee P., Curley S.A. Intracellular gold nanoparticles enhance non-invasive radiofrequency thermal destruction of human gastrointestinal cancer cells // *J. Nanobiotechnology.* 2008. V. 6. P. 2.

160. Simon-Deckers A., Brun E., Gouget B., Carrière M., Sicard-Roselli C. Impact of gold nanoparticles combined to X-Ray irradiation on bacteria // *Gold Bull.* 2008. V. 41. P. 187–194.

161. Sun L., Liu D., Wang Z. Functional gold nanoparticle-peptide complexes as cell-targeting agents // *Langmuir.* 2008. V. 24. P. 10293–10297.

162. Qu Y., Lü X. Aqueous synthesis of gold nanoparticles and their cytotoxicity in human dermal fibroblasts—Fetal // *Biomed. Mater.* 2009. V. 4. 025007.

163. Fan J.H., Hung W.I., Li W.T., Yeh J.M. Biocompatibility study of gold nanoparticles to human cells // In: *IFBME Proceedings,* V. 23 / Eds. Lim C.T., Goh J.C.H. New York: Springer. 2009. P. 870–873.

164. Alkilany A.M., Nagaria P.K., Hexel C.R., Shaw T.J., Murphy C.J., Wyatt M.D. Cellular uptake and cytotoxicity of gold nanorods: molecular origin of cytotoxicity and surface effects // *Small.* 2009. V. 5. P. 701–708.

165. Chen Y.-S., Hung Y.-C., Liau I., Huang G.S. Assessment of the *in vivo* toxicity of gold nanoparticles // *Nanoscale Res. Lett.* 2009. V. 4. P. 858–864.

166. Salmaso S., Caliceti P., Amendola V., Meneghetti M., Magnusson J.P., Pasparakis G., Alexander C. Cell up-take control of gold nanoparticles functionalized with a thermoresponsive polymer // *J. Mater. Chem.* 2009. V. 19. P. 1608–1615.

167. Li G., Li D., Zhang L., Zhai J., Wang E. One-step synthesis of folic acid protected gold nanoparticles and their receptor-mediated intracellular uptake // *Chem. Eur. J.* 2009 V. 15. P. 9868–9873.

168. Zhou M., Wang B., Rozynek Z., Xie Z., Fossum J.O., Yu X., Raaen S. Minute synthesis of extremely stable gold nanoparticles // *Nanotechnology.* 2009. V. 20. 505606.

169. Murawala P., Phadnis S.M., Bhonde R.R., Prasad B.L.V. In situ synthesis of water dispersible bovine serum albumin capped gold and silver nanoparticles and their cytocompatibility studies // *Coll. Surf. B.* 2009. V. 73. P. 224–228.

170. Pfaller T., Puntes V., Casals E., Duschl A., Oostingh G.J. *In vitro* investigation of immunomodulatory effects caused by engineered inorganic nanoparticles—The impact of experimental design and cell choice // *Nanotoxicology*. 2009. V. 3. P. 46–59.

171. Chen J., Irudayaraj J. Quantitative investigation of compartmentalized dynamics of ErbB2 targeting gold nanorods in live cells by single molecule spectroscopy // *ACS Nano*. 2009. V. 3. P. 4071–4079.

172. Ponti J., Colognato R., Franchini F., Gioria S., Simonelli F., Abbas K., Uboldi C., Kirkpatrick C.J., Holzwarth U., Rossi F. A quantitative *in vitro* approach to study the intracellular fate of gold nanoparticles: From synthesis to cytotoxicity // *Nanotoxicology*. 2009. V. 3. P. 296–306.

173. Villiers C.L., Freitas H., Couderc R., Villiers M.-B., Marche P.N. Analysis of the toxicity of gold nano particles on the immune system: Effect on dendritic cell functions // *J. Nanopart. Res.* 2010. V. 12. P. 55–60.

174. Singh S., D'Britto V., Prabhune A.A., Ramana C.V., Dhawan A., Prasad B.L.V. Cytotoxic and genotoxic assessment of glycolipid-reduced and -capped gold and silver nanoparticles // *N. J. Chem.* 2010. V. 34. P. 294–301.

175. Chen S., Ji Y., Lian Q., Wen Y., Shen H., Jia N. Gold nanorods coated with multilayer polyelectrolyte as intracellular delivery vector of antisense oligonucleotides // *Nano Biomed. Eng.* 2010. V. 2. P. 19–31.

176. Rayavarapu R.G., Petersen W., Hartsuiker L., Chin P., Janssen H., van Leeuwen F.W.B., Otto C., Manohar S., van Leeuwen T.G. *In vitro* toxicity studies of polymer-coated gold nanorods // *Nanotechnology*. 2010. V. 21. 145101.

177 Mironava T., Hadjiargyrou M., Simon M., Jurukovski V., Rafailovich M.H. Gold nanoparticles cellular toxicity and recovery: Effect of size, concentration and exposure time // *Nanotoxicology*. 2010. V. 4. P. 120–137.

178. Tarantola M., Pietuch A., Schneider D., Rother J., Sunnick E., Rosman C., Pierrat S., Sönnichsen C., Wegener J., Janshoff A. Toxicity of gold-nanoparticles: Synergistic effects of shape and surface functionalization on micromotility of epithelial cells // *Nanotoxicology*. 2011. V. 5. P. 254–268.

179. Choi S.Y., Jeong S., Jang S.H., Park J., Park J.H., Ock K.S., Lee S.Y., Joo S.-W. *In vitro* toxicity of serum protein-adsorbed citrate-reduced gold nanoparticles in human lung adenocarcinoma cells // *Toxicol. In Vitro*. 2012. V. 26. P. 229–237.

180. Soenen S.J., Manshian B., Montenegro J.M., Amin F., Meermann B., Thiron T., Cornelissen M., Vanhaecke F., Doak S., Parak W.J., De Smedt S., Braeckmans K. Cytotoxic effects of gold nanoparticles: A multiparametric study // *ACS Nano*. 2012. V. 6. P. 5767–5783.

181. Vetten M.A., Tlotleng N., Rascher D.T., Skepu A., Keter F.K., Boodhia K., Koekemoer L.-A., Andraos C., Tshikhudo R., Gulumian M. Label-free *in vitro* toxicity and uptake assessment of citrate stabilised gold nanoparticles in three cell lines // *Part. Fibre Toxicol.* 2013. V. 10. 50.

182. DeBrosse M.C., Comfort K.K., Untener E.A., Comfort D.A., Hussain S.M. High aspect ratio gold nanorods displayed augmented cellular internalization and surface chemistry mediated cytotoxicity // *Mater. Sci. Eng. C*. 2013. V. 33. P. 4094–4100.

183. Ng C.-T., Li J.J., Gurung R.L., Hande M.P., Ong C.-N., Bay B.-H., Yung L.-Y.L. Toxicological profile of small airway epithelial cells exposed to gold nanoparticles // *Exp. Biol. Med. (Maywood)*. 2013. V. 238. P. 1355–1361.

184. Mohan J.C., Praveen G., Chennazhi K.P., Jayakumar R., Nair S.V. Functionalised gold nanoparticles for selective induction of *in vitro* apoptosis among human cancer cell lines // *J. Exp. Nanosci.* 2013. V. 8. P. 32–45.

185. Chuang S.-M., Lee Y.-H., Liang R.-Y., Roam G.-D., Zeng Z.-M., Tu H.-F., Wang S.-K., Chueh P.J. Extensive evaluations of the cytotoxic effects of gold nanoparticles // *Biochim. Biophys. Acta*. 2013. V. 1830. P. 4960–4973.

186. Vijayakumara S., Ganesan S. Size-dependent *in vitro* cytotoxicity assay of gold nanoparticles // *Toxicol. Environ. Chem.* 2013. V. 95. P. 277–287.

187. Ding F., Li Y., Liu J., Liu L., Yu W., Wang Z., Ni H., Liu B., Chen P. Overendocytosis of gold nanoparticles increases autophagy and apoptosis in hypoxic human renal proximal tubular cells // *Int. J. Nanomedicine*. 2014. V. 9. P. 4317–4330.

188. Tatini F., Landini I., Scaletti F., Massai L., Centi S., Ratto F., Nobili S., Romano G., Fusi F., Messori L., Mini E., Pini R. Size dependent biological profiles of PEGylated gold nanorods // *J. Mater. Chem. B*. 2014. V. 2. P. 6072–6080.

189. Wahab R., Dwivedi S., Khan F., Mishra Y.K., Hwang I.H., Shin H.S., Musarrat J., Al-Khedhairy A.A. Statistical analysis of gold nanoparticle-induced oxidative stress and apoptosis in myoblast (C2C12) cells // *Colloids Surf. B*. 2014. V. 123. P. 664–672.

190. Bajak E., Fabbri M., Ponti J., Gioria S., Ojea-Jiménez I., Collotta A., Mariani V., Gilliland D., Rossi F., Gribaldo L. Changes in Caco-2 cells transcriptome profiles upon exposure to gold nanoparticles // *Toxicol. Lett.* 2015. V. 233. P. 187–199.

191. Mateo D., Morales P., Ávalos A., Haza A.I. Comparative cytotoxicity evaluation of different size gold nanoparticles in human dermal fibroblasts // *J. Exp. Nanosci.* 2015. V. 10. P. 1–17.

192. Tang Y., Shen Y., Huang L., Lv G., Lei C., Fan X., Lin F., Zhang Y., Wu L., Yang Y. *In vitro* cytotoxicity of gold nanorods in A549 cells // *Environ. Toxicol. Pharmacol.* 2015. V. 39. P. 871–878.

193. Senut M.-C., Zhang Y., Liu F., Sen A., Ruden D.M., Mao G. Size-dependent toxicity of gold nanoparticles on human embryonic stem cells and their neural derivatives // *Small*. 2016. V. 12. P. 631–646.

194. Cho W.-S., Kim S., Han B.S., Son W.C., Jeong J. Comparison of gene expression profiles in mice liver following intravenous injection of 4 and 100 nm-sized PEG-coated gold nanoparticles // *Toxicol. Lett.* 2009. V. 191. P. 96–102.

195. Browning L.M., Lee K.J., Huang T., Nallathamby P.D., Lowman J.E., Xu X.-H.N. Random walk of single gold nanoparticles in zebrafish embryos leading to stochastic toxic effects on embryonic developments // *Nanoscale*. 2009. V. 1. P. 138–152.

196. Pocheptsov A.Y., Mamulaishvili N.I., Babayeva Z.V., Velikorodnaya Y.I. Studying of gold nanoparticles distribution in organs and tissues of the experimental animals according to morphological investigation data // In: *Production and Application of Nanomaterials in Russia: Toxicological, Exposure and Regulatory Issues*. Moscow: T-Press. 2009. P. 62.

197. Lasagna-Reeves C., Gonzalez-Romero D., Barria M.A., Olmedo I., Clos A., Sadagopa Ramanujam V.M., Urayama A., Vergara L., Kogan M.J., Soto C. Bioaccumulation and toxicity of gold nanoparticles after repeated administration in mice // *Biochem. Biophys. Res. Commun.* 2010. V. 393. P. 649–655.

198. Abdelhalim M.A.K., Jarrar B.M. Gold nanoparticles induced cloudy swelling to hydropic degeneration, cytoplasmic hyaline vacuolation, polymorphism, binucleation, karyopyknosis, karyolysis, karyorrhexis and necrosis in the liver // *Lipids Health Dis.* 2011. V. 10. 166.

199. Zhang X.-D., Wu D., Shen X., Liu P.-X., Yang N., Zhao B., Zhang H., Sun Y.-M., Zhang L.-A., Fan F.-Y. Size-dependent *in vivo* toxicity of PEG-coated gold nanoparticles // *Int. J. Nanomedicine.* 2011. V. 6. P. 2071–2081.

200. Gad S.C., Sharp K.L., Montgomery C., Payne J.D., Goodrich G.P. Evaluation of the toxicity of intravenous delivery of auroshell particles (gold-silica nanoshells) // *Int. J. Toxicol.* 2012. V. 31. P. 584–594.

201. Jiang W., Kim B.Y.S., Rutka J.T., Chan W.C.W. Nanoparticle-mediated cellular response is size-dependent // *Nat. Nanotechnol.* 2008. V. 3. P. 145–150.

202. Tsoli M., Kuhn H., Brandau W., Esche H., Schmid G. Cellular uptake and toxicity of Au_{55} clusters // *Small.* 2005. V. 1. P. 841–844.

203. Ryan J.A., Overton K.W., Speight M.E., Oldenburg C.N., Loo L.N., Robarge W., Franzen S., Feldheim D.L. Cellular uptake of gold nanoparticles passivated with BSA-SV40 large T antigen conjugates // *Anal. Chem.* 2007. V. 79. P. 9150–9159.

204. Schaeublin N.M., Braydich-Stolle L.K., Schrand A.M., Miller J.M., Hutchison J., Schlager J.J., Hussain S.M. Surface charge of gold nanoparticles mediates mechanism of toxicity // *Nanoscale.* 2011. V. 3. P. 410–420.

205. Xia D.-L., Wang Y.-F., Bao N., He H., Li X.-d., Chen Y.-P., Gu H.-Y. Influence of reducing agents on biosafety and biocompatibility of gold nanoparticles // *Appl. Biochem. Biotechnol.* 2014. V. 174. P. 2458–2470.

206. Favi P.M., Gao M., Arango L.J.S., Ospina S.P., Morales M., Pavon J.J., Webster T.J. Shape and surface effects on the cytotoxicity of nanoparticles: Gold nanospheres versus gold nanostars // *J. Biomed. Mater. Res. A.* 2015. V. 103. P. 3449–3462.

207. Favi P.M., Valencia M.M., Elliott P.R., Restrepo A., Gao M., Huang H., Pavon J.J., Webster T.J. Shape and surface chemistry effects on the cytotoxicity and cellular uptake of metallic nanorods and nanospheres // *J. Biomed. Mater. Res. A.* 2015. V. 103. P. 3940–3955.

208. Jia H.Y., Liu Y., Zhang X.J., Han L., Du L.B., Tian Q., Xu Y.C. Potential oxidative stress of gold nanoparticles by induced-NO releasing in serum // *J. Am. Chem. Soc.* 2009. V. 131. P. 40–41.

209. Li J.J., Zou L., Hartono D., Ong C.-N., Bay B.-H., Lanry Yung L.-Y. Gold nanoparticles induce oxidative damage in lung fibroblasts *in vitro* // *Adv. Mater.* 2008. V. 20. P. 138–142.

210. Uboldi C., Bonacchi D., Lorenzi G., Hermanns M.I., Pohl C., Baldi G., Unger R.E., Kirkpatrick C.J. Gold nanoparticles induce cytotoxicity in the alveolar type-II cell lines A549 and NCIH441 // *Part Fibre Toxicol.* 2009. V. 6. 18.

211. Dobrovolskaia M.A., Patri A.K., Zheng J., Clogston J.D., Ayub N., Aggarwal P., Neun B.W., Hall J.B., McNeil S.E. Interaction of colloidal gold nanoparticles with human blood: Effects on particle size and analysis of plasma protein binding profiles // *Nanomedicine.* 2009. V. 5. P. 106–117.

212. Lacerda S.H.D.P., Park J.J., Meuse C., Pristinski D., Becker M.L., Karim A., Douglas J.F. Interaction of gold nanoparticles with common human blood proteins // *ACS Nano.* 2010. V. 4. P. 365–379.

213. Montes-Burgos I., Walczyk D., Hole P., Smith J., Lynch I., Dawson K. Characterisation of nanoparticle size and state prior to nanotoxicological studies // *J. Nanopart. Res.* 2010. V. 12. P. 47–53.

214. Staroverov S.A., Aksinenko N.M., Gabalov K.P., Vasilenko O.A., Vidyasheva I.V., Shchyogolev S.Y., Dykman L.A. Effect of gold nanoparticles on the respiratory activity of peritoneal macrophages // *Gold Bull.* 2009. V. 42. P. 153–156.

215. Yan L., Gu Z., Zhao Y. Chemical mechanisms of the toxicological properties of nanomaterials: Generation of intracellular reactive oxygen species // *Chem. Asian J.* 2013. V. 8. P. 2342–2353.

216. Cheung K.L., Chen H., Chen Q., Wang J., Ho H.P., Wong C.K., Kong S.K. CTAB-coated gold nanorods elicit allergic response through degranulation and cell death in human basophils // *Nanoscale.* 2012. V. 4. P. 4447–4449.

217. Huff T.B., Hansen M.N., Zhao Y., Cheng J.-X., Wei A. Controlling the cellular uptake of gold nanorods // *Langmuir.* 2007. V. 23. P. 1596–1599.

218. Parab H.J., Chen H.M., Lai T.-C., Huang J.H., Chen P.H., Liu R.-S., Hsiao M., Chen C.-H., Tsai D.-P., Hwu Y.-K. Biosensing, cytotoxicity, and cellular uptake studies of surface-modified gold nanorods // *J. Phys. Chem. C.* 2009. V. 113. P. 7574–7578.

219. Pissuwan D., Valenzuela S.M., Killingsworth M.C., Xu X., Cortie M.B. Targeted destruction of murine macrophage cells with bioconjugated gold nanorods // *J. Nanopart. Res.* 2007. V. 9. P. 1109–1124.

220. Wang L., Jiang X., Ji Y., Bai R., Zhao Y., Wu X., Chen C. Surface chemistry of gold nanorods: Origin of cell membrane damage and cytotoxicity // *Nanoscale.* 2013. V. 5. P. 8384–8391.

221. Choi B.-S., Iqbal M., Lee T., Kim Y.H., Tae G. Removal of cetyltrimethylammonium bromide to enhance the biocompatibility of Au nanorods synthesized by a modified seed mediated growth process. // *J. Nanosci. Nanotechnol.* 2008. V. 8. P. 4670–4674.

222. Grabinski C., Schaeublin N., Wijaya A., D'Couto H., Baxamusa S.H., Hamad-Schifferli K., Hussain S.M. Effect of gold nanorod surface chemistry on cellular response // *ACS Nano.* 2011. V. 5. P. 2870–2879.

223. Zeng Q., Zhang Y., Ji W., Ye W., Jiang Y., Song J. Inhibition of cellular toxicity of gold nanoparticles by surface encapsulation of silica shell for hepatocarcinoma cell application // *ACS Appl. Mater. Interfaces.* 2014. V. 6. P. 19327–19335.

224. Eghtedari M., Liopo A.V., Copland J.A., Oraevsky A.A., Motamedi M. Engineering of hetero-functional gold nanorods for the *in vivo* molecular targeting of breast cancer cells // *Nano Lett.* 2009. V. 9. P. 287–291.

225. Wan J., Wang J.-H., Liu T., Xie Z., Yu X.-F., Li W. Surface chemistry but not aspect ratio mediates the biological toxicity of gold nanorods *in vitro* and *in vivo* // *Sci. Rep.* 2015. V. 5. 11398.

226. Hirsch L.R., Stafford R.J., Bankson J.A., Sershen S.R., Rivera B., Price R.E., Hazle J.D., Halas N.J., West J.L. Nanoshell-mediated near-infrared thermal therapy of tumors under magnetic resonance guidance // *Proc. Natl. Acad. Sci. USA.* 2003. V. 100. P. 13549–13554.

227. Loo C., Lowery A., Halas N., West J., Drezek R. Immunotargeted nanoshells for integrated cancer imaging and therapy // *Nano Lett.* 2005. V. 5. P. 709–711.

228. Stern J.M, Stanfield J., Lotan Y., Park S., Hsieh J.-T., Cadeddu J.A. Efficacy of laser-activated gold nanoshells in ablating prostate cancer cells *in vitro* // *J. Endourol.* 2007. V. 21. P. 939–943.

229. Liu S.-Y., Liang Z.-S., Gao F., Luo S.-F., Lu G.-Q. *In vitro* photothermal study of gold nanoshells functionalized with small targeting peptides to liver cancer cells // *J. Mater. Sci. Mater. Med.* 2010. V. 21. P. 665–674.

230. Bar-Ilan O., Albrecht R.M., Fako V.E., Furgeson D.Y. Toxicity assessments of multisized gold and silver nanoparticles in zebrafish embryos // *Small.* 2009. V. 5. P. 1897–910.

231. Wiwanitkit V., Sereemaspun A., Rojanathanes R. Effect of gold nanoparticle on the microscopic morphology of white blood cell // *Cytopathology.* 2009. V. 20. P. 109–110.

232. Sereemaspun A., Rojanathanes R., Wiwanitkit V. Effect of gold nanoparticle on renal cell: An implication for exposure risk // *Ren. Fail.* 2008. V. 30. P. 323–325.

233. Wiwanitkit V., Sereemaspun A., Rojanathanes R. Effect of gold nanoparticles on spermatozoa: The first world report // *Fertil. Steril.* 2009. V. 91. P. 7–8.

234. Velikorodnaya Y.I., Pocheptsov A.Y., Sokolov O.I., Bogatyrev V.A., Dykman L.A. Effect of gold nanoparticles on proliferation and apoptosis during spermatogenesis in rats // *Nanotechnol. Russia.* 2015. V. 10. P. 814–819.

235. Khlebtsov N.G., Dykman L.A., Terentyuk G.S. Biodistribution and toxicity of engineered gold nanoparticles: State-of-the-art and prospects // In: *Medical Physics 2010, III Euroasian Congress on Medical Physics and Engineering.* Moscow: Moscow State Univ. Publ. 2010. V. 3. P. 209–211.

236. De Jong W.H., Borm P.J. Drug delivery and nanoparticles: Applications and hazards // *Int. J. Nanomedicine.* 2008. V. 3. P. 133–149.

237. Eck W., Nicholson A.I., Zentgraf H., Semmler W., Bartling S. Anti-CD4-targeted gold nanoparticles induce specific contrast enhancement of peripheral lymph nodes in X-ray computed tomography of live mice // *Nano Lett.* 2010. V. 10. P. 2318–2322.

238. Panessa-Warren B.J., Warren J.B., Maye M.M., Schiffer W. Nanoparticle interactions with living systems: *in vivo* and *in vitro* biocompatibility // In: *Nanoparticle and Nanodevices in Biological Applications* / Ed. Bellucci S. Heidelberg: Springer. 2009. V. 4. P. 1–45.

239. Alkilany A.M., Mahmoud N.N., Hashemi F., Hajipour M.J., Farvadi F., Mahmoudi M. Misinterpretation in nanotoxicology: A personal perspective // *Chem. Res. Toxicol.* 2016. V. 29. P. 943–948.

240. Falagan-Lotsch P., Grzincic E.M., Murphy C.J. One low-dose exposure of gold nanoparticles induces long-term changes in human cells // *Proc. Natl. Acad. Sci. USA.* 2016. V. 113. P. 13318–13323.

4

Uptake of Gold Nanoparticles into Mammalian Cells

4.1 Preliminary Remarks

In the past decade, mammalian cells have increasingly encountered gold nanoparticles (GNPs), owing to the active development of nanomedicine (see Chapter 2). Researchers employing GNPs *in vivo* inevitably have to deal with particle biodistribution over animal organs and tissues, pharmacokinetics, and assessment of possible particle cytotoxicity (see Chapter 3). In view of this, one of the hottest areas in current research is the interaction of GNPs with mammalian cells, routes of particle uptake into the cellular space, the fate of the particles inside the cell, and their elimination from the cell and the whole organism.

It is natural to suppose that the first cells that GNPs encounter on their way in the mammalian organism are those of the immune system, in particular its phagocytic link (neutrophils, monocytes, macrophages, dendritic cells, and mast cells). Indeed, as early as in the first attempts to investigate colloidal gold biodistribution, which were performed in the 1960s–1980s on rabbits [1], mice [2], and rats [3], it was found that after parenteral administration, colloidal gold particles are captured by liver cells, excreted through bile, and eliminated from the organism with feces. After injection, gold was identified mostly in Kupffer cells. Perhaps Scott *et al.* [1] were the first to note that the phagocytosis of GNPs is size dependent (maximal accumulation and clearance were observed for 40-nm-diameter particles). Besides Hardonk *et al.* [4], the important role of Kupffer cells in the elimination of GNPs was established by Sadauskas *et al.* [5], who injected 2- and 40-nm GNPs intravenously in mice. Electron microscopy showed that after injection, the GNPs accumulated in the macrophages of the liver (90%) and spleen (10%). The authors concluded that GNPs penetrate only phagocytes, primarily the Kupffer cells of the liver. In a subsequent study [6], Sadauskas *et al.* reported that 40-nm GNPs get localized in lysosomes (endosomes) of Kupffer cells and can be retained there for up to 6 months.

With recent achievements in the fabrication of functionalized GNPs, which have controllable properties and are loaded with targeting molecules, the endocytosis of such GNPs has become crucial for successful biomedical applications. Understanding the mechanisms by which GNPs are taken up and the particles' fate after cellular internalization may help to avoid endocytic metabolism and to retain the biological activity of delivered cargoes. Despite the notable progress in this field, real quantitative control of endocytosis kinetics and of the traffic of GNPs to their intracellular targets is still quite limited. The existing gaps in understanding the mechanisms of endocytosis, together with the potential nanotoxicity of GNPs, hamper the transfer of nanoparticle (NP)-based technologies to clinical practice.

By now, a large number of original research articles have addressed the mechanisms and critical parameters of GNP uptake into cells, and a large number of reviews covering these issues to this or that degree have been published [7–37]. However, as it usually happens at the initial stages in the study of a complex problem, it is often difficult to compare and organize the available data because of the many important factors, which differ in different studies. In particular, most work on cellular GNP uptake has employed various cell lines *in vitro*, making use of GNPs of various sizes, shapes, and structures (nanospheres [NSphs], nanorods [NRs], nanoshells [NSs], nanocages, nanostars, *etc.*). Finally, a vast diversity of molecules has been used to functionalize the surface of GNPs in studies of the mechanisms and effectiveness of intracellular uptake. So, there is a strong need to systematize the data available in the literature in a special review devoted entirely to the intracellular uptake of GNPs.

In this chapter, we consider recent progress in understanding how the size, shape, and surface properties of GNPs affect their uptake and intracellular fate. In particular, we discuss the selective penetration of GNPs into cancer and immune cells and the interaction of GNPs with immune cell receptors. We also analyze recent theoretical models for endocytosis of spherical and nonspherical particles in relation to experimental data on the effects of particle geometry, ligand density, and cell membrane properties. In accordance with our major goal, we focus on experiments *in vitro*, with those *in vivo* being omitted from discussion. There are two reasons for our choice. One is that most data on the mechanisms of GNP uptake by animal cells have come from the use of models *in vitro*. The other reason is that studies of cellular uptake *in vivo* depend inevitably on particle biodistribution, which itself relies on the chosen model and on the nature of the functionalized GNPs. Therefore, particle uptake into mammalian cells *in vivo* deserves separate consideration.

4.2 Dependence of Intracellular Uptake on GNP Size and Shape

4.2.1 Effects of GNP Size

The most important geometrical parameters of particles, which should be responsible for their cellular uptake, are size and shape. Pioneering experimental work on GNPs was done by Chan's group. In 2006 [38], Chan *et al.* studied HeLa cell uptake of citrate GNSphs with average diameters of 14, 30, 50, 74, and 100 nm and gold nanorods (GNRs) with dimensions (length × diameter) of 40×14 nm and 74×14 nm. The surface of the as-prepared NRs was coated with a bilayer of cetyltrimethylammonium bromide (CTAB) molecules. The uptake effectiveness was determined quantitatively with inductively coupled plasma atomic emission spectroscopy (ICP–AES) by the amount of gold present in a homogenate of cells treated with different particles. With knowledge of the amount of gold accumulated in the cells and the geometrical dimensions of the particles, the average number of particles in a cell was calculated. GNSphs of ~50 nm were best able to enter the cell interior (Figure 4.1a). The accumulation of NRs in the cells was three-fold and sixfold less for 40 × 14-nm and 74 × 14-nm NRs, respectively (Figure 4.1b). At first glance, these data seem unexpected, because the negatively charged citrate NSphs should have penetrated the negatively charged cells less effectively than the positively charged NRs did. However, they can be explained by the effect of coating of cell-incubated citrate NSphs with serum proteins from Dulbecco's modified Eagle's medium (DMEM), which lowered the effect of charge. Further, citrate GNPs coated with serum proteins penetrated the cells much better than did those conjugated with transferrin—a blood plasma protein that mediates transport of Fe ions into cells *via* the transferrin receptors at the cell surface (active targeting) [39]. Chan and coworkers attributed this difference in uptake to the fact that the cell surface has many more receptors for

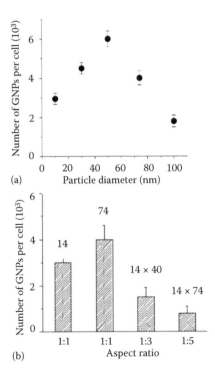

FIGURE 4.1 (a) Dependence of cellular uptake of spherical GNPs as a function of size. (b) Comparison of cellular uptake for rod-shaped NPs (with aspect ratios of 1:3 and 1:5) and spherical GNPs. Recalculated data of ICP–AES after 6-h incubation of HeLa cells with GNPs. (Adapted from Chithrani, B.D., Ghazani, A.A., Chan, W.C.W., *Nano Lett.*, 6, 662–668, 2006. With permission.)

serum proteins than it does for transferrin, and they concluded that the cellular uptake of GNPs is receptor-mediated endocytosis.

As is known, endocytosis [22,40] is a process by which materials are engulfed by a cell through the invagination of a portion of the plasma membrane, with subsequent formation inside the cell of vesicles containing the trapped materials. Endocytosis can arbitrarily be divided into phagocytosis (actin-dependent, "professional" endocytosis), macropinocytosis (receptor-independent), and receptor-mediated (specific) endocytosis [40]. In turn, receptor-mediated endocytosis can be clathrin, caveolin, and raft dependent. Clathrin and caveolin are proteins that ensure the curving of the plasma membrane and the formation of coated vesicles; rafts are specially organized domains of lipids in the plasma membrane. Receptor-mediated endocytosis is "turned on" for a rapid and controlled uptake of the corresponding ligand by a cell. Vesicles rapidly lose their coats and fuse to form larger compartments, known as endosomes, which then fuse with the primary lysosomes, forming secondary lysosomes [41–43]. Krpetić *et al.* [44], using transmission electron microscopy (TEM), showed that three stages are characteristic of the intracellular uptake of GNPs into macrophages: chemotaxis (including the formation of vacuoles at the cost of pseudopodia), phagocytosis (the formation of GNP-containing phagosomes and their merging with lysosomes), and digestion (the formation of phagolysosomes). Banerji and Hayes [45] demonstrated the impossibility of a nonendocytotic pathway for cellular GNP uptake through an intact lipid bilayer, although Taylor *et al.* [46] and Xia *et al.* [47] discussed the possibility in principle of nonendosomal GNP uptake. A detailed discussion of endocytotic pathways, trafficking, and endocytosis of NPs can be found in a recent review by Canton and Battaglia [22].

Further to their research [48,49], Chan and coworkers studied the mechanisms of intracellular uptake, accumulation, and elimination of transferrin-coated GNPs of different sizes and shapes. The cell lines STO, HeLa, and SNB19 served as models in this study. Uptake was evaluated by two methods: (i) laser confocal microscopy (LCM), by using Texas red-labeled transferrin as a fluorescent probe, and (ii) TEM. To confirm the suggestion made in the article [38], namely, that the endocytosis mechanism is receptor mediated, the authors investigated cellular GNP uptake at low temperature and after cell treatment with sodium azide, an inhibitor of electron transport. As a result, the level of uptake decreased by about 70% in both cases. Next, before GNPs were introduced, the cells had been treated with a hypertonic solution of sucrose or had been grown in a K^+-deficient medium to disrupt the formation of clathrin-coated vesicles. As a result, a considerable decrease in the level of GNP uptake was also shown. For this reason, the authors speculated that the process of GNP internalization is clathrin dependent. The best ability of spherical particles of ~50 nm to penetrate inside cells was explained by Chithrani and Chan [48] by the minimal time needed by the membrane to wrap around spheres, in agreement with the previously reported thermodynamic calculations [41].

Another interesting point reported in [48] is the time of GNP elimination from the cells (exocytosis). The smaller the NPs, the more rapidly they were removed from the cells. Specifically, 14-nm GNPs were eliminated two times faster than 74-nm particles, and this dependence was almost linear in the size range 10–100 nm. Transferrin-coated NRs were taken up much less effectively than spherical particles with the same surface coating. However, the fraction of exocytosed NRs was larger than that of spherical GNPs.

Continuing their work on size-dependent endocytosis, Chan's group studied the uptake of antibody-coated GNPs into tumor cells [50]. Specifically, they examined the interactions between herceptin and SK-BR-3 cells, which overexpress ErbB2 tyrosine kinase receptors. Herceptin is a recombinant humanized monoclonal G1 antibody that selectively interacts with the extracellular domain of the human epidermal growth factor receptor (EGFR) and blocks the proliferation of ErbB2-overexpressing human tumor cells. Most commonly, herceptin is used to treat breast cancer with overexpression of ErbB2 (early stages and metastatic cancer). GNP–herceptin complexes were first used for optoacoustic tomography imaging of deep tumors [51]. Chan and coworkers [50] conjugated herceptin to 2-, 10-, 25-, 40-, 50-, 70-, 80-, and 90-nm GNPs and assessed the degree of uptake by TEM and LCM. Similarly to the uptake of transferrin-coated GNPs [48], the uptake of GNP–herceptin into tumor cells was also found to be size dependent. Maximal uptake of GNPs, with localization in endosomes and lysosomes, was observed for 25–50-nm particles, whereas smaller or larger GNPs penetrated cells much worse and appeared mostly on the plasma membrane. This was explained by the optimal quantity of antibodies adsorbed on the surface of 25–50-nm GNPs and interacting with the cellular ErbB2 receptors. Using LCM and colocalization of ErbB2 and transferrin receptors (Tfr), Chan and coworkers [50] demonstrated that the GNP–Her conjugates and the ErbB2 receptors are internalized as a single complex and that this process is also size dependent (Figure 4.2). What is more, herceptin conjugated with 25–50-nm GNPs caused the most damage to the cells. Thus, the authors provided additional strong evidence for a receptor-mediated mechanism of GNP endocytosis.

In 2011, Chan's group reported on the effect of aggregation of GNPs on their intracellular uptake and cytotoxicity [52]. The amount of GNPs taken up by cells was evaluated

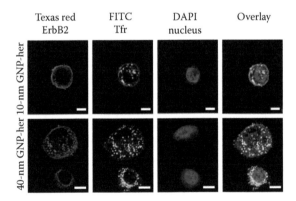

FIGURE 4.2 Size-dependent internalization of ErbB2 receptors. Laser confocal fluorescence microscopy of SK-BR-3 cells treated for 3 h with 10- and 40-nm GNPs conjugated to herceptin. Cells were then labeled with anti-ErbB2 (red) and antitransferrin receptor (Tfr) antibodies (green), and the nucleus was counterstained with DAPI (blue). Significant colocalization of ErbB2 and Tfr was observed in cells treated with 40-nm GNP–Her conjugates (yellow), whereas a fairly limited number of ErbB2 receptors were internalized in the case of 10-nm GNPs. Scale bars, 10 μm. (Adapted from Jiang, W., Kim, B.Y.S., Rutka, J.T., Chan, W.C.W., *Nat. Nanotechnol.*, 2008, 3, 145–150. With permission.)

by ICP–AES and TEM. The results showed that the amount of aggregated transferrin-coated GNPs taken up by HeLa cells was 25% smaller than that of nonaggregated particles. A similar picture was obtained for A549 cells as well. However, for cells of the line MDA-MB-435, the greatest capacity for being internalized was demonstrated by the largest aggregates (98 nm). Their accumulation was almost twofold higher than that of nonaggregated particles or small aggregates. This finding looks somewhat unexpected, because the surface of MDA-MB-435 cells contains the smallest number of receptors of transferrin CD71. Therefore, the authors proposed that in addition to receptor-mediated endocytosis, MDA-MB-435 cells might use another mechanism of GNP internalization. Possibly, the asymmetrical geometry of the aggregates may add to the complexity of their interaction with the cell membrane. Inhomogeneous aggregates may interact with the cell membrane through various polyvalent interactions by following the receptor-independent pathway. One also should bear in mind the results obtained by Kneipp *et al.* [53], who showed that aggregates may form directly in endosomes after GNPs have entered the cytoplasm.

The effect of the concentration of introduced GNPs on uptake by MC3T3-E1 cells was explored by Mustafa *et al.* [54] 12-nm GNPs at 160 µg/ml formed aggregates on the cell surface and effectively penetrated the cells by endocytosis. Yet, at 10 µg/ml, the GNPs were found on the membranes mostly as isolated particles, suggesting diffusion to be the predominant uptake pathway in this case.

Mironava *et al.* [55] compared intracellular uptake of 13- and 45-nm GNPs. From scanning electron microscopy (SEM) observations, 45-nm GNPs were taken up into the intracellular space of CF-31 cells five times more effectively than were 13-nm GNPs, and they were localized in large lysosomes, without entering the nuclei or mitochondria. To investigate the mechanism of GNP endocytosis, Mironava *et al.* pretreated the cells with phenylarsine oxide, an inhibitor of clathrin-dependent endocytosis. TEM analysis demonstrated that endocytosis was clathrin dependent only for 45-nm GNPs and was inhibited considerably by phenylarsine oxide. For 13-nm GNPs, there was no fundamental difference in uptake between inhibitor-treated and nontreated cells. In this case, in the authors' opinion, a receptor-independent mechanism of endocytosis (macropinocytosis) was possibly involved. An effective way to detect a receptor-independent mechanism is to study endocytosis at lowered temperature, because this mechanism depends strongly on temperature (as distinct from the receptor-dependent mechanism). Specifically, the data of Mironava *et al.* showed that reducing the temperature to 4°C brought about a 90% decrease in the endocytosis of 13-nm GNPs, whereas the endocytosis of 45-nm GNPs decreased only by 30%.

Sonavane *et al.* [56] examined the size effects of GNP uptake *in vitro* into cells of rat skin and intestine. They used GNPs with diameters of 15, 100, and 200 nm. The tissue content of GNPs was determined by TEM, ICP–mass spectrometry (MS), and energy-dispersive x-ray spectroscopy. The amount of GNPs taken up into the studied tissues was found to decrease in the order 15 > 100 > 200 nm.

Yen *et al.* [57] described the uptake of 3-, 6-, and 40-nm GNPs by murine macrophages (J774 A1). The tissue content of the GNPs was determined by TEM, ultraviolet–visible (UV–vis) spectrophotometry, and ICP–MS. Compared with 3- and 6-nm GNPs, 40-nm particles were internalized better and were less cytotoxic. Yet, the smaller particles were responsible for enhanced production of proinflammatory cytokines (interleukins-1 and -6 [IL-1 and IL-6] and tumor necrosis factor-α [TNF-α]). The internalized GNPs in the macrophage cytoplasm were shown by TEM to be located inside coated vesicles, a finding indicative of receptor-mediated endocytosis. The authors associated this mechanism with adsorption of serum proteins from the cell culture medium onto the GNPs. Similar data were recorded by Coradeghini *et al.* [58] for 5- and 15-nm GNPs and mouse fibroblasts.

Wang *et al.* [59] used dark-field microscopy to investigate the endocytosis of GNPs (Figure 4.3). The experiments used 45-, 70-, and 110-nm GNPs and the cell lines CL1-0 and HeLa. GNPs were functionalized with ssDNA—an aptamer to mucin, which is a cell surface glycoprotein overexpressed in tumor cells. The best uptake in both types of cells was obtained with 45-nm GNPs. Conversely, the cell cytoplasm contained almost no 110-nm GNPs, which, however, appeared on the cell membranes. To explain what had caused the revealed effect, Wang *et al.* put forward a thermodynamic model for the ligand–receptor interaction occurring in receptor-mediated endocytosis.

Rieznichenko *et al.* [60] presented data on the uptake of 10-, 20-, 30-, and 45-nm GNPs by U937 tumor cells. GNP internalization was detected by TEM and LCM. The most active cellular accumulation was found for 30-nm GNPs, with maximal uptake occurring very fast (at 3 to 5 min). It is interesting to note that for the same U937 cells,

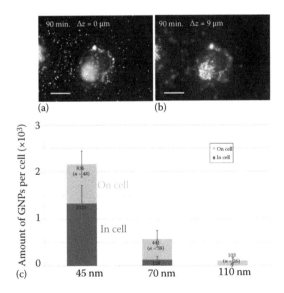

FIGURE 4.3 Dark-field images of a HeLa cell taken at the bottom (a) and top (b) of the cell at 90 min after interaction with 70-nm GNPs. The image (b) shows that many GNPs are localized on the cell membrane without being internalized. (c) Statistical data for 45-, 70-, and 110-nm GNPs localized on (soft pink) and in (ruby red) HeLa cells illustrate the maximal total amounts of 45-nm GNPs interacting with cells and the maximal percentage of 45-nm GNPs localized in cells. (Adapted from Wang, S.-H., Lee, C.-W., Chiou, A., Wei, P.-K., *J. Nanobiotechnol.*, 8, 33, 2010.)

the effectiveness of uptake of 15-nm GNPs, from ICP–AES data, was approximately an order of magnitude lower than was that for HeLa cells [61]. This observation, along with the data of Albanese and Chan [52], Coulter *et al.* [62], Freese *et al.* [63], Cui *et al.* [64], and Liu *et al.* [65], indicates that the intracellular uptake of GNPs depends also on the cell type. Wang *et al.* [66] reported also on the influence of cell size on cellular uptake of GNPs.

The same conclusion was drawn by Trono *et al.* [67] from a study of uptake of 5-, 10-, 20-, 30-, 40-, and 50-nm GNPs by PK-1, PK-45, and Panc-1 tumor cells. It was shown by atomic absorption spectrometry (AAS) that as distinct from HeLa cells, the pancreas cancer cells were best able to internalize 20-nm particles. In addition, uptake efficacy was influenced by the time of incubation, the concentration of introduced GNPs, and the temperature conditions for the experiment. Quite recently, Zhou *et al.* [68] showed that the thickness of the pericellular matrix of different cells is an important factor, which can enhance the retention and cellular uptake of NPs.

Size-dependent endocytosis of single GNPs was examined with HeLa cells by Shan *et al.* [69]. The GNP diameters used were 4, 12, and 17 nm. The force of GNP interaction with the cell membrane was measured by atomic force microscopy (AFM). It was found that the interaction force of a single GNP with the cell membrane increased with increasing particle diameter. Pretreatment of the cells with methyl-β-cyclodextrin inhibited the endocytosis. Because methyl-β-cyclodextrin lowers the content of membrane cholesterol and, accordingly, decreases the formation of lipid rafts, the mechanism of GNP endocytosis was concluded to be raft dependent. In their subsequent work [70], the authors determined the endocytosis of small (4 nm) GNPs to be caveolin dependent.

The endocytosis of variously sized GNPs that leads to lysosome impairment and to autophagosome accumulation was explored by Ma *et al.* [71]. They used GNPs of 10, 25, and 50 nm and cells of the rat kidney (NRK) as a model object. The cellular content of the GNPs was determined by TEM (Figure 4.4), light microscopy, and ICP–MS. Receptor-mediated endocytosis was found to increase with increasing particle diameter, reaching a maximum for 50-nm GNPs. Similar data were reported by Sobhan *et al.* [72] for 7-, 21-, and 31-nm GNPs and AR42J cells. Sabuncu *et al.* [73] demonstrated that 50-nm GNPs were taken up by tumor cells more effectively than were 25-nm GNPs and that uptake in Panc-1 cells was much more active than it was in Jurkat cells.

Different data were acquired by Huang *et al.* [74] in a TEM and ICP–MS study of uptake of 2-, 6-, and 15-nm tiopronin-functionalized GNPs by MCF-7 cells. The most active uptake was recorded for 2-nm GNPs (4×10^6 particles/cell); the uptake of 6- and

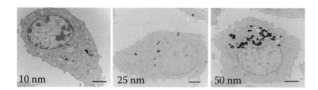

FIGURE 4.4 TEM images of rat kidney cells treated with 10-, 26-, and 50-nm GNPs. (Adapted from Ma, X., Wu, Y., Jin S., Tian, Y., Zhang, X., Zhao, Y., Yu, L., Liang, X.-J., *ACS Nano*, 5, 8629–8639, 2011. With permission.)

15-nm GNPs was approximately an order of magnitude less effective. Only 2- and 6-nm GNPs were present in the cell nuclei. However, in a subsequent study [75], the same group compared the uptake of 50- and 100-nm tiopronin-functionalized GNPs by MCF-7 cells and concluded that the smaller, 50-nm NPs demonstrated more advantages over the larger, 100-nm particles in the uptake and permeability in tumor cells and tissues. Yet, Arvizo *et al.* [76] demonstrated, by instrumental neutron activation analysis and TEM, that the uptake of 5- and 20-nm unmodified GNPs in SKOV3-ip, OVCAR5, and A2780 cells was higher, compared with 50- and 100-nm GNPs. Trickler *et al.* [77] examined blood–brain barrier (BBB) in response to variously sized GNPs (3, 5, 7, 10, 30, and 60 nm) *in vitro* using primary rat brain microvessel endothelial cells (rBMECs). According to spectrophotometric measurements at 500 nm, the smaller GNPs (3–7 nm) showed higher rBMEC accumulation compared to larger GNPs (10–60 nm). Even though slight changes in cell viability were observed with small GNPs, the rBMEC morphology appeared unaffected 24 h after exposure to GNPs, with only mild changes in fluorescein permeability, indicating that BBB integrity was unaltered.

A possible reason for the contradiction in the published data on size-dependent uptake can be related to the particle clusters formed prior to interaction with cells or during adsorption on the cell membrane. For example, Chithrani and Chan [48] observed single-particle uptake for 50-nm transferrin-coated GNPs, whereas 14-nm GNPs were taken up only as GNP clusters (at least six GNPs per cluster). By contrast, Lèvy *et al.* [78] observed single-particle entry of 10-nm peptide-capped GNPs in HeLa cells (see endosomes containing single 10-nm NPs in Figure 4.5). An obvious explanation for these apparently conflicting data may be the different surface functionalization of GNPs (transferrin versus peptide/poly(ethylene glycol) [PEG]). This example, together with the data of Albanese and Chan [52], clearly demonstrates the need for real-time single particle techniques in order to better understand the impact of clustering on GNP uptake.

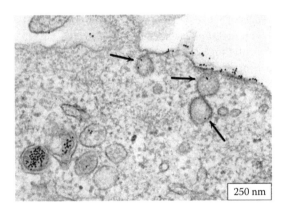

FIGURE 4.5 TEM image of single-particle uptake of 10-nm peptide-coated GNPs (6 nM) incubated with HeLa cells for 3 h in the presence of serum. The arrows indicate endosomes with a single and two GNPs. (Adapted from Lèvy, R., Thanh, N.T.K., Doty, R.C., Hussain, I., Nichols, R.J., Schiffrin, D.J., Brust, M., Fernig, D.G., *J. Am. Chem. Soc.*, 126, 10076–10084, 2004. With permission.)

4.2.2 Effects of GNP Shape

We now shift to discuss particle shape effects. The effects of GNP shape on intracellular uptake were addressed by Bartczak *et al.* [79]. They studied, by ICP–AES, the HUVEK cell uptake of four types of PEG-coated (PEGylated) GNPs, namely, NSphs (15 nm), NRs (47 × 17 nm), hollow NSs (91 nm, shell thickness of 9 nm), and NSs on silica cores (SiO$_2$ core of 43 nm, Au shell of 7 nm). In this study, the intracellular uptake of GNPs decreased in the order NSphs (~2400 GNPs/cell) > NRs (~2200 GNPs/cell) > NSs on silica cores (~400 GNPs/cell) > hollow NSs (~200 GNPs/cell). It should be stressed that these results do not reflect any effect of the size and shape of particles in pure form, because the negative charge of the particles and the quantity of stabilizer molecules associated with GNPs increased in the sequence from PEG-coated NSphs to hollow NSs. It is these two factors that are thought by the authors to be responsible for the obtained differences in uptake effectiveness.

Cho *et al.* [80] arrived at different conclusions in their UV–vis spectrophotometry study of the uptake of gold NSphs (15, 54, and 100 nm), nanocages (62 and 118 nm), and NRs (40 × 16 nm) into SK-BR-3, ATCC, and HTB-30 cells. After changing the experimental design to preclude any effects of GNP sedimentation, they showed that the effectiveness of intracellular uptake is affected by the rates of GNP diffusion and sedimentation much more than it is affected by the size and shape of NPs, the density of stabilizer coating, or the GNP concentration.

Schaeublin *et al.* [81] pointed out the differences in uptake of NSphs and NRs into HaCaT cells. On the whole, NRs were taken up less effectively; furthermore, TEM analysis revealed that NSphs inside the cells were present as aggregates and NRs as isolated particles. The authors concluded that the shape of GNPs not only affects intracellular uptake but also plays a crucial role in cellular response to particle introduction. Zhang *et al.* [82] demonstrated that the effectiveness of NR uptake by MDA-MB-231 cells, measured with UV–vis–near-infrared (NIR) adsorption spectroscopy, is affected by the time of incubation and by the concentration of GNPs introduced into a cell suspension. In agreement with the data of Schaeublin *et al.* [81], Tarantola *et al.* [83] showed that spherical GNPs (43 nm) with identical surface functionalization (CTAB) are generally more toxic and more efficiently ingested by MDCK II cells than rod-shaped particles (38 × 17 nm). Their experiment confirmed more efficient intracellular uptake of spherical functionalized GNPs as compared with NRs, with other experimental conditions being equal. Accordingly, an increased intracellular release of CTAB from spherical GNPs resulted in their increased toxicity.

Hutter *et al.* [84] compared the uptake of gold NSphs, NRs, and nano-sea-urchins (also named spiky NPs or nanostars) by phagocytic microglia cells and by nonphagocytic neurons. It was shown that gold NRs were mainly taken up by neurons, whereas spiky NPs were preferentially internalized by microglia cells, thus indicating the selectivity of cells with respect to particle shape. Chu *et al.* [85] found that NPs with sharp shapes, regardless of their surface chemistry, size, or composition, could pierce the membranes of endosomes that carried them into the cells and escape to the cytoplasm, which in turn significantly reduced the cellular excretion rate of the NPs.

Liu *et al.* [86] used dark-field microscopy and ICP–AES to study the uptake of PEGylated gold NSs on silica nanorattles by MCF-7 cells. Uptake efficacy was found to decrease in the particle-size: 84-nm–uptake > 142 > 315-nm–uptake. This conclusion agrees with the previously discussed data, as the actual particle shape was close to spherical.

Navarro *et al.* [87] investigated the uptake of PEGylated gold nanostars and bipyramids into melanoma B16-F10 cells. Dark-field microscopy showed that the biocompatible GNPs were easily internalized and most of them were localized within the cells. Avram *et al.* [88], using the same B16 melanoma cells but different particles (5-nm citrate-coated gold NSphs), presented similar data.

In a recent study *in vitro*, Plascencia-Villa *et al.* [89] showed gold nanostars to be biocompatible with murine macrophages. For precise control of the size, shape, and structure of the nanostars, these were prepared through a seed-mediated route by using 2-[4-(2-hydroxyethyl)-1-piperazinyl]ethanesulfonic acid, a zwitterionic buffering compound. Such NPs were efficiently adsorbed and internalized by the cells, as revealed by advanced field emission SEM and backscattered electron imaging of complete unstained uncoated cells.

Summing up the many experimental data reported in the cited articles (Table 4.1), we conclude that the endocytosis of GNPs is receptor mediated and size dependent [90]. The maximal effectiveness of the intracellular uptake of GNPs is observed for 30–50-nm particles, depending on the cell type. Gold NRs—particularly those with large axial ratios [91]—are taken up into cells much worse but can be eliminated faster from them. In the case of the receptor-independent mechanism, small particles are internalized the best. It should be noted that all GNPs listed in Table 4.1, regardless of their surface functionalization, size, and shape, were negatively charged (zeta potentials ranging from –20 to –70 mV).

4.3 Experimental Methods and Theoretical Models

One of the most important problems in understanding the mechanisms of GNP endocytosis is to achieve a reliable statistical estimation of particles that enter cells or cell compartments [92,93]. For instance, estimates of the amount of particles present in a cell are often based on the qualitative examination of TEM images and on the counting of the number of particles in various cellular structures. However, the statistical accuracy of such estimates is usually questionable if no special methods have been used for sampling [94]. The importance of quantitative judgment is well illustrated by the strong dependence of endocytosis on the modification of the NP surface. From many published data collected in this chapter, it follows that for the same type of cell and similar geometrical parameters of GNPs, replacing the surface modifier may cause the number of particles per cell to change by two or even three orders of magnitude. Such a strong difference in the effectiveness of endocytosis calls for a thorough analysis of the techniques used to generate quantitative data and of their statistical reliability.

It is also important to choose a quantitative parameter for assessing the effectiveness of endocytosis: the number, mass, or surface of the particles. For example, it is obvious that TEM images are better suited for estimating the number of particles in a sample and ICP–AES or ICP–MS for measuring the mass of particles per unit volume of that sample. For an ideal monodisperse ensemble of spherical particles of known size, the number of

TABLE 4.1 Size and Shape Effects of GNPs on Endocytosis

GNP Type and Size	Cell Line	Functionalization	Analysis Methods	Observed Effects	Type of Endocytosis	References
NSphs of 14, 30, 50, 74, and 100 nm NRs of 40 × 14 and 74 × 14 nm	HeLa	Transferrin	ICP–AES, TEM	Endocytosis depends on size; maximal uptake occurs for 50-nm NSphs	Receptor-mediated endocytosis	Chithrani et al. [38]
NSphs of 14, 30, 50, 74, and 100 nm NRs of 40 × 14 and 74 × 14 nm	STO, HeLa, SNB19	Transferrin	TEM, LCM	Endocytosis depends on size (linearly, inversely) and shape (NRs are eliminated faster than NSphs)	Clathrin-dependent receptor-mediated endocytosis	Chithrani and Chan [48]
NSphs of 2, 10, 25, 40, 50, 70, 80, and 90 nm	SK-BR-3	Herceptin	TEM, LCM	Endocytosis depends on size; maximal uptake occurs for 25–50-nm NSphs	Receptor-mediated endocytosis	Jiang et al. [50]
NSphs of 15, 30, 45, and 100 nm and their aggregates of 26, 49, and 98 nm	A549, MDA-MB-435, HeLa	Transferrin	ICP–AES, TEM	Endocytosis depends on cell phenotype, as well as on GNP size	Receptor-mediated and receptor-independent endocytosis	Albanese and Chan [52]
NSphs of 13 and 45 nm	CF-31	–	SEM, TEM	Endocytosis depends on size; maximal uptake occurs for 45-nm NSphs	Clathrin-dependent endocytosis for 45-nm GNPs and receptor-independent endocytosis for 13-nm GNPs	Mironava et al. [55]
NSphs of 3, 6, and 40 nm	J774 A1	–	TEM, UV-vis, ICP–MS	Endocytosis depends on size; maximal uptake occurs for 40-nm NSphs	Receptor-mediated endocytosis	Yen et al. [57]

(Continued)

TABLE 4.1 (CONTINUED) Size and Shape Effects of GNPs on Endocytosis

GNP Type and Size	Cell Line	Functionalization	Analysis Methods	Observed Effects	Type of Endocytosis	References
NSphs of 45, 70, and 110 nm	CL1-0, HeLa	Aptamer (ssDNA)	Dark-field microscopy	Endocytosis decreases in the order 45 > 70 > 100 nm	Receptor-mediated endocytosis	Wang *et al.* [59]
NSphs of 10, 20, 30, and 45 nm	U937	–	TEM, LCM	Endocytosis depends on size; maximal uptake occurs for 30-nm NSphs	Receptor-mediated endocytosis	Riezuichencko *et al.* [60]
NSphs of 5, 10, 20, 30, 40, and 50 nm	PK-1, PK-45, Panc-1	–	AAS	Endocytosis depends on size; maximal uptake occurs for 20-nm NSphs	Receptor-mediated endocytosis	Trono *et al.* [67]
NSphs of 4, 12, and 17 nm	HeLa	L-cysteine	AFM	Endocytosis increases with increasing particle diameter	Raft-dependent endocytosis	Shan *et al.* [69]
NSphs of 10, 25, and 50 nm	NRK	–	TEM, light microscopy, ICP-MS	Endocytosis increases with increasing particle diameter	Receptor-mediated endocytosis	Ma *et al.* [71]
NSphs of 7, 21, and 31 nm	AR42J	–	Dark-field microscopy, TEM, SEM	Endocytosis increases with increasing particle diameter	Clathrin-dependent endocytosis	Sobhan *et al.* [72]
NSphs of 25 and 50 nm	Panc-1, Jurkat	–	ICP-AES	Endocytosis increases with increasing particle diameter	Receptor-mediated endocytosis	Sabuncu *et al.* [73]
NSphs of 2, 6, and 15 nm	MCF-7	Tiopronin	ICP-MS, TEM	Maximal uptake occurs for 2-nm GNPs	Receptor-mediated endocytosis	Huang *et al.* [74]
NSphs of 50 and 100 nm	MCF-7	Tiopronin	ICP-MS, TEM	Maximal uptake occurs for 50-nm GNPs	Receptor-mediated endocytosis	Huo *et al.* [75]

(Continued)

TABLE 4.1 (CONTINUED) Size and Shape Effects of GNPs on Endocytosis

GNP Type and Size	Cell Line	Functionalization	Analysis Methods	Observed Effects	Type of Endocytosis	References
NSphs of 40, 60, and 100 nm	SH-SY5Y	–	TEM, surface-enhanced Raman spectroscopy imaging	Maximal uptake occurs for 60-nm GNPs	Receptor-mediated endocytosis	Huefner et al. [95]
NSphs of 40 nm NRs of 70 × 20 nm Nanocages of 60 nm	MCF-7, hMSC	Anti-HER2 antibody	Dark-field microscopy	Maximal uptake occurs for NSphs	Receptor-mediated endocytosis	Polat et al. [96]
NSphs of 2, 6, 10, and 16 nm	MCF-7	Tiopronin	ICP-MS, TEM	GNPs 2 and 6 nm could enter the nucleus while 10 and 16 nm were found only in the cytoplasm	Receptor-mediated endocytosis	Huo et al. [97]
NSphs of 15, 50, and 100 nm	Caco-2	–	ICP-MS, TEM	Maximal uptake occurs for 100-nm GNPs	Receptor-mediated endocytosis	Yao et al. [98]
NSphs of 20 and 70 nm	PMN	–	TEM	Uptake of the GNP of 70 nm required more time to enter the cells than GNP of 20 nm	Receptor-mediated endocytosis	Noël et al. [99]

particles per unit volume can easily be converted into their mass or surface. For actual samples, however, such conversion requires knowledge of particle distribution by size, shape, composition, *etc.*—information that is usually unavailable.

As it follows from the discussion of the experimental data presented in the two previous sections, the principal experimental methods to study intracellular uptake of GNPs *in vivo* and *in vitro* are TEM, LCM, dark-field microscopy, AFM, UV–vis spectrophotometry, ICP–AES, and ICP–MS. Other state-of-the-art methods have also been used for qualitative and quantitative analysis of cellular GNP uptake, including luminescence microscopy [100], scanning TEM [101], scanning transmission ion microscopy [102], differential interference contrast microscopy [103], photothermal heterodyne imaging [104], Rayleigh light scattering microscopy [105], fluorescence spectral imaging [106], laser desorption/ionization mass spectrometry [107], cell mass spectrometry [108], fluorescence correlation microscopy [109], anti-Stokes Raman scattering microscopy [110], confocal Raman microscopy [111], dynamic surface-enhanced Raman spectroscopy imaging [112], photothermal optical coherent tomography [113], high-resolution x-ray microscopy [114], x-ray computer tomography [115], synchrotron x-ray fluorescence microscopy [116], photoacoustic imaging [117], multiphoton imaging [118], instrumental neutron activation analysis [119], and combinations of different methods [120–122]. Elsaesser *et al.* [92] made a critical analysis of the basic methods used to determine NPs in biological samples, and they compared particle types, biosystems, applications, and the strong and weak points of such methods as light microscopy, TEM, radioactive labeling, mass spectrometry, magnetic NPs, and field-flow fractionation [92].

Despite the large number of the methods listed above, all of them fall into two major classes: destructive and nondestructive. Destructive methods are, in essence, analytical and permit only the total amount of gold in a sample to be determined, whereas nondestructive methods yield the distribution of GNPs over organs, cells, and cellular compartments. In addition, both classes of methods can take advantage of special labels for particle determination or estimate GNPs without the aid of any labels, by relying only on the properties of particles themselves (*e.g.*, by using their high TEM contrast or enhanced plasmonic scattering and absorption of light).

Among the great diversity of destructive techniques, ICP–MS and ICP–AES yield the most reliable data [123]. Both have exceptional sensitivity within the detection range from ppt to ppm and are commonly employed in trace analysis of various substances. Therefore, ICP can be recommended as a method of choice to estimate both the biodistribution of GNPs over organs and the total gold content per cell. An evident, but not the only, weak point of ICP is the destructive character of the analysis, which rules out GNP localization in various cellular compartments or organ sites [124]. Another weak point stems from problems in sample preparation for analysis. Whereas GNPs can be relatively simply digested and measured in suspensions, a more sophisticated sample preparation procedure is needed to determine them within complex biological matrices such as cells, blood, or tissues. It is common practice to use microwave digestion in closed pressure vials with high resistance to volatile compounds such as *aqua regia* or nitric acid. Because the sample preparation is complex, the results generated by ICP–AES/MS depend strongly on instrument calibration, and independent determinations of gold within one set of samples may show a large scatter of data. For this reason, the measured data should always be averaged over several samples, which makes the analysis even more labor-intensive.

The most reliable nondestructive method to estimate GNPs in cells or their compartments is TEM. As noted previously, the localization of GNPs in TEM images presents no problems owing to the strong contrast with respect to the biological components of the sample. Compared to even the most advanced light microscopy methods [125], TEM stands out because it provides an open view at high resolution on the cellular level by visualizing the inner structures in which GNPs are contained [94]. However, making statistically significant estimates of the average number of particles in a cell or its individual organelles is a nontrivial task. Usually, TEM images represent a selection of thin sections obtained after particle treatment of samples and their subsequent fixation and embedding in resins. For versatile quantification of the embedded NPs, several basic principles need to be applied [94]: (1) unbiased selection of specimens by multistage random sampling, (2) unbiased estimation of GNP number and compartment size, and (3) statistical testing of an appropriate null hypothesis. The work of Brandenberger *et al.* [126] is an instructive example of a thorough sampling to obtain reliable estimates of differences in the cellular uptake and trafficking of plain versus PEG-coated GNPs.

It should be stressed that TEM photos are actually a set of two-dimensional (2D) images; therefore, attaining statistically sound results in three dimensions (3D) requires combining TEM with the principles of stereology. Stereology is a set of tools used to estimate morphometric parameters of 3D objects, which is based on measurements made on 2D sections. For revealing the internal structure of cells or tissues at a sufficiently high level of lateral resolution, TEM images are usually obtained for thin or ultrathin (50–100 nm) sections. However, only structures with dimensions smaller than the section thickness will be seen without loss of dimensionality. For example a mitochondrion or endosome will appear on the section plane as one or more section profiles of a certain area. Thus, combining TEM with stereology offers a set of tools that allows unbiased and efficient quantification of GNPs in 3D samples (cells, tissues, and organs) and versatile comparison between experimental groups or between compartments within one group.

The presence of a size dependence of endocytosis has been established for many types of NPs and viruses [22] and has stimulated the development of various theoretical models, beginning with studies by Lipowsky and Döbereiner [127], Gao *et al.* [41], Bao and Bao [128], and Li *et al.* [129]. The attraction between the curved surface of an NP and the cell membrane deforms the membrane and causes the NP to be sucked up inside the membrane region being deformed. The attraction forces between NP and membrane, on the one hand, and the membrane rigidity, on the other, lead to the existence of a minimal radius for endocytosis [130]. Note that the attractive interaction and aggregation of small particles on the cell membrane will decrease the minimal size of particles, whereas nonspecific repulsive interactions will increase the minimal size for effective uptake [22]. The membrane deformation model has been extended to receptor-mediated interactions involved in endocytosis [130–132]. Decuzzi and Ferrari [133] generalized the original formulation by Gao *et al.* [41] to include the contribution of nonspecific interactions and a more realistic expression for the ligand–receptor binding energy.

In the initially developed theoretical models [41,128], the mechanisms of receptor-mediated endocytosis were treated from a kinetic point of view to elucidate the question of "how fast" a single NP can be transported into the cell. Zhang *et al.* [90] treated receptor-mediated endocytosis from a different standpoint by invoking thermodynamic arguments and considering a cell immersed in a solution with ligand-coated GNPs. The NP–cell system reaches a thermodynamic equilibrium at minimal free energy, which determines the number of bound receptors and the distribution of bound GNPs. Accordingly, this theory elucidates "how many" GNPs can be endocytosed in a sufficiently long period. Although the final results explained size-dependent endocytosis qualitatively, the minimal particle diameter was greater than 40 nm, in contrast to the experimental observations.

Yuan *et al.* [131] provided a phase diagram (Figure 4.6) that correlates uptake efficiency with particle size and with the density of the ligands expressed on the particle surface. The most efficient uptake was predicted for high ligand densities and for a particle radius of about 25–30 nm (the right-bottom part of phase II). The ligand-shortage phase corresponds to a critical ligand density, for which the cellular uptake vanishes, because the NP is too short of ligands (phase I in Figure 4.6). On the other hand, at critically large ligand density or particle size, the receptor density can reach an entropic

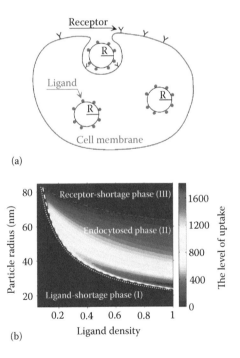

(a)

(b)

FIGURE 4.6 (a) Scheme of wrapped and endocytosed NPs covered by ligands to cell-membrane receptors. (b) Dependence of cellular uptake on the particle radius R of spherical particles and the ligand density on their surface. The dashed and dash-dotted lines are the lower and upper transition boundaries between effective endocytosis phase II and the other two phases, in which uptake is impossible. (Adapted from Yuan, H., Li, J., Bao, G., Zhang, S., *Phys. Rev. Lett.*, 105, 138101, 2010. With permission.)

limit at which adhesion becomes insufficient to overcome the membrane deformation cost. This entropic limit gives rise to a regime noted as "receptor shortage" (phase III in Figure 4.6) [131].

The endocytosis of nonspherical particles, unlike that of spherical particles, has been the subject of few theoretical studies. Decuzzi and Ferrari [134] showed that only circular cylinders and those with moderate aspect ratios can be internalized, whereas particles with small and high aspect ratios cannot be taken up effectively.

Recently [135], Li developed a thermodynamic theoretical model to explain the size and shape effects of cigar-like NPs (Figure 4.7) on endocytosis. In agreement with the previous theoretical considerations [22], it was found that endocytosis needs to overcome a thermodynamic energy barrier and that there exists a minimal NP radius for endocytosis. On the basis of the "diffusion length of receptors" concept [135], Li obtained the following analytical expression for the minimal diameter of cigar-like particles: $R_{min} = [A_0\kappa(8 + a)/2(2 + a)(\mu + \ln \xi_0]^{1/2}$, where κ is the membrane bending modulus, $a = L/R$ is the aspect ratio of the length of the cylindrical part to the radius of the hemispherical parts, A_0 is the cross-sectional area of the receptor, μ is the released chemical energy caused by a ligand-receptor chemical binding, and ξ_0 is the initial density of receptors on the membrane surface. Note that the particle surface area $S = 2\pi R^2 (2 + a)/A_0$ is measured in terms of A_0, which is taken as unit area. Below the minimal size, the particle cannot be internalized through receptor-mediated endocytosis. For spherical particles, the aspect ratio $a = 0$, and the minimal size reduces to

$$R_{min} = [2A_0\kappa/(\mu + \ln \xi_0)]^{1/2}, \tag{4.1}$$

which is close to the previously reported expressions of Gao *et al.* [41] and other authors [22]. As the wrapping of NPs needs continuous binding of ligands to

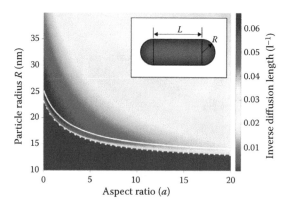

FIGURE 4.7   The endocytosis phase diagram in the space of the particle size R and aspect ratio a, as calculated by Equation 4.2. The solid white line represents the theoretical optimal NP size for endocytosis according to Equation 4.3, and the dotted white line is the threshold NP size for endocytosis according to Equation 4.1. In the blue region, endocytosis is impossible. The color bar represents uptake efficiency in terms of the dimensionless inverse diffusion length, expressed in $A_0^{-1/2}$ units. (Adapted from Li, X., *J. Appl. Phys.*, 111, 024702, 2012. With permission.)

receptors diffused from the vicinity of the NPs, there exists a diffusion length for full wrapping:

$$l_{diff} = \left[s\left(\xi_b^* - \xi_0 \right) \Big/ \pi\left(\xi_f^* - \xi_0 \right) \right]^{1/2},$$ (4.2)

where ξ_b^* and ξ_f^* are the densities of bound and free receptors, respectively, as defined in [135]. Physically, the diffusion length is the distance from the NP to the farthest receptors that diffuse to it. Therefore, a small diffusion length corresponds to fast endocytosis. Accordingly, by minimizing the diffusion length with respect to the surface area, one can obtain the optimal size [135],

$$R_{opt} = \left\{ \frac{\kappa A_0(8+a)}{2(2+a)[W(\xi_0 e^{\mu+1})-1]} \right\}^{1/2},$$ (4.3)

where the Lambert $W(x)$ function is defined as the inverse function $f(W) = W \exp(W)$. Using Equations 4.1–4.3, Li calculated a phase diagram (Figure 4.7) clarifying the relation between the geometry of NPs and the rate of their endocytosis.

In general, all theoretical models confirm that there exists an optimal diameter (40–60 nm) for effective intracellular uptake and of a threshold size (depending on the model's details), below which uptake is impossible. To conclude this section, we point to a recent theoretical study by Dobay *et al.* [136] in which a stochastic approach was applied to simulate NP uptake and intracellular distribution. These simulations were based on experimental TEM data obtained with human bronchial epithelial cells (Beas-2B) and with 4-nm GNPs.

4.4 Effect of GNP Functionalization on Intracellular Uptake

4.4.1 Formation of a GNP–Protein Corona in a Biological Environment

Functionalization of GNPs with surface molecules is aimed at fabricating multifunctional NP bioconjugates possessing various modalities, such as active biosensing, enhanced imaging contrast, drug delivery, and tumor targeting. The type of the bond between GNPs and functional molecules is, in particular, responsible for the release of the target substance inside a cell [137]. Thus, the presence on the GNP surface of polymers and biological macromolecules that differ in chemical makeup and properties leads to differences in the cellular uptake and exocytosis of functionalized particles [138].

Immediately on contact of GNPs with blood, lymph, gastric juice, or any other biological liquid *in vivo*, as well as on contact with a culture medium *in vitro*, the interaction between GNPs and solvable proteins and other biomolecules results in the formation of a protein "corona" [139–144]. Similarly to the concept of functionalized GNPs, the concept of a GNP–protein corona is important in tuning the surface physicochemical properties of GNPs, such as charge, hydrodynamic size, and colloidal stability. In fact, it is the GNP–protein corona that forms the first nano–bio interface and determines the first interactions of GNPs with/or within living cells. This is because the GNP–protein corona is a dynamic biopolymer layer that can strongly affect cellular uptake owing to modification of the particle properties (the overall size, charge, *etc.*). Furthermore, owing to the GNP–protein interaction, the adsorbed proteins can change their conformations. It is a kind of change that exposes new functional groups and alters the protein functions, avidity effects, and so on (Figure 4.8) [140,145]. In general, the corona structure and properties depend on the prior history of particles, as the GNP surface may already be covered by various ligands coming from fabrication and stabilization procedures [140].

Although as much as 69 plasma proteins can bind to the GNP surface [146,147], only some of them, such as albumin, apolipoprotein, immunoglobulin, complement, and fibrinogen, are the most abundantly bound proteins forming the GNP–protein corona. After intravenous injection, the coating of GNPs by these proteins largely determines the particles' fate in the body—biodistribution over organs, tissues, and cells, the efficiency of cellular uptake and clearance, and so on. Oh and Park demonstrate that the native surface chemistries of GNPs and their subsequent opsonization by serum proteins play critical roles in the exocytosis patterns in cells [148].

Thus, the corona is a complex mixture of proteins adsorbed on the surface of NPs. These proteins play an important role in determining what surface is actually presented to cells that take the nanostructure up and activate signaling pathways. The protein corona is composed of an inner layer of selected proteins with a lifetime of several hours in slow exchange with the environment (the hard corona) and an outer layer of weakly bound proteins that are characterized by a faster exchange rate with the free proteins (the soft corona). The biological impact of protein-coated NPs is mainly related to the hard corona and their specificity and suitable orientation for a particular receptor. Although low-affinity, high-abundance proteins may initially adsorb to the surface of NPs, proteins with lower abundance but higher affinity quickly replace them [149].

The corona formation process depends mainly on GNP size, charge, preliminary surface functionalization, particle curvature, and surface roughness. In turn, these factors determine the main physicochemical contributions to corona formation, namely, hydrodynamic, electrodynamic, electrostatic, solvent, steric, and polymer–bridging interactions [140]. The dynamic composition of the GNP–protein corona depends on association/dissociation constants, the concurrent binding processes, steric hindrance (which prevents binding), and the composition of the biological liquid surrounding GNPs [150]. Accordingly, Arvizo *et al.* [151] pointed out that the composition and properties of the GNP–protein corona should be considered in designing new GNP-based therapeutic targets.

Recently, Cheng *et al.* [152] showed that protein corona significantly decreases the internalization of GNPs in a particle size- and cell type-dependent manner. Protein corona exhibits much more significant inhibition on the uptake of large-sized GNPs by phagocytic cell than that of small-sized GNPs by nonphagocytic cell. The presence of

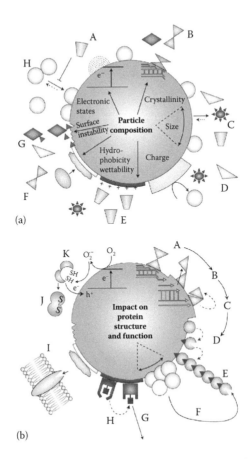

FIGURE 4.8 Effects of a GNP–protein corona. (a) The initial material characteristics of as-prepared GNPs contribute to the formation of the corona in a biological environment. The corona can change when particles move from one biological compartment to another. Symbol designations: A—competitive binding interactions depend on the protein concentration and the body fluid composition; B—binding interactions release surface free energy, leading to surface reconstruction; C—the available surface area, surface coverage, and angle of curvature determine the adsorption profiles; D—steric hindrance prevents binding; E—charge (*e.g.*, cationic binding); F—hydrophilic or hydrophobic interactions; G—protein binding accelerates the dissolution of some materials; H—characteristic protein on/off rates depend on the material type and the protein characteristics. (b) Possible changes in corona structure and functions owing to interaction with the GNP surface. The colored symbols represent various types of proteins, including charged, lipophilic, and conformationally flexible proteins; catalytic enzymes with sensitive thiol groups; and proteins that crowd together or interact to form fibrils. Symbol designations: A—surface reconstruction; B—release of surface free energy; C—protein conformation, functional changes; D—protein fibrillation; E—amyloid fiber; F—protein crowding, layering, nucleation; G—immune recognition; H—exposure to cryptic epitopes; I—surface opsonization/liganding allows interaction with additional nano–bio interfaces; J—protein conformational changes cause loss of enzyme activity; K—electron–hole pairs lead to oxidative damage. (Adapted from Nel, A.E., Mädler, L., Velegol, D., Xia, T., Hoek, E.M.V., Somasundaran, P., Klaessig, F., Castranova, V., Thompson, M., *Nat. Mater.*, 8, 543–557, 2009. With permission.)

a protein corona has been shown to have a protective effect, and the hemolytic potential of GNPs featuring both hydrophobic and hydrophilic surface functionalization was reduced [153].

4.4.2 Intracellular Uptake of PEGylated GNPs

In current nanotechnologies, extensive use has been made of GNPs functionalized with PEGs having different polymer-chain lengths, density, and diverse functional groups [154,155]. This is due to a number of factors: PEGylated GNPs are worse "recognized" by cells of the immune system because of the lower opsonization by serum proteins (stealth properties) and are much more tolerant of salt aggregation than citrate- or CTAB-coated GNPs, which makes them able to circulate in the bloodstream for a longer time [156–158]. *In vivo*, PEGylated NPs preferentially accumulate in tumor tissue owing to the increased permeability of the tumor vessels and are retained in it owing to reduced lymph outflow [159]. PEG hydrophilizes the surface of NPs and can serve as a ligand for the attachment of drugs or genes to deliver to target cells [160,161]. Finally, PEGylation of GNPs is affected on their cellular uptake and cytotoxicity [162,163].

Shenoy *et al.* [164] used 10-nm GNPs coated with thiolated PEG1500 with attached coumarin, a fluorescent dye. This is an example of a "hetero-bifunctional" PEG, which has a thiogroup for binding to GNPs at one end and a mobile spacer for binding to a dye or to other ligands at the other. With the help of this structure, the authors investigated by LCM the uptake of GNPs in MDA-MB-231 cells. They showed that GNPs modified with coumarin–PEG thiol (PEG-SH) are taken up effectively by cells within 1 h of incubation and localize mostly in the perinuclear space. Endocytosis proper begins as early as at 5 min, and at 30 min, GNPs appear in endosomes. Despite the high rate of endocytosis, no GNPs could be found in the cell nuclei even after a day's incubation. It was concluded that with the aid of PEG-coated GNPs, it is possible to perform intracellular delivery of various ligands (dyes, antibodies, drugs, *etc.*).

Bergen *et al.* [165] explored the liver cell uptake *in vivo* of 50-, 80-, 100-, and 150-nm GNPs modified with PEG5000-SH and galactose–PEG5000-SH. The cellular concentration of gold was determined by neutron activation. The time of GNP circulation in blood and the amount of GNPs in liver cells were found to be substantially affected by the size and functionalization (the presence of galactose in the ligand) of the particles (Figures 4.9 and 4.10). The increased accumulation of GNPs coated with galactose–PEG5000-SH was attributed to receptor-mediated endocytosis of such GNPs by the Kupffer cells of the liver through galactose receptors or by hepatocytes through asialoglycoprotein receptors on their surface.

Cho *et al.* [166] made a TEM study of the toxicity and pharmacokinetics of 13-nm GNPs modified with PEG5000. In complete agreement with the data obtained by Sadauskas *et al.* [5,6] for nonfunctionalized GNPs, the PEGylated particles accumulated mostly in Kupffer cells of the liver and in macrophages of the spleen. In addition, the results of TEM demonstrated that inside a cell, GNPs are found in endosomes and lysosomes and are not associated with the nucleus, mitochondria, or the Golgi apparatus (Figure 4.11). Similar data were reported by Cho *et al.* for 4- and 100-nm PEGylated GNPs [167]. Liu *et al.* [168] also showed that 5-nm GNPs modified with PEG5000 were

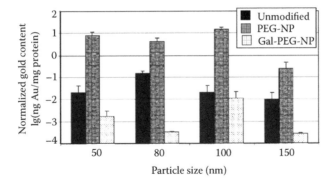

FIGURE 4.9 The normalized gold content in blood, in terms of lg(ng gold/mg protein from four mice/group), at 2 h postinjection. (Adapted from Dykman, L.A., Khlebtsov, N.G., *Chem. Rev.*, 114, 1258–1288, 2014. With permission.)

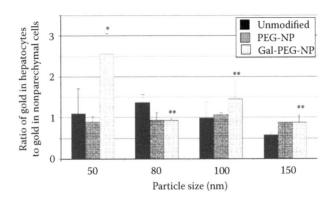

FIGURE 4.10 Liver-cell biodistribution of GNPs at 2 h postinjection. Values presented are the mean ratio of gold in hepatocytes to gold in nonparechymal cells from four mice/group. *, significant statistical difference compared to the PEG–GNP control ($p < 0.05$). **, no significant statistical difference compared to the PEG–GNP control ($p > 0.05$). (Adapted from Dykman, L.A., Khlebtsov, N.G., *Chem. Rev.*, 114, 1258–1288, 2014. With permission.)

localized in the cytoplasm of CT26 cells and were not found in the nuclei. Zhang *et al.* [169] reported that the most effective uptake *in vivo* into cells of the liver and spleen was demonstrated by 20-nm GNPs modified with PEG5000 through thioctic acid.

Different data were gathered by Gu *et al.* [170], who found, by TEM, LCM, and ICP-MS, that 3.7-nm GNPs successively coated with mercaptopropionic acid and PEG2000 can penetrate the nuclei of HeLa cells (Figure 4.12) [171]. Thus, as Gu *et al.* [170] conclude, PEGylated GNPs can be used for the delivery of drugs or genetic material directly to the nuclei of cells.

Brandenberger *et al.* [172] made a comparative study of the uptake of 15-nm PEGylated (PEG5000) and plain GNPs by A549 cells. Quantitative analysis of GNPs in the cells was done by TEM. Both types of particles were found to be present in the cytoplasm as part of variously sized vesicles: <150 nm (primary endosomes), 150–1000 nm (endosomes),

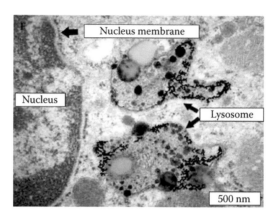

FIGURE 4.11 Thin-section TEM image of mouse spleen macrophages at 7 days after intravenous injection of PEG-modified GNPs. Magnification, 50,000×. (Adapted from Cho, W.-S., Cho, M., Jeong, J., Choi, M., Cho, H.-Y., Han, B.S., Kim, S.H., Kim, H.O., Lim, Y.T., Chung, B.H., Jeong, J., *Toxicol. Appl. Pharmacol.*, 236, 16–24, 2009. With permission.)

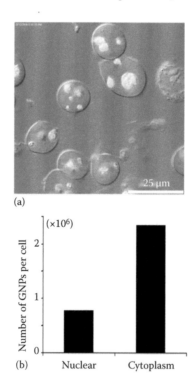

FIGURE 4.12 Confocal image of isolated nuclei of HeLa cells incubated with GNPs–PEG–FITC (a). The GNP concentration in the nuclei and cytoplasm of HeLa cells (data by ICP–MS) (b). (Adapted from Gu, Y.-J., Cheng, J., Lin, C.-C., Lam, Y.W., Cheng, S.H., Wong, W.-T., *Toxicol. Appl. Pharmacol.*, 237, 196–204, 2009. With permission.)

and >1000 nm (phagosomes and lysosomes). In the other cell compartments—the nucleus, mitochondria, endoplasmic reticulum, and Golgi apparatus—no GNPs were observed. After 4 h of incubation, the per-cell number of plain GNPs was 3500 and that of PEGylated GNPs was 1000. When the cells were treated with methyl-β-cyclodextrin, an inhibitor of caveolin- and clathrin-dependent endocytosis, the per-cell number of GNPs decreased particularly strongly for PEGylated GNPs—by 95% (for plain GNPs, the decrease was 50%). It was concluded that the uptake of non-PEGylated GNPs is effected both through macropinocytosis and through caveolin- and clathrin-dependent endocytosis and that the uptake of PEGylated GNPs occurs only *via* receptor-mediated endocytosis.

Lund *et al.* [173] presented data on the tumor cell (HCT-116, HT 29, LS154T, and SW640) uptake of two types of 2-nm GNPs: (1) those coated with PEG and (2) those coated with 50% PEG/50% glucose. Particles of the latter type were found to be taken up into cells much more effectively. The authors inferred that the major role in GNP internalization is played by the structural organization of ligands on the NP surface, rather than by the charge of functional groups. In addition, the authors suggested that apart from receptor-mediated endocytosis, an important role in the uptake of small GNPs is played by passive transport of GNPs through the pores of the plasma membrane.

An in-depth study of the uptake of PEGylated NSs by macrophages (RAW264.7) *in vitro* was described by Kah *et al.* [174]. They synthesized gold NSs on SiO_2 cores, with diameters of 79, 100, 140, 162, 181, and 196 nm and a gold shell thickness of ~23 nm. For functionalization, they used different concentrations of methoxy-PEG-SH with different molecular masses, including 750, 2,000, 5,000, 10,000, and 20,000 Da. For visualization of gold NSs in the macrophages, dark-field microscopy and LCM were employed. The results showed that the intracellular uptake of NSs by macrophages was affected by the amount of PEG added. The highest uptake was exhibited by non-PEGylated NSs (80%). The percentage of phagocytosed NSs decreased to 40% with 0.05 mmol/L PEG, to 30% with 0.25 mmol/L PEG, to 20% with 0.5 mmol/L PEG, and to ~10% with 2.5–20 mmol/L PEG. It was also found that the phagocytosis of PEGylated NSs depends on the molecular mass of adsorbed PEG (2.5 mmol/L). The uptake of NSs (140/30) in the macrophages, in percentage of the amount of administered particles (2×10^{13} particles/ml), was 10.2% for PEG750, 0.6% for PEG2000, 0.9% for PEG5000, 2.1% for PEG10000, and 48.2% for PEG20000. Furthermore, the process of phagocytosis depended on the NP size: the uptake of NS–PEG2000 was 8.1% for NSs of 79/23 nm, 1.7% for NSs of 100/23 nm, 0.6% for NSs of 140/23 nm, 0.9% for NSs of 162/23 nm, 1.4% for NSs of 181/23 nm, and 1.2% for NSs of 196/23 nm. Thus, the best ability to be taken up was displayed by NSs with a core diameter of 79 nm and NSs functionalized with PEG20000. The data from this study are summarized in Table 4.2.

Unlike gold NSphs and NSs, gold NRs obtained by the seed-mediated method are coated with the cationic surfactant CTAB. CTAB-coated gold NRs actively and irreversibly penetrate into mammalian cells, and they are markedly cytotoxic. The cytotoxic effect is often reduced by replacing CTAB with inert polymers, specifically with PEG [156]. Such replacement not only sharply decreases cytotoxicity but also appreciably lowers, according to several authors [84,175–178], the level of nonspecific uptake of gold NRs in various cells.

TABLE 4.2 Effect of the Concentration and Molecular Mass of PEG and the Diameter of NSs on the Phagocytosis of NSs

PEG Concentration (mmol/L)	Uptake in Macrophages (%)[a]	Size of SiO₂/ Au NSs (nm)	Uptake in Macrophages (%)[b]	Molecular Mass of PEG (Da)	Uptake in Macrophages (%)[c]
0	80	79/23	8.1	750	10.2
0.05	40	100/23	1.7	2,000	0.6
0.25	30	140/23	0.6	5,000	0.9
0.5	20	162/23	0.9	10,000	2.1
2.5	10	181/23	1.4	20,000	48.2
5.0	10	196/23	1.2		
10	10				
20	10				

Source: Kah, J.C., Wong, K.Y., Neoh, K.G., Song, J.H., Fu, J.W., Mhaisalkar, S., Olivo, M., Sheppard, C.J., *J. Drug Target*, 17, 181–193, 2009.

[a] NSs of 181/23 nm, PEG2000.
[b] PEG2000, PEG concentration of 2.5 mmol/L.
[c] NSs of 140/23 nm, PEG concentration of 2.5 mmol/L.

Arnida *et al.* [179,180] compared the intracellular uptake of PEGylated and plain gold NSphs and PEGylated gold NRs by PC-3 and RAW264.7 cells. They used NSphs with diameters of 30, 50, and 90 nm and NRs with dimensions of 35 × 40 and 45 × 10 nm. TEM and ICP–MS analysis showed that 50-nm non-PEGylated NSphs were taken up best. Modification of the surface with PEG led to a considerable decline in uptake, as also did the presence of serum proteins. PEGylated NRs (especially the shorter ones) were taken up by the cells much worse than NSphs.

Somewhat different data were recorded by Cho *et al.* [181], who compared the uptake by SK-BR-3 cells of 17-nm citrate-capped and PEGylated NSphs and CTAB- and PEG5000-coated NRs (50 × 20 nm). They showed by UV–vis spectrophotometry that the levels of uptake of citrate-capped NSphs and NR–CTAB were approximately the same, constituting 8×10^3 GNPs/cell. After PEGylation, the uptake level decreased approximately fourfold (2×10^3 GNPs/cell) for NSphs and twofold (4×10^3 GNPs/cell) for NRs. Of interest is the fact that when the cells were incubated with a mixture of NSphs and NRs, the latter were found to be present in large amounts in the intracellular space regardless of the modification of the surface. Hence, as the authors concluded, it is surface functionalization of NPs, rather than their shape, that has a greater influence on the intracellular uptake of GNPs. Similar data were recorded by Pietuch *et al.* [182]. In contrast to this conclusion, Puvanakrishnan *et al.* [183] showed that PEGylated NRs penetrated tumor cells 12-fold more actively than did PEGylated NSs and that the accumulation of these particles in the liver cells was approximately of the same level.

Although PEG functionalization ensures stealth properties of GNPs in a biological environment, this coating may be destroyed by some concurrent binding. For example, Larson *et al.* [184] showed that the physiological concentration of cysteine and cystine can displace methoxy-PEG-thiol molecules from the GNP surface. This displacement is accompanied by plasma protein coatings and by enlargement of the protein corona; as a result, the modified GNPs demonstrate enhanced cellular uptake. To avoid such loss of stealth properties and to greatly reduce the cellular uptake, the authors incorporated a small hydrophobic shield (alkyl linker) in between the GNP core and the hydrophilic PEG shell [184].

Thus, the existing data point to a combined effect of the shape and functionalization of the particles used in comparative experiments to study the effectiveness of intracellular uptake. Therefore, further work is needed to identify the effects of shape in pure form, without interference from the effects of surface molecules.

4.4.3 Uptake of GNPs Functionalized with Synthetic Polymers and Proteins

Apart from PEG, other polymer molecules are also frequently employed to reduce the cytotoxicity of gold NRs. This changes the level of uptake of such GNPs by mammalian cells. More specifically, Takahashi *et al.* [185] modified the surface of NRs (65 × 11 nm) with phosphatidylcholine and polyethyleneimine (PEI). It is known that conjugating GNPs to PEI improves the effectiveness of transfection of plasmid DNA into cells [186]. According to the data of Takahashi *et al.* [185], PEI-modified GNRs were taken up by HeLa cells much better than were NR–phosphatidylcholine conjugates. Thus, for nonspherical particle shape, the effect of surface functionalization with PEI is evident. Pyshnaya *et al.* [187] showed that PEI-GNRs and PEI-NSphs demonstrated fast and active penetration into cells by caveolin-dependent and lipid raft-mediated endocytosis and accumulated in endosomes and lysosomes, while bovine serum albumin (BSA)-modified GNPs showed prolonged flotation and a significant delay in cell penetration.

Probably the most thorough analysis of the effect of functionalization with various polymers on the intracellular uptake of NRs was reported by Chan's group [188]. To modify the surface of NRs (40 × 18 nm), they used poly(4-styrenesulfonic acid) (PSS1), poly(diallyldimethylammonium chloride) (PDDAC), poly(allylamine hydrochloride) (PAH), and PSS–PDDAC–PSS (PSS2). NRs coated with CTAB (the initial ones), PDDAC, and PAH were charged positively, and those capped with PSS1 and PSS2 were charged negatively. Uptake was studied with HeLA cells by using TEM and ICP–AES. NRs (150 µM) were added to cells in DMEM containing fetal bovine serum and also to cells in DMEM serum-free medium. It was found that in serum-free medium, PSS1- and PAH-coated NRs were taken up approximately twofold worse than CTAB-, PDDAC-, and PSS2-coated NRs were (~50,000 GNPs/cell, as compared with ~10,0000). The membrane of HeLa cells is charged negatively, and a possible explanation for the higher uptake of positively charged NRs could be the electrostatic interactions between the cell and the NR surface. In addition, the chemical nature of the coating polymer is probably important for the uptake of NRs. Specifically, NRs coated with quaternary amines (CTAB and PDDAC) were taken up by the cells much more effectively than were NRs coated with a primary amine (PAH). Experiments with cells in serum medium yielded somewhat different results. The highest uptake was demonstrated by NRs coated with PDDAC (~150,000 GNPs/cell). The average value yielded by CTAB-, PAH-, and PSS2-functionalized NRs was 25,000 GNPs/cell. PSS1-coated NRs displayed a fairly low uptake level—<10,000 GNPs/cell. This effect was explained by differences in the adsorption of serum proteins onto the functionalized NRs and by the enhanced receptor-mediated endocytosis of GNPs covered with serum proteins [38].

Slightly different results for the same type of particle were produced by Alkilany *et al.* [189]. They used TEM and ICP–MS to investigate the uptake of gold NRs (λ_{max} = 840 nm, aspect ratio of 4.1) by HT-29 cells. The particles used were positively charged NRs, coated with CTAB and PAH, and negatively charged NRs, coated with polyacrylic acid (PAA). The best cellular uptake (in serum medium) was shown by PAH-coated NRs (final concentration of 0.2 nM; ~2500 GNPs/cell); much worse uptake, by PAA-coated NRs (~300 GNPs/cell); and minimal uptake, by CTAB-coated NRs (~50 GNPs/cell). This is in disagreement with the data of Hauck *et al.* [188] for CTAB-coated NRs and for the average values for the taken-up particles. It should be noted that after incubation of NRs in a medium containing serum proteins, the charge of all NRs leveled off, becoming approximately –20 mV. It is quite possible that the differences between the data of Hauk *et al.* [188] and those of Alkilany *et al.* [189] have to do with the different types of cells, used amounts of NRs added, geometrical parameters of the NRs, and incubation times (6 and 24 h, respectively). Also in conflict with the data of Hauk *et al.* is a closely related study by Parab *et al.* [190], who used TEM to compare the uptake of CTAB- and PSS-coated NRs (λ_{max} = 700 nm, aspect ratio of 2–3) by S-G and TW 2.6 cells. PSS-coated NRs were found to be taken up into the cell cytoplasm better than their CTAB-coated counterparts.

Fan *et al.* [191] reported that the uptake of PEG-coated NRs (65 × 12 nm) in MEF-1 and MRC-5 cells was much less effective than it was when the particles were functionalized with PSS and PDDAC. Goh *et al.* [192] demonstrated that gold NRs encapsulated in Pluronic triblock copolymers more effectively penetrated OSCC cells and were much less toxic than CTAB-coated NRs.

An in-depth study of the effect of NR geometry and surface functionalization on intracellular uptake was made by Qiu *et al.* [193]. They used NRs with dimensions of 33 × 30 (1), 40 × 21 (2), 50 × 17 (3), and 55 × 14 (4) nm, which were coated with CTAB (the initial NRs, 1–4), PSS (1 and 4), and PDDAC (1 and 4). The uptake was examined with MCF-7 cells grown on Roswell Park Memorial Institute (RPMI) medium containing fetal bovine serum, and the methods used were TEM and ICP–MS. The results showed that out of the initial NRs, the best uptake was observed for GNPs with smaller aspect ratios (almost nanospherical), whereas the NRs with the maximal aspect ratio exhibited the worst uptake. However, the effect of the geometrical parameters on uptake effectiveness was less strong—9 × 10³ GNPs/cell for NR-1–CTAB and 5 × 10³ for NR-4–CTAB. A much stronger effect was produced by surface modification: 9 × 10³ GNPs/cell for NR-1–CTAB, 4 × 10³ GNPs/cell for NR-1–PSS, 1.2 × 10⁴ GNPs/cell for NR-1–PDDAC, 5 × 10³ GNPs/cell for NR-4–CTAB, 3 × 10³ GNPs/cell for NR-4–PSS, and 9 × 10⁴ GNPs/cell for NR-4–PDDAC (Figure 4.13).

As indicated by TEM, some of the CTAB-coated NRs inside the cells were localized in lysosomes and some were found in mitochondria (Figure 4.14a and b). It follows from the images (c and d) that the CTAB molecules in cells can cause swelling and stimulate the production of mitochondria, thus indicating a cellular inflammatory response.

On the basis of the results of their work, the authors reasoned that it is surface functionalization, rather than geometrical parameters, that has a major effect on the cellular

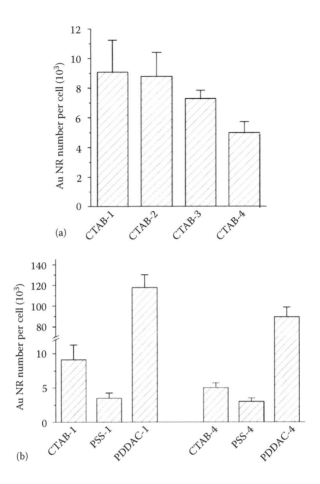

FIGURE 4.13 Effect of shape and functionalization on the intracellular uptake of NRs. The cellular amounts of CTAB-coated (a) and CTAB-, PSS-, and PDDAC-coated (b) NRs with four different aspect ratios after cell incubation with 70 pM NRs for 24 h. (Adapted from Qiu, Y., Liu, Y., Wang, L., Xu, L., Bai, R., Ji, Y., Wu, X., Zhao, Y., Li, Y., Chen, C., *Biomaterials*, 31, 7606–7619, 2010. With permission.)

uptake of NRs. They also proposed a scheme for the mechanism of intracellular uptake and cytotoxicity of NRs (Figure 4.15). In the authors' opinion, NRs are rapidly covered by serum proteins, which mediate endocytosis. Once taken up, the NR–protein complex is transported by vesicles to lysosomes, where the proteins are digested and CTAB is released. The NR aggregates are delivered to mitochondria, where they accumulate. The released CTAB molecules damage the mitochondria and induce cell apoptosis and death. If NRs are coated with an inert polymer, the lysosomal enzymes are unable to digest it; therefore, CTAB is not released and, consequently, the cytotoxic effect of NRs is relieved. Qiu *et al.* [193] believe that PDDAC-complexed NRs with an aspect ratio of about 4 make the optimal nanocomposites for medical applications.

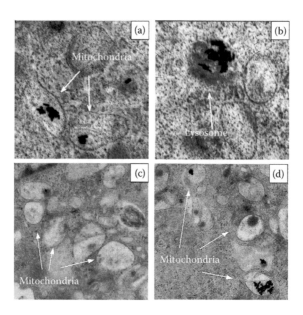

FIGURE 4.14 TEM images of cells incubated with CTAB-coated NRs. Some NRs are in mitochondria (a), whereas others are found in lysosomes (b). The CTAB molecules in cells can cause swelling and stimulate the production of mitochondria, which is an ordinary pathological indicator for cellular inflammatory responses (c, d). (Adapted from Qiu, Y., Liu, Y., Wang, L., Xu, L., Bai, R., Ji, Y., Wu, X., Zhao, Y., Li, Y., Chen, C., *Biomaterials*, 31, 7606–7619, 2010. With permission.)

In a subsequent report from the same group [194], the study of uptake of gold NRs into various types of cells was continued by using A549 carcinoma cells, 16HBE normal bronchial epithelial cells, and mesenchymal stem cells. The intracellular uptake of 55 × 13-nm NRs functionalized with fetal bovine serum was examined by ICP–MS and TEM. It was found that uptake was more effective in the carcinoma and stem cells and was much worse in the normal cells. Exocytosis was more active in the stem cells. To investigate the mechanism of endocytosis, the authors used a variety of inhibitors. Sodium azide, 2-deoxy-D-glucose, and low temperature sharply decreased the endocytosis (to as low as 80%) and noticeably decreased ATP synthesis—that is, the endocytosis was energy demanding. The use of chlorpromazine and a hypertonic solution of sucrose as inhibitors reduced NR uptake by 75 and 89%, respectively. It follows that the principal mechanism of endocytosis was clathrin dependent. If nystatin, methyl-β-cyclodextrin, and dynamin were served as inhibitors, the endocytosis decreased by 54%, 48%, and 42%, respectively, attesting that a raft-dependent endocytosis route is also possible. TEM data indicated that the intracellular localization of NRs was different in the three types of cells: only in the carcinoma cells were NRs present in mitochondria as well as in lysosomes. The authors associated the toxic effect of NRs on the cancer cells with mitochondrial damage by CTAB, which is released in the lysosomes of cancerous cells. Thus, cell viability is affected not only by endocytosis routes but also (and primarily) by intracellular localization.

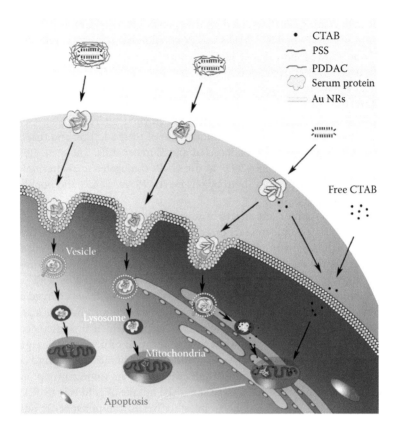

FIGURE 4.15 Scheme for the pathway of intracellular uptake and the mechanism of cytotoxicity of NRs. (Adapted from Qiu, Y., Liu, Y., Wang, L., Xu, L., Bai, R., Ji, Y., Wu, X., Zhao, Y., Li, Y., Chen, C., *Biomaterials*, 31, 7606–7619, 2010. With permission.)

Very intriguing data on the influence of surface modification of NRs on their intracellular uptake were presented recently by Vigderman *et al.* [195]. Using TEM, SEM, and ICP–AES, they demonstrated that replacing CTAB on the surface of NRs (42 × 10 nm) by its analog, (16-mercaptohexadecyl)trimethylammonium bromide, increases drastically (approximately 200-fold) the effectiveness of NR uptake by MCF-7 cells. In addition, these NRs were very stable and caused no toxicity.

Li *et al.* [196] examined the intracellular uptake of transferrin-functionalized and PEG-capped GNPs (25 nm) in tumor (Hs578T) and normal (3T3) cells. Dark-field microscopy and LCM showed that transferrin-coated GNPs were taken up by tumor cells *via* transferrin-receptor endocytosis sixfold more effectively than were PEGylated GNPs. Blocking of transferrin receptors by cell pretreatment with native transferrin also brought about a sixfold decrease in uptake effectiveness. In addition, the uptake of transferrin-capped GNPs by normal cells was 75% lower than that by tumor cells owing to the small number of transferrin receptors on the surface of normal cells.

The effect of coating GNPs with a thermoresponsive polyacrylamide polymer on the uptake of 18-nm GNPs by MCF-7 cells was explored by Salmaso *et al.* [197]. They used ICP–AES and TEM to show that at 40°C, polymer-capped GNPs were taken up 80-fold more effectively (12,000 particles/cell) than they were at 34°C (140 particles/cell). The intracellular uptake of uncapped GNPs was 6000 particles/cell and did not depend on temperature.

Dragoni *et al.* [198] reported that 5-nm polyvinylpyrrolidone-coated GNPs were found in endosomes of hepatocytes within 30 min of their addition and that after 2 h, they appeared in endosomes of endothelial and Kuppfer cells. Liu *et al.* [199] found that as compared with GNP–PEG2000, 16-nm GNPs conjugated to zwitterionic surfactants were taken up much worse both by phagocytic (RAW 264.7) and nonphagocytic (HUVEC and HepG2) cells.

One of the most important effects of functionalization is the change of particle charge, because electrostatic interactions can control cellular uptake of particles much stronger than can, *e.g.*, hydrophobic or Van der Waals interactions. The effect of surface-polymer charge on the intracellular uptake of spherical GNPs was investigated by several authors [200–207]. It was demonstrated that GNPs coated with positively charged polymers were taken up best, as compared with negatively charged and neutral NSphs. These results are in harmony with the data obtained by Hauck *et al.* [188] and by Zhu *et al.* [208] for GNRs. Cho *et al.* [200] explained the enhanced uptake of positively charged GNPs by the enhanced adsorption of positively charged GNPs onto the negatively charged cell surface, facilitating the higher uptake into cells. Yet, Liang *et al.* [209] successfully demonstrated the use of an anionic group, meso-2,3-dimercaptosuccinic acid (DMSA), to enhance the uptake of gold NSs by RAW 265.7, A543, and BEL-7402 cells. The reason possibly lies in the nonspecific adsorption of serum protein on the DMSA-modified gold NSs, which enhanced the cellular uptake. Arvizo *et al.* [210] demonstrated that the surface charge of functionalized GNPs plays a prominent role in modulating the membrane potential of different malignant and normal cells and in subsequent intracellular uptake of GNPs. In particular, positively charged GNPs depolarized the membrane to the greatest extent, whereas NPs of other charges had negligible effect. Such membrane potential perturbations resulted in an increased intracellular Ca^{2+} concentration, which in turn inhibited the proliferation of normal cells, whereas malignant cells remained unaffected. Furthermore, as shown by Lin *et al.* [211], the charge of coating polymers influences the mechanism of endocytosis, in particular the involvement of tubulin microtubules in this process.

Researchers who study the influence of various factors on the effectiveness of cellular GNP uptake should take into consideration the components of the cell growth medium they are using. Just like in the case of NRs, the composition of the cell growth medium may be profound for cellular uptake of gold NSphs. For instance, Maiorano *et al.* [212], using ICP–AES, reported that the uptake of 15-nm GNPs by HeLa cells grown on RPMI medium was approximately threefold more effective than the uptake by the same cells grown on DMEM medium. The authors' explanation for this finding is that in RPMI medium, the diameter of the protein corona formed on GNPs is twofold greater than that observed in DMEM medium.

An interesting research direction is the use of hybrid particles, combining GNPs with components of liposomes, which are traditionally considered promising carriers for intracellular delivery. Pal *et al.* [213] and Boca *et al.* [214] found that GNPs incorporated in liposomes or capped with chitosan were taken up by CHO cells better than native GNPs and also were less toxic. The data of Chithrani *et al.* [215] indicate that 1.4-nm GNPs incorporated in liposomes were taken up in HeLa cells 1000-fold more effectively than were native GNPs. In addition, the biocompatibility of GNPs can be improved by the use of, *e.g.*, gellan gum [216], collagen [217], or mesoporous silica [218]. According to LCM data, 14-nm GNPs coated with this polysaccharide entered LN220 tumor cells much more effectively than they entered NIH-3T3 normal fibroblasts. It was also proposed that biocompatibility can be increased with lysozyme-conjugated GNPs, and such conjugates were found to be effective at entering the cells and nuclei of an NIH-3T3 culture by clathrin-dependent endocytosis [219]. Hao *et al.* [220] showed that phospholipid-coated GNPs were taken up by MCF-7 cells more effectively than were PEGylated GNPs. Wang and Petersen [221,222] also showed that lipid-coated GNPs were readily taken up by A549 cells. Similar data were published by Yang *et al.* [223] for high-density-lipoprotein–coated GNPs and lymphoma cells. Amarnath *et al.* [224] demonstrated that glutathione- or lipoic-acid-coated GNPs entered HBL-100 tumor cells much more effectively than nonconjugated GNPs did.

4.4.4 Uptake of GNPs Conjugated with Antibodies

The use of GNP–antibody conjugates was first suggested by Faulk and Taylor [225] for TEM identification of bacterial antigens. Currently, such conjugates are widely used in immunochemistry. In recent years, several publications have appeared describing the penetration of GNP–antibody conjugates into cancer cells, and in those studies, antibodies were employed mainly as target molecules.

Hu *et al.* [226] compared the uptake by Panc-1 and MiaPaCa cells of gold NRs (40 × 14 nm) capped with PDDAC, transferrin, and antibodies to the cell surface antigens claudin 4 and mesothelin. Dark-field microscopy and TEM indicated that bioconjugates of NRs with transferrin and with the antibodies were taken up better than NR–PDDAC, a finding that corroborates the specific receptor-mediated nature of GNP endocytosis.

Rejiya *et al.* [227] reported similar results. They examined, by luminescence microscopy and ICP–AES, the intracellular uptake by A431 cells of NRs (42 × 10 nm) modified with antibodies to the EGFR. The amount of nonconjugated (CTAB-capped) NRs that had entered the cellular space within 3 h proved to be twofold smaller than that of NR–antibody conjugates. Thus, modification of the NR surface with antibodies specific for cell-surface antigens improves the effectiveness of intracellular GNP uptake, which may prove to be useful both for targeted drug delivery and for photothermal therapy [228].

Lapotko *et al.* [229] demonstrated that 10- and 30-nm NSphs conjugated with antibodies to the CD33 receptor on the surface of K562 and AML cells accumulate in the cells more effectively than do nonconjugated GNPs. A possible explanation [229] is that immunospecific conjugates form cell-surface clusters of approximately 20 particles, conducing to more active endocytosis. Furthermore, GNP clustering improves the effectiveness of photothermal therapy in the NIR region, because the maximum of the clusters' plasmon resonance shifts to the longer wavelength region of the spectrum, as compared with the resonance of isolated GNPs.

Tumor cells are characterized by overexpression of EGFRs. Antibodies to these receptors also can serve for the selective entry of GNPs into tumor cells [50]. Specifically, Melancon *et al.* [230] demonstrated that 30-nm hollow NSs complexed with anti-EGFR antibodies were effectively taken up by A431 cells *in vitro* and by cells of this tumor transplanted in mice, *in vivo*. In a further work [231], this group successfully used the melanocyte-stimulating hormone, bound by the melacortin receptors of tumor cells, as a selective ligand. Marega *et al.* [232] demonstrated that 5-nm GNPs coated with plasma-polymerized allylamine can be produced through plasma vapor deposition and can be conjugated with a monoclonal antibody to EGFR. The resulting nanoconjugates displayed an antibody loading of about 1.7 nmol/mg and efficiently targeted EGFR-overexpressing cell lines, as ascertained by enzyme-linked immunosorbent assay and Western blot assays. Similar results were reported by Raoof *et al.* [233] for 10-nm GNPs coated with anti-EGFR antibodies. Such conjugates penetrated the endosomes of tumor cells, and this effect could be used in noninvasive radiofrequency-based cancer therapy [233]. A new and rapidly growing field is the fabrication of multifunctional nanocarriers for anticancer drugs and the synthesis of plasmonic imaging probes to label specifically targeted cancer cells. For example, Song *et al.* [234] fabricated GNPs conjugated with both anti-EGFR antibodies and doxorubicin.

Liu *et al.* [235] aimed to demonstrate the potential of lymphotropic NP contrast agents designed to bind with high affinity to lymphoid cells overexpressing the CD45 antigen. To this end, 18-nm GNPs were prepared and conjugated with anti-CD45 antibodies through an optimized PEG coating to protect the particles from aggregation. By using a murine macrophage cell line as a model, the high binding affinity of the anti-CD45 NPs for lymphoid cells was demonstrated *in vitro*. In contrast, unconjugated and nonspecific antibody-conjugated particles showed minimal nonspecific binding to the macrophage owing to the dense PEG coating. Similarly to spherical GNPs, gold NRs functionalized with anti-EGFR antibodies were taken up by CAL 27 cells much more effectively, as compared to NRs coated with chitosan oligosaccharides, PEG, PAA, PSS, or CTAB [236].

In the huge arsenal of NPs being synthesized currently, gold nanocages are of special interest owing to their optimal size (about 40–50 nm) for uptake into cells and also owing to their plasmon resonance being in the biological tissue optical window (about 800 nm). Au *et al.* [237] focused on the intracellular uptake of gold nanocages. They examined the uptake into U87MGwtEGFR cells of gold nanocages functionalized with PEG and with antibodies to the EGFR, by using two-photon luminescence and ICP–MS. By combining these two methods, they showed that nanocage–antibody conjugates entered the cell interior by receptor-mediated endocytosis fourfold better than did PEGylated GNPs. The endocytosis also depended on the time and temperature of incubation, the size of nanocages (35 > 50 > 90 nm), and the amount of antibody molecules immobilized on each GNP. The same group concluded [238] that although both the size and shape of NPs do affect intracellular uptake—15-nm NSphs were taken up more effectively than 45-nm NSphs, 33-nm nanocages more effectively than 55-nm nanocages, and NSphs more effectively than nanocages—the major contribution to internalization is made by functionalization of the GNP surface. Using SK-BR-3 cells as an example, the authors showed by ICP–MS that the effectiveness of intracellular uptake of both types of functionalized GNPs increased in the order PAA >> antibodies > PEG. Another interesting fact they observed [239] was that as shown by two-photon microscopy, U87-MG cells did not have equal amounts of nanocages at division.

4.5 Use of Penetrating Peptides for the Delivery of GNPs into Cells

Penetrating peptides are short peptides that facilitate the cellular uptake of various molecular cargoes, from small chemical molecules and nanosized particles to large DNA fragments. The cargoes are associated with the peptides either through covalent bonds or through noncovalent interactions.

As a rule, penetrating peptides have relatively high contents of positively charged amino acids, such as lysine or arginine, or they contain sequences of polar (charged) amino acids interspersed with nonpolar (hydrophobic) ones. These two types of structures are named polycationic and amphipathic, respectively [240,241]. Peptides capable of entering cells have been isolated from proteins of various organisms—from viruses (HIV-1, SV40, adenovirus, herpes virus, and influenza virus) to vertebrates [242–244]. The penetrating peptides most commonly used in complex with GNPs are listed in Table 4.3.

Most often, penetrating peptides are employed for the delivery of GNPs loaded with target molecules *via* the cytoplasm to the cell nucleus. Pioneering work in this area was performed by Feldheim and colleagues [245]. Specifically, they described the delivery into the nuclei of HepG2 cells of 20-nm GNPs coated with BSA, which was conjugated to four penetrating peptides (CGGGPKKKRKVGG, CGGFSTSLRARKA, CKKKKKKSEDEYPYVPN, and CKKKKKKKSEDEYPYVPNFSTSLRARKA). Peptide 3 facilitates receptor-mediated endocytosis, peptides 1 and 2 interact with the nucleopore complex, and peptide 4 has both these mechanisms of entry. The entry of GNPs into the cells and nuclei was monitored by TEM and differential interference contrast

TABLE 4.3 Most Commonly Used Penetrating Peptides

Amino Acid Composition	Source or Function
CGGGPKKKRKVGG	SV40 large T, NLS (nuclear localization signal peptide)
AAVALLPAVLLALLAP	SV40 large T, RME (receptor-mediated endocytosis peptide)
PKKKRKVAAVALLPAVLLALLAP	SV40 large T, NLS, and RME
CGGFSTSLRARKA	Adenoviral, NLS
CKKKKKKSEDEYPYVPN	Adenoviral, RME
CKKKKKKKSEDEYPYVPNFSTSLRARKA	Adenoviral fiber protein, NLS, and RME
CGGRKKRRQRRRAP	HIV 1 TAT (transactivator of transcription) protein, NLS
GRQIKIWFQNRRMKWKK	pANTN from Antennapedia protein from *Drosophila*, NLS
CKKKKKKGGRGDMFG	Integrin binding domain and oligolysine
CALNN (CALNNR$_8$, CALND, CALNS)	GNP-stabilizing peptides

microscopy. GNP–BSA complexes conjugated to peptides 1 and 3 entered the cells and were found in endosomes, those conjugated to peptide 2 did not enter the cells, and only those conjugated to peptide 4 entered the nuclei. However, GNPs carrying both peptides 2 and 3 were best able to enter the nuclei. Similar data were reported by Mandal *et al.* [246] for 5-nm GNPs and TE85 cells. Thus, in order to achieve efficient nuclear penetration for GNPs, it is necessary to use their conjugates with peptides (a single peptide or a complex of peptides) able to penetrate both through membrane receptors and through nuclear pores. In a subsequent study [247], Feldheim's group showed that the entry of GNPs complexed with penetrating peptides depends on the type of cell, the nature of peptide (adenovirus, SV40, or HIV-1 origin), the time of incubation, and the temperature conditions for the experiment.

Later [248], the same group demonstrated that the intranuclear entry of GNPs depends also on the size of particles and on the number of peptide molecules on the surface of GNP–BSA. By ICP-AES, the authors found that uptake into HeLa cells was highest with 20-nm GNPs, lower with 15-nm GNPs, and still lower with 10-nm GNPs. Moreover, the effectiveness of nuclear penetration depended almost directly on the number of peptide molecules on the surface of GNPs (Table 4.4). Similar results were also attained for GNP–PEG conjugated to penetrating peptides [249]. Liu and Franzen [250] showed that it is possible to use GNP-conjugated penetrating peptides for the delivery of oligonucleotides to the nuclei of HeLa cells. Ghosh *et al.* [251] reported on the capacity of 3-nm GNPs complexed with a penetrating short cationic peptide, HKRK, for intracellular delivery of large protein (β-galactosidase) molecules, which under usual conditions are membrane impermeable.

De la Fuente and Berry [252] compared the intracellular entry of GNPs with that of GNPs conjugated to transactivator of transcription (TAT) peptide. It was shown by TEM that both types of 3-nm GNPs entered the cytoplasm of hTERT-BJ1 cells but that only GNPs functionalized with the penetrating peptide appeared in the cell nucleus. A similar result came from Oh *et al.* [253]; in addition, the authors, using LCM, demonstrated that 2.4-nm GNP–PEG–TAT complexes were localized in the nucleus; 5.5- and 8.2-nm complexes, in the perinuclear space; and 16-nm complexes, in the cytoplasm of COS-1 cells.

TABLE 4.4 Number of GNPs per Cell, Depending on the Amount of Peptide Used

Number of Peptide Molecules per GNP	Number of GNPs per Cell	Number of GNPs per Cytosolic Fraction	Number of GNPs per Nuclear Fraction	% of GNPs in the Nucleus
0 (native BSA)	3.48×10^4	1.33×10^4	8.18×10^3	37.5
30	3.52×10^4	5.22×10^3	9.23×10^3	62.5
70	1.21×10^5	1.54×10^4	7.04×10^3	82.4
80	4.97×10^5	3.21×10^4	1.73×10^5	84.4
110	1.16×10^6	4.74×10^4	9.61×10^5	95.3
130	1.80×10^6	8.97×10^4	1.47×10^6	94.2
150	5.44×10^6	1.20×10^5	3.11×10^6	96.3

Source: Ryan, J., Overton, K.W., Speight, M.E., Oldenburg, C.N., Loo, L., Robarge, W., Franzen, S., Feldheim D.L., *Anal. Chem.*, 79, 9150–9159, 2007.

Krpetić *et al.* [254] found that modifying GNPs with the cell-penetrating peptide TAT enhances HeLa cell uptake and leads to an unusual distribution pattern, by which particles are initially found in the cytosol, the nucleus, and the mitochondria and later within densely filled vesicles, from which they can be released again. Once inside the cell, these particles appear to overcome intracellular membrane barriers quite freely, including the possibility of direct membrane transfer. Ultimately, in the absence of extracellular NPs, the gold is completely cleared from the cells.

Combined conjugation of 13-nm GNPs with penetrating and lysosomal sorting peptides was suggested by Dekiwadia *et al.* [255] in order to minimize cytotoxic effects and ensure selective delivery of NPs into cell lysosomes.

Several researchers have employed the CALNN pentapeptide and its derivatives as a penetrating peptide to conjugate with GNPs. This pentapeptide transforms citrate-stabilized GNPs to GNPs that are extremely stable in aqueous solution and have some chemical properties that are analogous to those of proteins [78]. Nativo *et al.* [256] provided TEM and ICP–AES data on the uptake of 16-nm CALNN-stabilized GNPs by HeLa cells. Citrate- and CALNN-stabilized GNPs effectively entered the cell cytoplasm and localized in endosomes, whereas GNP–PEG complexes were absent from the cells. However, when penetrating peptides (transactivator of transcription [TAT], Pntn) were attached to GNP–PEG, such complexes were detected not only in the cytoplasm but also in the nuclei of the cells.

Sun *et al.* [257] explored the HeLa cell uptake of 30-nm GNPs coated with an arginine derivative of the CALNN peptide—CALNNRRRRRRRR (CALNNR$_8$). As indicated by dark-field microscopy and ICP–AES, GNPs complexed with CALNNR$_8$ entered the cell cytoplasm much more effectively than did those complexed with CALNN. However, the best nuclear penetration was shown by GNPs conjugated to a 1:9 CALNNR$_8$/CALNN mixture.

Finally, it was reported recently [258] that as compared with citrate-capped GNPs, GNP–CALNN conjugates were much more effective at penetration *in vivo* of the endosomes of rat liver Kupffer cells.

Penetrating peptides serve to modify not only NSphs but also NRs, NSs, and nanostars. Specifically, Oyelere *et al.* [259] modified NRs with a virus (SV40) nuclear localization signal (NLS) peptide. The results by dark-field microscopy and Raman spectroscopy demonstrated that such NRs entered both the cytoplasm and the nucleus of the cells much better than did CTAB-capped NRs. Furthermore, they entered the nuclei of tumor cells (HSC) much more effectively than they entered normal cells (HaCat). This fact was explained by disruption of the normal cellular processes and by damage to the nuclear membranes in the tumor cells, and it was concluded that it is possible to use NRs modified with penetrating peptides for the diagnosis of tumors and for the delivery of target substances to the nuclei of cancer cells. And, as early as in their subsequent work [260], the authors demonstrated that the NR–peptide conjugates that have entered the nuclei can lead to DNA damage, arrest of cytokinesis and cell division, and, ultimately, apoptosis.

Yuan *et al.* [261] showed that TAT-peptide-functionalized gold nanostars enter cells significantly better than bare or PEGylated nanostars. Their major cellular uptake mechanism involves actin-driven lipid raft-mediated macropinocytosis, in which particles primarily accumulate in macropinosomes but may also leak out into the cytoplasm.

Liu *et al.* [262] investigated the uptake by normal and tumorous liver cells of gold NSs conjugated to the A54 (AGKGTPSLETTP) peptide, which specifically adheres to liver tumor cells. TEM and ICP–AES data suggested that the A54-functionalized NSs were efficiently taken up into tumor cells (BEL-7404 and BEL-7402) and were not taken up into normal ones (HL-7702).

The rate of GNP uptake can also be determined by the terminal amino acid residues in the ligand shell [263]. For example, by replacing only 5%–10% of the terminal amino acids of the peptide ligands from glutamic acid to tryptophan or serine, one can dramatically increase or decrease the GNP uptake [263].

Wang *et al.* [264] showed that 20-nm KDEL-peptide-covered GNPs penetrated Sol8 cells very rapidly (within 5–15 min) and were localized only in the endoplasmic reticulum. These data indicate that the GNP–KDEL nanoconstructs are internalized *via* a clathrin-mediated pathway and are trafficked to the endoplasmic reticulum *via* a retrograde transport pathway, bypassing the lysosomal degradation pathway. Thus, this novel approach to developing nanoconstruct-based drug delivery has the potential to evade intracellular degradation, enhancing drug efficacy.

The use of cyclic peptide-coated GNPs containing tryptophan and arginine residues for intracellular delivery of antitumor drugs was proposed by Shirazi *et al.* [265]. Their complexes exhibited approximately 12 and 15 times higher cellular uptake than that of antitumor drugs alone in CCRF-CEM cells and SK-OV-3 cells, respectively. Recently, Yao *et al.* [266] demonstrated a significant enhancement of tumor cell (HeLa-GFP) uptake of GNPs conjugated with pH (low) insertion peptide.

Yang *et al.* [267] presented an uptake and removal of GNPs functionalized with three peptides. The first peptide (RGD peptide) enhanced the uptake, the second peptide (NLS peptide) enhanced the nuclear delivery, while the third one (pentapeptide) covered the rest of the surface and protected it from the binding of serum proteins onto the GNP surface. This peptide-capped GNP showed a fivefold increase in GNP uptake followed by effective nuclear localization. Enhanced uptake and prolonged intracellular retention of peptide-capped GNPs could allow GNPs to perform their desired applications more efficiently in cells.

Todorova *et al.* [268] investigated the effects of TAT peptide concentration and arrangement in solution on functionalized NPs' efficacy for membrane permeation. The authors found that cell internalization correlates with the positive charge distribution achieved prior to NP encountering interactions with membrane.

Lin and Alexander-Katz [269], using coarse-grained molecular dynamics simulations, showed that a model cell membrane generates a nanoscale hole to assist the spontaneous translocation of GNPs as well as HIV-1 TAT peptides to the cytoplasm side under a transmembrane potential.

It is interesting that some peptides (*e.g.*, EKEKEKE-PPPPC-Am) cannot increase, but reduce, the penetration of GNPs into cells (stealth peptides) [270].

Thus, the use of penetrating peptides conjugated to variously sized and shaped GNPs can enhance the entry of GNPs into the cell cytoplasm and nucleus. This may become a basis for the development of diagnostic and therapeutic nanoplatforms to actively deliver target substances inside cells [271]. However, it should be considered that serum proteins can affect the transport function of penetrating peptides, decreasing the effectiveness of uptake [272].

4.6 Intracellular Uptake of Oligonucleotide-Coated GNPs

Currently, oligonucleotide-conjugated NPs enjoy active use in molecular diagnostics, gene therapy, and vaccination methods. Therefore, the study of the mechanisms responsible for the effective delivery of genes to cells is of major importance. The study of intracellular uptake of GNPs functionalized with nucleic acids is associated, first and foremost, with the work of Mirkin's group addressed to intracellular gene regulation [273]. Specifically, Giljohann *et al.* [274] reported on the uptake of GNPs modified with antisense oligonucleotides (28 bases) into C-166, HeLa, and A594 cells. With ICP–MS, they examined the cellular uptake of 13-nm GNPs functionalized with different numbers of oligonucleotide chains (from 0 to 80 per particle). It turned out that uptake of the conjugates was best in HeLa cells—3×10^7 GNPs/cell (for comparison, uptake in C-166 cells was 1×10^7 GNPs/cell and that in A594 cells was 1×10^6 GNPs/cell). The uptake depended strongly on the density of the oligonucleotide coating on GNPs, increasing from 1×10^3 GNPs/cell for the least laden GNPs to 1.3×10^6 GNPs/cell at a load of 60–80 molecules/particle (for A594 cells). In addition, uptake efficacy increased with increasing amount of GNP-adsorbed serum proteins in the cell growth medium and with increasing concentration of added GNPs. In another report from the same group [275], it was shown that when GNPs are coated with a combination of oligonucleotides and penetrating peptides, the effectiveness of intracellular uptake increases.

Subsequently [276], Mirkin's group demonstrated that GNP uptake into HeLa cells is affected by the nature of the adsorbate, as well as by the previously mentioned factors influencing the internalization of GNPs conjugated with different nucleic acids. As measured by ICP–MS, the highest uptake was shown by GNPs conjugated to double-stranded RNA; somewhat lower uptake, by GNPs conjugated to single-stranded DNA; and the lowest uptake, by BSA-conjugated GNPs (Figure 4.16).

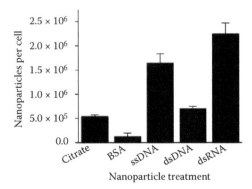

FIGURE 4.16 Quantification of HeLa cell uptake of NPs after 24-h treatment with 10-nM GNPs functionalized with various surface ligands. The mean and STD values were determined from three separate experiments. (Adapted from Massich, M.D., Giljohann, D.A., Schmucker, A.L., Patel, P.C., Mirkin, C.A., *ACS Nano*, 4, 5641–5646, 2010. With permission.)

These NPs are called nanoflares or spherical nucleic acids [277,278]. Despite their high negative charge, these nanostructures are naturally taken up by cells, without the need for positively charged polymeric cocarriers, and are highly effective as gene regulation agents in both siRNA and antisense gene regulation pathways [279,280].

Jewell *et al.* [281] found that the level of GNP–oligonucleotide uptake is also affected by the methods used to attach oligonucleotides to the particle surface. They used thiol-modified oligonucleotides for conjugation with GNPs, but they also modified the GNP surface with the linkers mercapto-1-undecanesulfonate (MUS) and MUS-1-octanethiol (MUS–OT). These ligands were reported to substantially increase the effectiveness of GNP–oligonucleotide uptake by B16-F0 tumor cells (5-fold for GNP–MUS–oligonucleotide and 10-fold for GNP–MUS–OT–oligonucleotide).

Bonoiu *et al.* [282] presented data for the uptake of siRNA-modified NRs by DAN cells. Dark-field microscopy and LCM indicated that such complexes were taken up by the cells much better than were native siRNAs. A similar result was obtained by Crew *et al.* [283] for the transfection of 13-nm GNPs conjugated to microRNA into MM.1S cells. Guo *et al.* [284] reported more effective intracellular uptake of siRNA conjugated to GNPs and polyelectrolytes, as compared to that of siRNA conjugated to lipofectamine, a commercial transfection agent. A similar finding was reported by Ghosh *et al.* [285] for the uptake of 13-nm cysteamine-functionalized GNPs conjugated to microRNA into neuroblastoma (NGP, SH-SY5Y) and ovarian cancer (HEYA8, OVCAR8) cells.

Braun *et al.* [286] developed a GNS functionalized with a TAT-lipid (TAT-peptide-lipid cell internalizing agent) layer for transfection and selective release of siRNA. The TAT-lipid coating mediated the cellular uptake of the nanomaterial, whereas the release of the siRNA was dependent on NIR laser pulses.

Apart from that, DNA or RNA transfection into cells is enhanced with GNPs complexed with PEI [186,287,288], cationic lipids [289,290], bacterial toxins [14], poly-L-lysine [291], and hyaluronic acid [292] or it can be enhanced with functionalized GNPs in combination with conventional transfection reagents [293].

Elbakry *et al.* [294] studied the size-dependent uptake of GNPs (20, 30, 50, and 80 nm), coated with a layer-by-layer approach with nucleic acid and PEI, into a variety of mammalian cell lines. In contrast to other studies, the optimal particle diameter for cellular uptake and the number of therapeutic cargo molecules per cell were determined. It was found that 20-nm GNPs, with diameters of about 32 nm after the coating process and about 88 nm (including the protein corona) after incubation in cell culture medium, yield the largest number of NPs and therapeutic DNA molecules per cell.

4.7 Selective Internalization of Engineered GNPs into Cancer Cells

The GNP-internalization mechanisms discussed previously are based mostly on the nonspecific interaction of NPs with the cell surface and on the subsequent endocytosis of GNPs. Of major importance in current biomedicine, however, is the selective recognition of cells at the cost of interaction of functionalized GNPs with specific receptors of cells, in particular tumorous cells. Such selective interaction will allow one, *e.g.*, to perform targeted delivery of drugs or to treat GNP-containing cells with various types of irradiation [295,296].

Specific receptors, in particular, include folate receptors, which are high-affinity cell receptors for folic acid and some of its derivatives. In humans, four types of receptors from this family have been described. Folate receptors ensure the delivery to cells of 5-tetrahydrofolate, a cofactor necessary for cellular proliferation. For this reason, folate receptors are targeted by anticancer therapy, since blocking of folate transport to cancer cells prevents their further proliferation. Consequently, the use of folic acid as part of nanocomposites seems very promising for the selective interaction with tumor cells [297].

Dixit *et al.* [298] explored the uptake of 10-nm PEGylated GNPs functionalized with folic acid into KB tumor cells with overexpression of folate receptors. Using TEM, they showed that folic acid-conjugated GNP–PEG1500 complexes were taken up effectively by KB cells through receptor-mediated endocytosis. The uptake effectiveness decreased substantially when cell lines with low expression of folate receptors were used, when the receptors were inhibited with native folic acid, and when GNP–PEG conjugates were folate uncoated. Similar results emerged from a study by Li *et al.* [299], who used HeLa cells and 18-nm non-PEGylated GNPs capped with folate, and also from that by Tong *et al.* [300] with folic acid–NR (46 × 12 nm) conjugates.

Bhattacharya *et al.* [301] investigated the uptake of GNP–folate conjugates by cells of seven tumor lines. They showed that the level of uptake depended on the number of folate molecules on GNPs. In turn, this number depended on the molecular mass of the PEG conjugated to GNPs. For the conjugates used, the uptake level decreased in the sequence folate–GNP–PEG20000 > folate–GNP–PEG10000 > folate–GNP–PEG2000, in agreement with the data of Kah *et al.* [174] data for NSs.

The use of folate-coated GNPs for intracellular delivery of the antitumor drug 6-mercaptopurine was proposed by Park *et al.* [302]. Their method enabled tumor cell (HeLa and KB) death to be increased by 20%, as compared with other drug delivery techniques.

Paciotti *et al.* [303] proposed the use of TNF in complex with 26-nm GNPs for selective uptake in tumor cells *in vivo*. Shao *et al.* [304] demonstrated that maximal uptake of TNF–GNP conjugates was in cancer cells (SCK), compared to no or low uptake in mouse red blood cells and white blood cells. For selective delivery to cancerous cells, Li *et al.* [100] employed 20-nm GNPs conjugated to anti-EGFR aptamers, and Wong *et al.* [59] used GNPs conjugated to aptamers specific for mucin, which was overexpressed on the surface of cancer cells.

An interesting approach to delivering drugs to the tumor cell nucleus was put forward by Dam *et al.* [305]. Gold nanostars were conjugated with an aptamer to nucleolin, a protein overexpressed in cancer cells and found both on the cell surface and in the cell interior. Owing to the aptamer–protein interaction, GNPs are delivered to the cell nucleus, after which the DNA aptamer is released from the nanostars and begins to act as a drug, causing cell death. As well as making it possible to load a large drug amount, the nanostars shape helps to concentrate light at the tops of the beams in photothermal therapy, facilitating drug release in these regions.

Kumar *et al.* [306,307] suggested the use of multifunctionalized GNPs (conjugated with a combination of specific antibodies, penetrating peptides, and PEG) for intracellular delivery of GNPs for the purpose of bioimaging. Lukianova-Hleb *et al.* [308] enhanced therapeutic effectiveness simultaneously with (i) gold NSs attached to anti-EGFR and (ii) liposomal doxorubicin.

Yet another important target for GNP interactions with the tumor cell surface is vascular endothelial growth factor receptors, which play a central role in angiogenesis. Mukherjee *et al.* [309] presented TEM data for the preferential internalization by CLL-B cells of GNPs complexed with antibodies to vascular endothelial growth factor receptors, as compared with that of nonconjugated 5-nm GNPs. Such internalization resulted in an enhancement of tumor cell apoptosis. Kalishwaralal *et al.* [310] showed that after 50-nm GNPs had been internalized by BREC cells, they could be detected by TEM in multivesicular bodies. If the cells were incubated with GNPs conjugated to vascular endothelial growth factor, the particles were found mostly in the membrane, with only a small number entering the cytoplasm. GNPs were also found to inhibit angiogenesis.

Wang *et al.* [311] proposed the use of gold nanocages conjugated to the low-molecular-weight ligand SV119, which specifically interacts with sigma-2 receptors on the surface of tumor cells, as a platform for a theranostic agent. The resultant conjugate effectively entered MDA-MB-231 and PC-3 cancer cells overexpressing sigma-2 receptors. Yet another ligand that selectively interacts with tumor cells is 5-aminovaleric acid. Using LCM, Krpetic *et al.* [312] demonstrated that 7-nm GNPs coated with this acid selectively penetrated into K562 cancer cells and almost did not penetrate into normal epithelial cells.

Kasten *et al.* [313] fabricated 5-nm GNPs coated with a prostate-specific membrane antigen inhibitor (CTT54). These conjugates exhibited selective, significantly higher uptake into LNCaP prostate tumor cells, as compared to the nontargeted control GNPs.

Huang *et al.* [314] studied the biodistribution and localization of gold NRs labeled with three types of probing molecules: (1) an scFv fragment of antibodies to the EGFR, (2) an amino terminal fragment (ATF) peptide that recognizes the urokinase plasminogen activator receptor, and (3) a cyclic RGD peptide that recognizes the $\alpha_v\beta_3$ integrin receptor. By dark-field microscopy and ICP–MS, the authors showed that the effectiveness of A549 cell uptake decreased in the order NR–RGD (~9000 NRs/cell) > NR–ATF (~6500 NRs/cell) > NR–scFv (~2500 NRs/cell) > NR–PEG (~100 NRs/cell). By contrast, the biodistribution data from xenograft animal models did not reveal any significant effect of active targeting on the total tumor uptake of long-circulating gold NRs, although their localization was quite different for different targeting ligands. Ali *et al.* [315] showed that rifampicin-conjugated GNPs can greatly enhance the rate as well as efficiency of endocytosis of GNPs and, hence, their concentration inside the cancer cell. Rifampicin-loaded GNRs can enhance the rate and efficacy of endocytosis of paclitaxel-carrying GNRs into the cancer cells.

In a study by Song *et al.* [316], HeLa and MCF-7 cancer cells took up more glucose-capped GNPs than they took up naked GNPs and the uptake curve showed size- and cell-dependent uptake. The glucose-capped GNPs were mainly located in the cytoplasm, and endocytosis was concluded to be the mechanism behind the internalization of both naked and glucose-capped GNPs.

Interesting data came from a study by Li *et al.* [317]. Five-nm GNPs conjugated with a dual ligand—folic acid and glucose—penetrated KB cells 3.9- and 12.7-fold more effectively than did GNP–folate and GNP–glucose, respectively. Thus, a definite synergetic effect of the two ligands was found. Bhattacharyya *et al.* [318] also used two ligands, conjugated to 50-nm GNPs: antibodies to folate receptors and anti-EGFR antibodies (cetuximab). Evaluation of conjugate uptake into OVCAR-5 and Skov3-ip tumor cells by using TEM and neutron activation indicated that GNPs complexed with the dual ligand were internalized best, as compared with those complexed with antibodies to one receptor. The same group found [319,320] that in the process of tumor cell endocytosis of GNPs conjugated with anti-EGFR antibodies, an important role is played by the intracellular GTFases dynamin (caveolin-dependent endocytosis) and Cdc42 (pinocytosis/phagocytosis).

The use of multifunctionalized GNPs was also addressed by Hosta-Rigau *et al.* [321]. They suggested the conjugation of 20-nm GNPs simultaneously with an analog of the peptide Bombesin (interacting with gastrin-releasing receptors, which are overexpressed on the cancer cell surface) and the therapeutic antitumor peptide RAF. Such complexes were found to possess high efficacy and selectivity of uptake and also marked therapeutic activity. Suresh *et al.* [322] showed that bombesin conjugated gold nanocages uptake in human prostate tumor cells is mediated by clathrin mediated endocytosis.

Kumar *et al.* [323] conjugated 2-nm GNPs also with two peptides: the peptide CRGDK, selectively binding to neuropilin-1 receptors, and the therapeutic peptide PMI (p12). As found by ICP–MS and MTT assay, the GNPs coated with the two peptides were effectively taken up by MDA-MB-321 cells, exerting a pronounced cytotoxic effect.

Heo *et al.* [324] reported on effective and selective tumor uptake of GNPs conjugated to the antitumor drug paclitaxel and to biotin. Biotin interacts with the biotin receptors, which are overexpressed on the tumor cell surface. As a result, uptake of functionalized GNPs by tumor cells (HeLa, A549, and MG63) was more effective than that by normal cells (NIH3T3), and correspondingly, cytotoxicity increased because of the larger amount of paclitaxel that had entered the cells.

Several authors have described the tumor uptake, *via* specific receptor-mediated endocytosis, of GNPs functionalized with antitumor drugs (daunorubicin [325], herceptin [109,326], tamoxifen [327], doxorubicin [328,329], prospidine [330], chloroquine [331], topotecan [332], and docetaxel [333]). The ultimate goal of such studies is to increase the effectiveness of drugs already used in oncology through conjugation to GNPs. Additionally, Comenge *et al.* [334] demonstrated that GNPs not only act as carriers but also protect the drug from deactivation by plasma proteins until conjugates are internalized in cells and cisplatin is released. For better drug penetration of the cell nuclei, it was suggested using small-sized (2–3 nm) [335] or dextran-capped GNPs [336].

4.8 Concluding Remarks

Without doubt, the use of various NPs, including GNPs, to deliver genetic, medicinal, or other materials into cells is a leading trend in current nanobiotechnology and its biomedical applications. The data available on the cellular uptake of GNPs are still insufficient to gain a complete understanding of the physical chemistry and biology of endocytosis and its dependence on particle parameters and cell type. Nevertheless, the information presented in this chapter leads to several conclusions, which have been confirmed in independent experiments by different research groups and also in existing theoretical models.

1. The cellular uptake of spherical GNPs is a receptor-mediated process, the effectiveness of which depends on the size of particles and on the density of ligand coating. Most experimental data accrued for colloidal gold particles confirm the existence of an optimal diameter range (30–50 nm), whereas the specific optimal size for uptake may depend on cell type. It seems likely that the inner structure of spherical gold particles (and, in particular, their average density) has no large role; therefore, gold NSs are amenable to the same regularities concerning size dependence as the usual GNPs. Specifically, 60-nm NSs are taken up by cells better than larger, 110-nm particles, and among NSs with outer diameters of 130–250 nm, the smallest particles are the best to take up. It is remarkable that in all experimental cases, optimally sized NSphs enter cells more effectively than large NSs do. The effectiveness of GNP endocytosis also depends on the time and temperature of incubation and on the concentration of GNPs used [337]. According to current theoretical models, the universal character of endocytosis can be expressed in terms of phase diagrams, which relate uptake efficiency to the particle size and aspect ratio and to the ligand surface density (Figures 4.6 and 4.7).

2. With an increase in the particle aspect ratio, the effectiveness of GNP uptake into cells decreases; the exocytosis time may also decrease. The theoretical data for thick NRs with diameters of 50–60 nm are in harmony with experimental data; for thinner NRs, however, theory predicts a more complex dependence on particle shape. Because the size, shape, and surface functionalization of nonspherical particles can change cellular uptake, additional experimental and theoretical studies are required to elucidate the dependence of endocytosis on particle shape only, with all other factors being equal. The importance of such work is determined also by the specific peculiarity of uptake of CTAB-coated NRs into tumor cell mitochondria followed by cell apoptosis, which has not been observed for gold NSphs.

3. Functionalization of gold NSphs with PEG molecules reduces the effectiveness of endocytosis, possibly because of the switching of GNP uptake from receptor-mediated pathways to other mechanisms. A similar effect was recorded for NSs and NRs as well, with uptake effectiveness diminishing with increasing surface density of PEG. The dependence of GNP uptake on the molecular mass of PEG is nonmonotonic and calls for further studies. Evidence shows that PEGylated NRs enter tumor cells an order of magnitude more actively than do PEGylated NSs, but both particle types accumulate in liver cells in approximately the same amounts. This property is important for the use of GNPs in photothermal therapy.

4. Although the effect of surface functionalization of GNPs on their intracellular uptake has been studied for a wide variety of molecules, it seems that in one way or another, all data reflect the influence of the principal factor—the surface charge of particles. On the whole, it should be considered proven that the best uptake is observed for particles coated with positively charged polymers, as compared to those coated with negatively charged and neutral polymers. Apart from the positive charge value itself, the chemical nature of the coating polymer may also be

important, as shown for uptake of NRs capped with quaternary amines (CTAB and PDDAC, effective uptake) and primary amines (PAH, less effective uptake).

5. In further work with functionalized GNPs, it is important to account for the role of the protein components of the medium used for cell culturing or of blood serum. It was shown that surface hydrophobicity is a critical factor for controlling serum protein binding, which in turn decreases the cellular uptake of GNPs [338]. Protein adsorption may strongly interfere with the influence of the polymers used to coat functionalized particles. Specifically, culture-medium serum proteins may affect the transport function of penetrating peptides, reducing the uptake effectiveness. Overall, though, the existing data attest to elevated uptake of particles coated with penetrating peptides into the cell cytoplasm and nucleus. For NP functionalization with oligonucleotides, an opposite effect was observed: the uptake increased with increasing serum protein concentration in the medium. On the whole, however, the basic parameter for conjugates of this type is the density of the surface oligonucleotide coating, a change in which may cause uptake effectiveness to change by three orders of magnitude.

6. Functionalization of GNPs with molecular vectors to tumor cell receptors leads to a marked enhancement of uptake into target cells and is undoubtedly a promising direction for the delivery of antitumor drugs inside cells. However, this question deserves additional studies, as recent reexamination of biodistribution data from xenograft animal models did not reveal any significant effect of active targeting on the total tumor uptake of long-circulating gold NRs, although their localization was quite different for different targeting ligands [314]. It should also be admitted that in this area, GNPs compete strongly with biodegradable and other biocompatible carriers.

7. As justly noted by Canton and Battaglia [22], it is hardly probable that there will emerge an all-purpose nanocarrier for all types of cargoes. Rather, the type of carrier should be optimized for both the payload to be carried and the biological target. In particular, for biosensorics and bioimaging, the optical properties of GNPs and the valence properties of surface ligands are most important. For therapy and drug delivery, the stability of conjugates in the blood stream and their weak interaction with nontarget and immune cells should be combined with effective uptake and accumulation in targets. Perhaps, these goals could be achieved with nanoscale systems, whose properties can be switched and activated dynamically on demand. For example, one can expect rapid progress in the development of multifunctional nanocomposites, combining controllable physical properties (magnetic, optical, photodynamic, radioactive, *etc.*) with advantages in the molecular surface targeting to meet the emerging theranostic demands.

8. Sensitive and reliable determination of GNPs in cells or tissues is key when GNP uptake and biodistribution over different organs, cells, or cellular compartments are assessed. Although various techniques with labeled and label-free GNPs have been developed, tested, and applied to GNP quantification, only a few of them can be recommended as best compromises between statistically sounding and unbiased estimations on the one hand and preparative and measuring complexity on the other. Specifically, ICP–MS and ICP–AES are most suitable for sensitive

determination of the total gold content in a cell or a tissue sample, provided that the sampling procedures are properly optimized. For example, simply mixing GNPs with cell suspensions may lead to significant errors owing to aggregation of GNPs before their intracellular uptake. At present, only TEM in combination with stereology sampling principles enables statistically reliable estimation of GNPs with high resolution at the cell compartment level.

9. Although the ICP–MC technique is sensitive and reliable, there are serious problems associated with extrapolating from ICP–MS data to estimate GNP number per cell. On the one hand, such extrapolation depends on the particle size and shape distribution model assumptions. On the other hand, assessing the number of cells in which the estimated number of GNP resides is a difficult task. Clearly, both factors can lead to an unwanted bias in the ultimate results for individual cellular uptake. It should be emphasized that some of the estimates of GNP number per cell presented in this Chapter (see, *e.g.*, Table 4.4, Figure 4.16) are several orders higher than those shown in Figures 4.1 and 4.3, which are more consistent with the results obtained by unbiased TEM/stereology approaches. One may assume that the potential systematic bias of any particular method is the same across all experimental groups. Therefore, it would be reasonable to use the GNP-per-cell data in a comparative manner within a particular study, as the experimental conditions can be quite different. For example, the data of Figure 4.1 were obtained after 6-h incubation of cells with GNPs [38], whereas the data shown in Figure 4.3 [59] correspond to 1.5-h incubation. Similarly, Mirkin and coworkers [276,339] measured GNP uptake after 24-h incubation, whereas Ryan *et al.* [248] (Table 4.4) used 6-h incubation in combination with penetrating peptides. In any case, this important question deserves further study.

References

1. Scott G.B., Williams H.S., Marriott P.M. The phagocytosis of colloidal particles of different sizes // *Br. J. Exp. Pathol.* 1967. V. 48. P. 411–416.
2. Singer J.M., Adlersberg L., Sadek M. Long-term observation of intravenously injected colloidal gold in mice // *J. Reticuloendothel. Soc.* 1972. V. 12. P. 658–671.
3. Renaud G., Hamilton R.L., Havel R. Hepatic metabolism of colloidal gold-low-density lipoprotein complexes in the rat: Evidence for bulk excretion of lysosomal contents into bile // *Hepatology.* 1989. V. 9. P. 380–392.
4. Hardonk M.J., Harms G., Koudstaal J. Zonal heterogeneity of rat hepatocytes in the *in vivo* uptake of 17 nm colloidal gold granules // *Histochemistry.* 1985. V. 83. P. 473–477.
5. Sadauskas E., Wallin H., Stoltenberg M., Vogel U., Doering P., Larsen A., Danscher G. Kupffer cells are central in the removal of nanoparticles from the organism // *Part. Fibre Toxicol.* 2007. V. 4. 10.
6. Sadauskas E., Danscher G., Stoltenberg M., Vogel U., Larsen A., Wallin H. Protracted elimination of gold nanoparticles from mouse liver // *Nanomedicine.* 2009. V. 5. P. 162–169.

7. Xu Z.P., Zeng Q.H., Lu G.Q., Yu A.B. Inorganic nanoparticles as carriers for efficient cellular delivery // *Chem. Eng. Sci.* 2006. V. 61. P. 1027–1040.

8. Unfried K., Albrecht C., Klotz L.-O., von Mikecz A., Grether-Beck S., Schins R.P.F. Cellular responses to nanoparticles: Target structures and mechanisms // *Nanotoxicology.* 2007. V. 1. P. 52–71.

9. Sperling R.A., Gil P.R., Zhang F., Zanella M., Parak W.J. Biological applications of gold nanoparticles // *Chem. Soc. Rev.* 2008. V. 37. P. 1896–1908.

10. Roca M., Haes A.J. Probing cells with noble metal nanoparticle aggregates // *Nanomedicine (Lond.).* 2008. V. 3. P. 555–565.

11. Boisselier E., Astruc D. Gold nanoparticles in nanomedicine: Preparations, imaging, diagnostics, therapies and toxicity // *Chem. Soc. Rev.* 2009. V. 38. P. 1759–1782.

12. Giljohann D.A., Seferos D.S., Daniel W.L., Massich M.D., Patel P.C., Mirkin C.A. Gold nanoparticles for biology and medicine // *Angew. Chem. Int. Ed.* 2010. V. 49. P. 3280–3294.

13. Chithrani B.D. Intracellular uptake, transport, and processing of gold nanostructures // *Mol. Membr. Biol.* 2010. V. 27. P. 299–311.

14. Lévy R., Shaheen U., Cesbron Y., Sée V. Gold nanoparticles delivery in mammalian live cells: A critical review // *Nano Rev.* 2010. V. 1. 4889.

15. Verma A., Stellacci F. Effect of surface properties on nanoparticle–cell interactions // *Small.* 2010. V. 6. P. 12–21.

16. Johnston H.J., Hutchison G., Christensen F.M., Peters S., Hankin S., Stone V. A review of the *in vivo* and *in vitro* toxicity of silver and gold particulates: Particle attributes and biological mechanisms responsible for the observed toxicity // *Crit. Rev. Toxicol.* 2010. V. 40. P. 328–346.

17. Chithrani B.D. Optimization of bio-nano interface using gold nanostructures as a model nanoparticle system // *Insciences J.* 2011. V. 1. P. 115–135.

18. Jiang X.-M., Wang L.-M., Chen C.-Y. Cellular uptake, intracellular trafficking and biological responses of gold nanoparticles // *J. Chin. Chem. Soc.* 2011. V. 58. P. 273–281.

19. Cobley C.M., Chen J., Cho E.C., Wang L.V., Xia Y. Gold nanostructures: A class of multifunctional materials for biomedical applications // *Chem. Soc. Rev.* 2011. V. 40. P. 44–56.

20. Chou L.Y.T., Ming K., Chan W.C.W. Strategies for the intracellular delivery of nanoparticles // *Chem. Soc. Rev.* 2011. V. 40. P. 233–245.

21. Zhao F., Zhao Y., Liu Y., Chang X., Chen C., Zhao Y. Cellular uptake, intracellular trafficking, and cytotoxicity of nanomaterials // *Small.* 2011. V. 7. P. 1322–1337.

22. Canton I., Battaglia G. Endocytosis at the nanoscale // *Chem. Soc. Rev.* 2012. V. 41. P. 2718–2739.

23. Podkolodnaya O.A., Ignatieva E.V., Podkolodnyy N.L., Kolchanov N.A. Routes of nanoparticle uptake into mammalian organisms, their biocompatibility and cellular effects // *Biol. Bull. Rev.* 2012. V. 2. P. 279–289.

24. Albanese A., Tang P.S., Chan W.C.W. The effect of nanoparticle size, shape, and surface chemistry on biological systems // *Annu. Rev. Biomed. Eng.* 2012. V. 14. P. 1–16.

25. Zhu M., Nie G., Meng H., Xia T., Nel A., Zhao Y. Physicochemical properties determine nanomaterial cellular uptake, transport, and fate // *Acc. Chem. Res.* *2013.* V. 46. P. 622–631.

26. Cheng L.-C., Jiang X., Wang J., Chen C., Liu R.-S. Nano-bio effects: Interaction of nanomaterials with cells // *Nanoscale.* 2013. V. 5. P. 3547–3569.

27. Mao Z., Zhou X., Gao C. Influence of structure and properties of colloidal biomaterials on cellular uptake and cell functions // *Biomater. Sci.* 2013. V. 1. P. 896–911.

28. Ma N., Ma C., Li C., Wang T., Tang Y., Wang H., Mou X., Chen Z., He N. Influence of nanoparticle shape, size, and surface functionalization on cellular uptake // *J. Nanosci. Nanotechnol.* 2013. V. 13. P. 6485–6498.

29. Mu Q., Jiang G., Chen L., Zhou H., Fourches D., Tropsha A., Yan B. Chemical basis of interactions between engineered nanoparticles and biological systems // *Chem. Rev.* 2014. V. 114. P. 7740–7781.

30. Kettler K., Veltman K., van de Meent D., van Wezel A., Hendriks A.J. Cellular uptake of nanoparticles as determined by particle properties, experimental conditions, and cell type // *Environ. Toxicol. Chem.* 2014. V. 33. P. 481–492.

31. Nazarenus M., Zhang Q., Soliman M.G., del Pino P., Pelaz B., Carregal-Romero S., Rejman J., Rothen-Rutishauser B., Clift M.J.D., Zellner R., Nienhaus G.U., Delehanty J.B., Medintz I.L., Parak W.J. *In vitro* interaction of colloidal nanoparticles with mammalian cells: What have we learned thus far? // *Beilstein J. Nanotechnol.* 2014. V. 5. P. 1477–1490.

32. Martens T.F., Remaut K., Demeester J., De Smedt S.C., Braeckmans K. Intracellular delivery of nanomaterials: How to catch endosomal escape in the act // *Nano Today.* 2014. V. 9. P. 344–364.

33. Oh N., Park J.-H. Endocytosis and exocytosis of nanoparticles in mammalian cells // *Int. J. Nanomedicine.* 2014. V. 9. Suppl. 1. P. 51–63.

34. England C.G., Gobin A.M., Frieboes H.B. Evaluation of uptake and distribution of gold nanoparticles in solid tumors // *Eur. Phys. J. Plus.* 2015. V. 130. 231.

35. Gustafson H.H., Holt-Casper D., Grainger D.W., Ghandehari H. Nanoparticle uptake: The phagocyte problem // *Nano Today.* 2015. V. 10. P. 487–510.

36. Gong N., Chen S., Jin S., Zhang J., Wang P.C., Liang X.-J. Effects of the physicochemical properties of gold nanostructures on cellular internalization // *Regen. Biomater.* 2015. V. 2. P. 273–280.

37. Wilhelm S., Tavares A.J., Dai Q., Ohta S., Audet J., Dvorak H.F., Chan W.C.W. Analysis of nanoparticle delivery to tumours // *Nat. Rev. Mater.* 2016 V. 1. 16014.

38. Chithrani B.D., Ghazani A.A., Chan W.C.W. Determining the size and shape dependence of gold nanoparticle uptake into mammalian cells // *Nano Lett.* 2006. V. 6. P. 662–668.

39. Yang P.-H., Sun X., Chiu J.-F., Sun H., He Q.-Y. Transferrin-mediated gold nanoparticle cellular uptake // *Bioconjug. Chem.* 2005. V. 16. P. 494–496.

40. Mukherjee S., Ghosh R.N., Maxfield F.R. Endocytosis // *Physiol. Rev.* 1997. V. 77. P. 759–803.

41. Gao H., Shi W., Freund L.B. Mechanics of receptor-mediated endocytosis // *Proc. Natl. Acad. Sci. USA.* 2005. V. 102. P. 9469–9474.

42. Boraschi D., Costantino L., Italiani P. Interaction of nanoparticles with immunocompetent cells: Nanosafety considerations // *Nanomedicine (Lond)*. 2012. V. 7. P. 121–131.

43. Chithrani B.D., Stewart J., Allen C., Jaffray D.A. Intracellular uptake, transport, and processing of nanostructures in cancer cells // *Nanomedicine*. 2009. V. 5. P. 118–127.

44. Krpetić Z., Porta F., Caneva E., Dal Santo V., Scarì G. Phagocytosis of biocompatible gold nanoparticles // *Langmuir*. 2010. V. 26. P. 14799–14805.

45. Banerji S.K., Hayes M.A. Examination of nonendocytotic bulk transport of nanoparticles across phospholipid membranes // *Langmuir*. 2007. V. 23. P. 3305–3313.

46. Taylor U., Klein S., Petersen S., Kues W., Barcikowski S., Rath D. Nonendosomal cellular uptake of ligand-free, positively charged gold nanoparticles // *Cytometry A*. 2010. V. 77. P. 439–446.

47. Xia T., Rome L., Nel A. Particles slip cell security // *Nat. Mater*. 2008. V. 7. P. 519–520.

48. Chithrani B.D., Chan W.C.W. Elucidating the mechanism of cellular uptake and removal of protein-coated gold nanoparticles of different sizes and shapes // *Nano Lett*. 2007. V. 7. P. 1542–1550.

49. Sykes E.A., Chen J., Zheng G., Chan W.C.W. Investigating the impact of nanoparticle size on active and passive tumor targeting efficiency // *ACS Nano*. 2014. V. 8. P. 5696–5706.

50. Jiang W., Kim B.Y.S., Rutka J.T., Chan W.C.W. Nanoparticle-mediated cellular response is size-dependent // *Nat. Nanotechnol*. 2008. V. 3. P. 145–150.

51. Copland J.A., Eghtedari M., Popov V.L., Kotov N., Mamedova N., Motamedi M., Oraevsky A.A. Bioconjugated gold nanoparticles as a molecular based contrast agent: Implications for imaging of deep tumors using optoacoustic tomography // *Mol. Imaging Biol*. 2004. V. 6. P. 341–349.

52. Albanese A., Chan W.C.W. Effect of gold nanoparticle aggregation on cell uptake and toxicity // *ACS Nano*. 2011. V. 5. P. 5478–5489.

53. Kneipp J., Kneipp H., McLaughlin M., Brown D., Kneipp K. *In vivo* molecular probing of cellular compartments with gold nanoparticles and nanoaggregates // *Nano Lett*. 2006. V. 6. P. 2225–2231.

54. Mustafa T., Watanabe F., Monroe W., Mahmood M., Xu Y., Saeed L.M., Karmakar A., Casciano D., Ali S., Biris A.S. Impact of gold nanoparticle concentration on their cellular uptake by MC3T3-E1 mouse osteoblastic cells as analyzed by transmission electron microscopy // *J. Nanomedic. Nanotechnol*. 2011. V. 2. 118.

55. Mironava T., Hadjiargyrou M., Simon M., Jurukovski V., Rafailovich M.H. Gold nanoparticles cellular toxicity and recovery: Effect of size, concentration and exposure time // *Nanotoxicology*. 2010. V. 4. P. 120–137.

56. Sonavane G., Tomoda K., Sano A., Ohshima H., Terada H., Makino K. *In vitro* permeation of gold nanoparticles through rat skin and rat intestine: Effect of particle size // *Colloids Surf. B*. 2008. V. 65. P. 1–10.

57. Yen H.-J., Hsu S.-h., Tsai C.-L. Cytotoxicity and immunological response of gold and silver nanoparticles of different sizes // *Small*. 2009. V. 5. P. 1553–1561.

58. Coradeghini R., Gioria S., García C.P., Nativo P., Franchini F., Gilliland D., Ponti J., Rossi F. Size-dependent toxicity and cell interaction mechanisms of gold nanoparticles on mouse fibroblasts // *Toxicol. Lett.* 2013. V. 217. P. 205–216.

59. Wang S.-H., Lee C.-W., Chiou A., Wei P.-K. Size-dependent endocytosis of gold nanoparticles studied by three-dimensional mapping of plasmonic scattering images // *J. Nanobiotechnology.* 2010. V. 8. 33.

60. Rieznichenko L.S., Shpyleva S.I., Gruzina T.G., Dybkova S.N., Ulberg Z.R., Chekhun V.F. Cancer cells–gold nanoparticles contact interaction determined by their size and concentration // *Reports of the National Academy of Sciences of Ukraine.* 2010. No. 2. P. 170–174.

61. Sabella S., Brunetti V., Vecchio G., Galeone A., Maiorano G., Cingolani R., Pompa P.P. Toxicity of citrate-capped AuNPs: An *in vitro* and *in vivo* assessment // *J. Nanopart. Res.* 2011. V. 13. P. 6821–6835.

62. Coulter J.A., Jain S., Butterworth K.T., Taggart L.E., Dickson G.R., McMahon S.J., Hyland W.B., Muir M.F., Trainor C., Hounsell A.R., O'Sullivan J.M., Schettino G., Currell F.J., Hirst D.G., Prise K.M. Cell type-dependent uptake, localization, and cytotoxicity of 1.9 nm gold nanoparticles // *Int. J. Nanomedicine.* 2012. V. 7. P. 2673–2685.

63. Freese C., Uboldi C., Gibson M.I., Unger R.E., Weksler B.B., Romero I.A., Couraud P.-O., Kirkpatrick C.J. Uptake and cytotoxicity of citrate-coated gold nanospheres: Comparative studies on human endothelial and epithelial cells // *Part. Fibre Toxicol.* 2012. V. 9. 23.

64. Cui L., Zahedi P., Saraceno J., Bristow R., Jaffray D., Allen C. Neoplastic cell response to tiopronin-coated gold nanoparticles // *Nanomedicine.* 2013. V. 9. P. 264–273.

65. Liu Z., Wu Y., Guo Z., Liu Y., Shen Y., Zhou P., Lu X. Effects of internalized gold nanoparticles with respect to cytotoxicity and invasion activity in lung cancer cells // *PLoS One.* 2014. V. 9. e99175.

66. Wang X., Hu X., Li J., Russe A.C.M., Kawazoe N., Yang Y., Chen G. Influence of cell size on cellular uptake of gold nanoparticles // *Biomater. Sci.* 2016. V. 4. P. 970–978.

67. Trono J.D., Mizuno K., Yusa N., Matsukawa T., Yokoyama K., Uesaka M. Size, concentration and incubation time dependence of gold nanoparticle uptake into pancreas cancer cells and its future application to X-ray drug delivery system // *J. Radiat. Res.* 2011. V. 52. P. 103–109.

68. Zhou R., Zhou H., Xiong B., He Y., Yeung E.S. Pericellular matrix enhances retention and cellular uptake of nanoparticles // *J. Am. Chem. Soc.* 2012. V. 134. P. 13404–13409.

69. Shan Y., Ma S., Nie L., Shang X., Hao X., Tang Z., Wang H. Size-dependent endocytosis of single gold nanoparticles // *Chem. Commun.* 2011. V. 47. P. 8091–8093.

70. Hao X., Wu J., Shan Y., Cai M., Shang X., Jiang J., Wang H. Caveolae-mediated endocytosis of biocompatible gold nanoparticles in living HeLa cells // *J. Phys. Condens. Matter.* 2012. V. 24. 164207.

71. Ma X., Wu Y., Jin S., Tian Y., Zhang X., Zhao Y., Yu L., Liang X.-J. Gold nanoparticles induce autophagosome accumulation through size-dependent nanoparticle uptake and lysosome impairment // *ACS Nano.* 2011. V. 5. P. 8629–8639.

72. Sobhan M.A., Sreenivasan V.K.A., Withford M.J., Goldys E.M. Non-specific internalization of laser ablated pure gold nanoparticles in pancreatic tumor cell // *Colloids Surf. B Biointerfaces*. 2012. V. 92. P. 190–195.

73. Sabuncu A.C., Grubbs J., Qian S., Abdel-Fattah T.M., Stacey M.W., Beskok A. Probing nanoparticle interactions in cell culture media // *Colloids Surf. B*. 2012. V. 95. P. 96–102.

74. Huang K., Ma H., Liu J., Huo S., Kumar A., Wei T., Zhang X., Jin S., Gan Y., Wang P.C., He S., Zhang X., Liang X.-J. Size-dependent localization and penetration of ultrasmall gold nanoparticles in cancer cells, multicellular spheroids, and tumors *in vivo* // *ACS Nano*. 2012. V. 6. P. 4483–4493.

75. Huo S., Ma H., Huang K., Liu J., Wei T., Jin S., Zhang J., He S., Liang X.-J. Superior penetration and retention behavior of 50 nm gold nanoparticles in tumors // *Cancer Res*. 2013. V. 73. P. 319–330.

76. Arvizo R.R., Saha S., Wang E., Robertson J.D., Bhattacharya R., Mukherjee P. Inhibition of tumor growth and metastasis by a self-therapeutic nanoparticle // *Proc. Natl. Acad. Sci. USA*. 2013. V. 110. P. 6700–6705.

77. Trickler W.J., Lantz S.M., Murdock R.C., Schrand A.M., Robinson B.L., Newport G.D., Schlager J.J., Oldenburg S.J., Paule M.G., Slikker W. Jr., Hussain S.M., Ali S.F. Brain microvessel endothelial cells responses to gold nanoparticles: In vitro proinflammatory mediators and permeability.// *Nanotoxicology*. 2011. V. 5. P. 479–492.

78. Lèvy R., Thanh N.T.K., Doty R.C., Hussain I., Nichols R.J., Schiffrin D.J., Brust M., Fernig D.G. Rational and combinatorial design of peptide capping ligands for gold nanoparticles // *J. Am. Chem. Soc*. 2004. V. 126. P. 10076–10084.

79. Bartczak D., Muskens O.L., Nitti S., Sanchez-Elsner T., Millar T.M., Kanaras A.G. Interactions of human endothelial cells with gold nanoparticles of different morphologies // *Small*. 2012. V. 8. P. 122–130.

80. Cho E.C., Zhang Q., Xia Y. The effect of sedimentation and diffusion on cellular uptake of gold nanoparticles // *Nat. Nanotechnol*. 2011. V. 6. P. 385–391.

81. Schaeublin N.M., Braydich-Stolle L.K., Maurer E.I., Park K., MacCuspie R.I., Afrooz A.R.M.N., Vaia R.A., Saleh N.B., Hussain S.M. Does shape matter? Bioeffects of gold nanomaterials in a human skin cell model // *Langmuir*. 2012. V. 28. P. 3248–3258.

82. Zhang W., Ji Y., Meng J., Wu X., Xu H. Probing the behaviors of gold nanorods in metastatic breast cancer cells based on UV-vis-NIR absorption spectroscopy // *PLoS One*. 2012. V. 7. e31957.

83. Tarantola M., Pietuch A., Schneider D., Rother J., Sunnick E., Rosman C., Pierrat S., Sönnichsen C., Wegener J., Janshoff A. Toxicity of gold-nanoparticles: Synergistic effects of shape and surface functionalization on micromotility of epithelial cells // *Nanotoxicology*. 2011. V. 5. P. 254–268.

84. Hutter E., Boridy S., Labrecque S., Lalancette-Hébert M., Kriz J., Winnik F.M., Maysinger D. Microglial response to gold nanoparticles // *ACS Nano*. 2010. V. 4. P. 2595–2606.

85. Chu Z., Zhang S., Zhang B., Zhang C., Fang C.-Y., Rehor I., Cigler P., Chang H.-C., Lin G., Liu R., Li Q. Unambiguous observation of shape effects on cellular fate of nanoparticles // *Sci. Rep*. 2014. V. 4. 4495.

86. Liu H., Liu T., Li L., Hao N., Tan L., Meng X., Ren J., Chen D., Tang F. Size dependent cellular uptake, *in vivo* fate and light–heat conversion efficiency of gold nanoshells on silica nanorattles // *Nanoscale*. 2012. V. 4. P. 3523–3529.

87. Navarro J.R.G., Manchon D., Lerouge F., Blanchard N.P., Marotte S., Leverrier Y., Marvel J., Chaput F., Micouin G., Gabudean A.-M., Mosset A., Cottancin E., Baldeck P.L., Kamada K., Parola S. Synthesis of PEGylated gold nanostars and bipyramids for intracellular uptake // *Nanotechnology*. 2012. V. 23. 465602.

88. Avram M., Bălan C.M., Petrescu I., Schiopu V., Mărculescu C., Avram A. Gold nanoparticle uptake by tumour cells of B16 mouse melanoma // *Plasmonics*. 2012. V. 7. P. 717–724.

89. Plascencia-Villa G., Bahena D., Rodríguez A.R., Ponce A., José-Yacamán M. Advanced microscopy of star-shaped gold nanoparticles and their adsorption-uptake by macrophages // *Metallomics*. 2013. V. 5. P. 242–250.

90. Zhang S., Li J., Lykotrafitis G., Bao G., Suresh S. Size dependent endocytosis of nanoparticles // *Adv. Mater.* 2009. V. 21. P. 419–424.

91. Kuo C.-W., Lai J.-J., Wei K.H., Chen P. Studies of surface-modified gold nanowires inside living cells // *Adv. Funct. Mater.* 2007. V. 17. P. 3707–3714.

92. Elsaesser A., Barnes C.A., McKerr G., Salvati A., Lynch I., Dawson K.A., Howard C.V. Quantification of nanoparticle uptake by cells using an unbiased sampling method and electron microscopy // *Nanomedicine (Lond.)*. 2011. V. 6. P. 1189–1198.

93. Elsaesser A., Taylor A., de Yanes G.S., McKerr G., Kim E.-M., O'Hare E., Howard C.V. Quantification of nanoparticle uptake by cells using microscopical and analytical techniques // *Nanomedicine (Lond.)*. 2010. V. 5. P. 1447–1457.

94. Mayhew T.M., Mühlfeld C., Vanhecke D., Ochs M. A review of recent methods for efficiently quantifying immunogold and other nanoparticles using TEM sections through cells, tissues and organs // *Ann. Anat.* 2009. V. 191. P. 153–170.

95. Huefner A., Septiadi D., Wilts B.D., Patel I.I., Kuan W.-L., Fragniere A., Barker R.A., Mahajan S. Gold nanoparticles explore cells: Cellular uptake and their use as intracellular probes // *Methods*. 2014. V. 68. P. 354–363.

96. Polat O., Karagoz A., Isık S., Ozturk R. Influence of gold nanoparticle architecture on in vitro bioimaging and cellular uptake // *Nanopart. Res.* 2014. V. 16. 2725.

97. Huo S., Jin S., Xiaowei M., Xue X., Yang K., Kumar A., Wang P.C., Zhang J., Hu Z., Liang X.-J. Ultra-small gold nanoparticles as carriers for nucleus-based gene therapy due to size-dependent nuclear entry // *ACS Nano*. 2014. V. 8. P. 5852–5862.

98. Yao M., He L., McClements D.J., Xiao H. Uptake of gold nanoparticles by intestinal epithelial cells: Impact of particle size on their absorption, accumulation, and toxicity // *J. Agric. Food Chem.* 2015. V. 63. P. 8044–8049.

99. Noël C., Simard J.-C., Girard D. Gold nanoparticles induce apoptosis, endoplasmic reticulum stress events and cleavage of cytoskeletal proteins in human neutrophils // *Toxicol. In Vitro*. 2016. V. 31. P. 12–22.

100. Li N., Larson T., Nguyen H.H., Sokolov K.V., Ellington A.D. Directed evolution of gold nanoparticle delivery to cells // *Chem. Commun.* 2010. V. 46. P. 392–394.

101. Peckys D.B., de Jonge N. Visualizing gold nanoparticle uptake in live cells with liquid scanning transmission electron microscopy // *Nano Lett.* 2011. V. 11. P. 1733–1738.

102. Chen X., Chen C.-B., Udalagama C.N.B., Ren M., Fong K.E., Yung L.Y.L., Giorgia P., Bettiol A.A., Watt F. High-resolution 3D imaging and quantification of gold nanoparticles in a whole cell using scanning transmission ion microscopy // *Biophys. J.* 2013. V. 104. P. 1419–1425.

103. Gu Y., Sun W., Wang G., Fang N. Single particle orientation and rotation tracking discloses distinctive rotational dynamics of drug delivery vectors on live cell membranes // *J. Am. Chem. Soc.* 2011. V. 133. P. 5720–5723.

104. Berciaud S., Cognet L., Tamarat P., Lounis B. Observation of intrinsic size effects in the optical response of individual gold nanoparticles // *Nano Lett.* 2005. V. 5. P. 515–518.

105. Louit G., Asahi T., Tanaka G., Uwada T., Masuhara H. Spectral and 3-dimensional tracking of single gold nanoparticles in living cells studied by Rayleigh light scattering microscopy // *J. Phys. Chem. C.* 2009. V. 113. P. 11766–11772.

106. Hartsuiker L., Petersen W., Rayavarapu R.G., Lenferink A., Poot A.A., Terstappen L.W.M.M., van Leeuwen T.G., Manohar S., Otto C. Raman and fluorescence spectral imaging of live breast cancer cells incubated with PEGylated gold nanorods // *Appl. Spectrosc.* 2012. V. 66. P. 66–74.

107. Zhu Z.-J., Ghosh P.S., Miranda O.R., Vachet R.W., Rotello V.M. Multiplexed screening of cellular uptake of gold nanoparticles using laser desorption/ionization mass spectrometry // *J. Am. Chem. Soc.* 2008. V. 130. P. 14139–14143.

108. Lin H.-C., Lin H.-H., Kao C.-Y., Yu A.L., Peng W.-P., Chen C.-H. Quantitative measurement of nano-/microparticle endocytosis by cell mass spectrometry // *Angew. Chem. Int. Ed.* 2010. V. 49. P. 3460–3464.

109. Chen J., Irudayaraj J. Quantitative investigation of compartmentalized dynamics of ErbB2 targeting gold nanorods in live cells by single molecule spectroscopy // *ACS Nano.* 2009. V. 3. P. 4071–4079.

110. Rago G., Bauer B., Svedberg F., Gunnarsson L., Ericson M.B., Bonn M., Enejder A. Uptake of gold nanoparticles in healthy and tumor cells visualized by nonlinear optical microscopy // *J. Phys. Chem. B.* 2011. V. 115. P. 5008–5016.

111. Shah N.B., Dong J., Bischof J.C. Cellular uptake and nanoscale localization of gold nanoparticles in cancer using label-free confocal Raman microscopy // *Mol. Pharm.* 2011. V. 8. P. 176–184.

112. Ando J., Fujita K., Smith N.I., Kawata S. Dynamic SERS imaging of cellular transport pathways with endocytosed gold nanoparticles // *Nano Lett.* 2011. V. 11. P. 5344–5348.

113. Jung Y., Reif R., Zeng Y., Wang R.K. Three-dimensional high-resolution imaging of gold nanorods uptake in sentinel lymph nodes // *Nano Lett.* 2011. V. 11. P. 2938–2943.

114. Chen H.-H., Chien C.-C., Petibois C., Wang C.-L., Chu Y.S., Lai S.-F., Hua T.-E., Chen Y.-Y., Cai X., Kempson I.M., Hwu Y., Margaritondo G. Quantitative analysis of nanoparticle internalization in mammalian cells by high resolution X-ray microscopy // *J. Nanobiotechnology.* 2011. V. 9. 14.

115. Menk R.H., Schültke E., Hall C., Arfelli F., Astolfo A., Rigon L., Round A., Ataelmannan K., MacDonald S.R., Juurlink B.H.J. Gold nanoparticle labeling of cells is a sensitive method to investigate cell distribution and migration in animal models of human disease // *Nanomedicine.* 2011. V. 7. P. 647–654.

116. Liu T., Kempson I., de Jonge M., Howard D.L., Thierry B. Quantitative synchrotron X-ray fluorescence study of the penetration of transferrin-conjugated gold nanoparticle inside model tumour tissues // *Nanoscale.* 2014. V. 6. P. 9774–9782.

117. Nam S.Y., Ricles L.M., Suggs L.J., Emelianov S.Y. Nonlinear photoacoustic signal increase from endocytosis of gold nanoparticles // *Opt. Lett.* 2012. V. 37. P. 4708–4710.

118. Lia K., Schneider M. Quantitative evaluation and visualization of size effect on cellular uptake of gold nanoparticles by multiphoton imaging-UV/Vis spectroscopic analysis // *J. Biomed. Opt.* 2014. V. 19. 101505.

119. Bancos S., Tyner K.M. Evaluating the effect of assay preparation on the uptake of gold nanoparticles by RAW264.7 cells // *J. Nanobiotechnology.* 2014. V. 12. 45.

120. Park J.-H., Park J., Dembereldorj U., Cho K., Lee K., Yang S.I., Lee S.Y., Joo S.-W. Raman detection of localized transferrin-coated gold nanoparticles inside a single cell // *Anal. Bioanal. Chem.* 2011. V. 401. P. 1631–1639.

121. Rosman C., Pierrat S., Henkel A., Tarantola M., Schneider D., Sunnick E., Janshoff A., Sönnichsen C. A new approach to assess gold nanoparticle uptake by mammalian cells: Combining optical dark-field and transmission electron microscopy // *Small.* 2012. V. 8. P. 3683–3690.

122. Black K.C.L., Wang Y., Luehmann H.P., Cai X., Xing W., Pang B., Zhao Y., Cutler C.S., Wang L.V., Liu Y., Xia Y. Radioactive 198Au-doped nanostructures with different shapes for *in vivo* analyses of their biodistribution, tumor uptake, and intratumoral distribution // *ACS Nano.* 2014. V. 8. P. 4385–4394.

123. Tan S.H., Horlick G. Background spectral features in inductively coupled plasma/mass spectrometry // *Appl. Spectrosc.* 1986. V. 40. P. 445–460.

124. Tanner S.D., Baranov V.I. A dynamic reaction cell for inductively coupled plasma mass spectrometry (ICP-DRC-MS). II. Reduction of interferences produced within the cell // *J. Am. Soc. Mass Spectrom.* 1999. V. 10. P. 1083–1094.

125. Stender A.S., Marchuk K., Liu C., Sander S., Meyer M.W., Smith E.A., Neupane B., Wang G., Li J., Cheng J-X., Huang B., Fang N. Single cell optical imaging and spectroscopy // *Chem. Rev.* 2013. V. 113. P. 2469–2527.

126. Brandenberger C., Mühlfeld C., Ali Z., Lenz A.-G., Schmid O., Parak W.J., Gehr P., Rothen-Rutishauser B. Quantitative evaluation of cellular uptake and trafficking of plain and polyethylene glycol-coated gold nanoparticles // *Small.* 2010. V. 6. P. 1669–1678.

127. Lipowsky R., Döbereiner H.G. Vesicles in contact with nanoparticles and colloids // *Europhys. Lett.* 1998. V. 43. P. 219–225.

128. Bao G., Bao X.R. Shedding light on the dynamics of endocytosis and viral budding // *Proc. Natl. Acad. Sci. USA.* 2005. V. 102. P. 9997–9998.

129. Li Y., Krögerb M., Liu W.K. Shape effect in cellular uptake of PEGylated nanoparticles: Comparison between sphere, rod, cube and disk // *Nanoscale.* 2015. V. 7. P. 16631–16646.

130. Chaudhuri A., Battaglia G., Golestanian R. The effect of interactions on the cellular uptake of nanoparticles // *Phys. Biol.* 2011. V. 8. 046002.

131. Yuan H., Li J., Bao G., Zhang S. Variable nanoparticle-cell adhesion strength regulates cellular uptake // *Phys. Rev. Lett.* 2010. V. 105. 138101.

132. Yi X., Shi X., Gao H. Cellular uptake of elastic nanoparticles // *Phys. Rev. Lett.* 2011. V. 107. 098101.

133. Decuzzi P., Ferrari M. The role of specific and non-specific interactions in receptor-mediated endocytosis of nanoparticles // *Biomaterials*. 2007. V. 28. P. 2916–2922.

134. Decuzzi P., Ferrari M. The receptor-mediated endocytosis of nonspherical particles // *Biophys. J.* 2008. V. 94. P. 3790–3797.

135. Li X. Size and shape effects on receptor-mediated endocytosis of nanoparticles // *J. Appl. Phys.* 2012. V. 111. 024702.

136. Dobay M.P.D., Piera Alberola A., Mendoza E.R., Rädler J.O.J. Modeling nanoparticle uptake and intracellular distribution using stochastic process algebras // *Nanopart. Res.* 2012. V. 14. 821.

137. Cheng Y., Samia A.C., Li J., Kenney M.E., Resnick A., Burda C. Delivery and efficacy of a cancer drug as a function of the bond to the gold nanoparticle surface // *Langmuir*. 2010. V. 26. P. 2248–2255.

138. Kim C.S., Le N.D.B., Xing Y., Yan B., Tonga G.Y., Kim C., Vachet R.W., Rotello V.M. The role of surface functionality in nanoparticle exocytosis // *Adv. Healthcare Mater*. 2014. V. 3. P. 1200–1202.

139. Lynch I., Dawson K.A. Protein–nanoparticle interactions // *Nano Today*. 2008. V. 3. P. 40–47.

140. Nel A.E., Mädler L., Velegol D., Xia T., Hoek E.M.V., Somasundaran P., Klaessig F., Castranova V., Thompson M. Understanding biophysicochemical interactions at the nano-bio interface // *Nat. Mater*. 2009. V. 8. P. 543–557.

141. Fleischer C.C., Payne C.K. Nanoparticle–cell interactions: Molecular structure of the protein corona and cellular outcomes // *Acc. Chem. Res*. 2014. V. 47. P. 2651–2659.

142. Dobrovolskaia M.A., Neun B.W., Man S., Ye X., Hansen M., Patri A.K., Crist R.M., McNeil S.E. Protein corona composition does not accurately predict hematocompatibility of colloidal gold nanoparticles // *Nanomedicine*. 2014. V. 10. P. 1453–1463.

143. Hamad-Schifferli K. Exploiting the novel properties of protein coronas: Emerging applications in nanomedicine // *Nanomedicine (Lond.)*. 2015. V. 10. P. 1663–1674.

144. Braun N.J., DeBrosse M.C., Hussain S.M., Comfort K.K. Modification of the protein corona–nanoparticle complex by physiological factors // *Mater. Sci. Eng. C*. 2016. V. 64. P. 34–42.

145. Tsai Y.-S., Chen Y.-H., Cheng P.-C., Tsai H.-T., Shiau A.-L., Tzai T.-S., Wu C.-L. TGF-β1 conjugated to gold nanoparticles results in protein conformational changes and attenuates the biological function // *Small*. 2013. V. 9. P. 2119–2128.

146. Dobrovolskaia M.A., Patri A.K., Zheng J., Clogston J.D., Ayub N., Aggarwal P., Neun B.W., Hall J.B., McNeil S.E. Interaction of colloidal gold nanoparticles with human blood: Effects on particle size and analysis of plasma protein binding profiles // *Nanomedicine*. 2009. V. 5. P. 106–117.

147. Lacerda S.H.D.P., Park J.J., Meuse C., Pristinski D., Becker M.L., Karim A., Douglas J.F. Interaction of gold nanoparticles with common human blood proteins // *ACS Nano*. 2010. V. 4. P. 365–379.

148. Oh N., Park J.-H. Surface chemistry of gold nanoparticles mediates their exocytosis in macrophages // *ACS Nano*. 2014. V. 8. P. 6232–6241.

149. Mahmoudi M., Lynch I., Ejtehadi M.R., Monopoli M.P., Bombelli F.B., Laurent S. Protein–nanoparticle interactions: Opportunities and challenges // *Chem. Rev.* 2011. V. 111. P. 5610–5637.

150. Goy-López S., Juárez J., Alatorre-Meda M., Casals E., Puntes V.F., Taboada P., Mosquera V. Physicochemical characteristics of protein–NP bioconjugates: The role of particle curvature and solution conditions on human serum albumin conformation and fibrillogenesis inhibition // *Langmuir.* 2012. V. 28. P. 9113–9126.

151. Arvizo R.R., Giri K., Moyano D., Miranda O.R., Madden B., McCormick D.J., Bhattacharya R., Rotello V.M., Kocher J.-P., Mukherjee P. Identifying new therapeutic targets via modulation of protein corona formation by engineered nanoparticles // *PLoS One.* 2012. V. 7. e33650.

152. Cheng X., Tian X., Wu A., Li J., Tian J., Chong Y., Chai Z., Zhao Y., Chen C., Ge C. Protein corona influences cellular uptake of gold nanoparticles by phagocytic and nonphagocytic cells in a size-dependent manner // *ACS Appl. Mater. Interfaces.* 2015. V. 7. P. 20568–20575.

153. Saha K., Moyano D.F., Rotello V.M. Protein coronas suppress the hemolytic activity of hydrophilic and hydrophobic nanoparticles // *Mater. Horiz.* 2014. V. 1. P. 102–105.

154. Cruje C., Chithrani D.B. Polyethylene glycol density and length affects nanoparticle uptake by cancer cells // *J. Nanomed. Res.* 2014. V. 1. 00006.

155. Pelaz B., del Pino P., Maffre P., Hartmann R., Gallego M., Rivera-Fernández S., de la Fuente J.M., Nienhaus G.U., Parak W.J. Surface functionalization of nanoparticles with polyethylene glycol (PEG): Effects on protein adsorption and cellular uptake // *ACS Nano.* 2015. V. 9. P. 6996–7008.

156. Niidome T., Yamagata M., Okamoto Y., Akiyama Y., Takahashi H., Kawano T., Katayama Y., Niidome Y. PEG-modified gold nanorods with a stealth character for *in vivo* applications // *J. Control. Release.* 2006. V. 114. P. 343–347.

157. Jokerst J.V., Lobovkina T., Zare R.N., Gambhir S.S. Nanoparticle PEGylation for imaging and therapy // *Nanomedicine (Lond.).* 2011. V. 6. P. 715–728.

158. Karakoti A.S., Das S., Thevuthasan S., Seal S. PEGylated inorganic nanoparticles // *Angew. Chem. Int. Ed.* 2011. V. 50. P. 1980–1994.

159. Jain R.K., Booth M.F. What brings pericytes to tumor vessels? // *J. Clin. Invest.* 2003. V. 112. P. 1134–1136.

160. Otsuka H., Nagasaki Y., Kataoka K. PEGylated nanoparticles for biological and pharmaceutical applications // *Adv. Drug Deliv. Rev.* 2003. V. 55. P. 403–419.

161. Sperling R.A., Parak W.J. Surface modification, functionalization and bioconjugation of colloidal inorganic nanoparticles // *Phil. Trans. R. Soc. A.* 2010. V. 368. P. 1333–1383.

162. Soenen S.J., Manshian B.B., Abdelmonem A.M., Montenegro J.-M., Tan S., Balcaen L., Vanhaecke F., Brisson A.R., Parak W.J., De Smedt S.C., Braeckmans K. The cellular interactions of PEGylated gold nanoparticles: Effect of PEGylation on cellular uptake and cytotoxicity // *Part. Part. Syst. Charact.* 2014. V. 31. P. 794–800.

163. Song H., Xu Q., Di H., Guo T., Qi Z., Zhao S. Time evolution and dynamic cellular uptake of PEGYlated gold nanorods // *RSC Adv.* 2016. V. 6. P. 8089–8092.

164. Shenoy D., Fu W., Li J., Crasto C., Jones G., Dimarzio C., Sridhar S., Amiji M. Surface functionalization of gold nanoparticles using hetero-bifunctional poly(ethylene glycol) spacer for intracellular tracking and delivery // *Int. J. Nanomedicine.* 2006. V. 1. P. 51–58.

165. Bergen J.M., von Recum H.A., Goodman T.T., Massey A.P., Pun S.H. Gold nanoparticles as a versatile platform for optimizing physicochemical parameters for targeted drug delivery // *Macromol. Biosci.* 2006. V. 6. P. 506–516.

166. Cho W.-S., Cho M., Jeong J., Choi M., Cho H.-Y., Han B.S., Kim S.H., Kim H.O., Lim Y.T., Chung B.H., Jeong J. Acute toxicity and pharmacokinetics of 13 nm-sized PEG-coated gold nanoparticles // *Toxicol. Appl. Pharmacol.* 2009. V. 236. P. 16–24.

167. Cho W.-S., Cho M., Jeong J., Choi M., Han B.S., Shin H.-S., Hong J., Chung B.H., Jeong J., Cho M.-H. Size-dependent tissue kinetics of PEG-coated gold nanoparticles // *Toxicol. Appl. Pharmacol.* 2010. V. 245. P. 116–123.

168. Liu C.J., Wang C.H., Chien C.C., Yang T.Y., Chen S.T., Leng W.H., Lee C.F., Lee K.H., Hwu Y., Lee Y.C., Cheng C.L., Yang C.S., Chen Y.J., Je J.H., Margaritondo G. Enhanced X-ray irradiation-induced cancer cell damage by gold nanoparticles treated by a new synthesis method of polyethylene glycol modification // *Nanotechnology.* 2008. V. 19. 295104.

169. Zhang G., Yang Z., Lu W., Zhang R., Huang Q., Tian M., Li L., Liang D., Li C. Influence of anchoring ligands and particle size on the colloidal stability and *in vivo* biodistribution of polyethylene glycol-coated gold nanoparticles in tumor-xenografted mice // *Biomaterials.* 2009. V. 30. P. 1928–1936.

170. Gu Y.-J., Cheng J., Lin C.-C., Lam Y.W., Cheng S.H., Wong W.-T. Nuclear penetration of surface functionalized gold nanoparticles // *Toxicol. Appl. Pharmacol.* 2009. V. 237. P. 196–204.

171. Dykman L.A., Khlebtsov N.G. Uptake of engineered gold nanoparticles into mammalian cells // *Chem. Rev.* 2014. V. 114. P. 1258–1288.

172. Brandenberger C., Mühlfeld C., Ali Z., Lenz A.-G., Schmid O., Parak W.J., Gehr P., Rothen-Rutishauser B. Quantitative evaluation of cellular uptake and trafficking of plain and polyethylene glycol-coated gold nanoparticles // *Small.* 2010. V. 6. P. 1669–1678.

173. Lund T., Callaghan M.F., Williams P., Turmaine M., Bachmann C., Rademacher T., Roitt I.M., Bayford R. The influence of ligand organization on the rate of uptake of gold nanoparticles by colorectal cancer cells // *Biomaterials.* 2011. V. 32. P. 9776–9784.

174. Kah J.C., Wong K.Y., Neoh K.G., Song J.H., Fu J.W., Mhaisalkar S., Olivo M., Sheppard C.J. Critical parameters in the pegylation of gold nanoshells for bio-medical applications: An *in vitro* macrophage study // *J. Drug Target.* 2009. V. 17. P. 181–193.

175. Huff T.B., Hansen M.N., Zhao Y., Cheng J.-X., Wei A. Controlling the cellular uptake of gold nanorods // *Langmuir.* 2007. V. 23. P. 1596–1599.

176. Rayavarapu R.G., Petersen W., Hartsuiker L., Chin P., Janssen H., van Leeuwen F.W.B., Otto C., Manohar S., van Leeuwen T.G. *In vitro* toxicity studies of polymer-coated gold nanorods // *Nanotechnology.* 2010. V. 21. 145101.

177. Grabinski C., Schaeublin N., Wijaya A., D'Couto H., Baxamusa S.H., Hamad-Schifferli K., Hussain S.M. Effect of gold nanorod surface chemistry on cellular response // *ACS Nano.* 2011. V. 5. P. 2870–2879.
178. Alkilany A.M., Shatanawi A., Kurtz T., Caldwell R.B., Caldwell R.W. Toxicity and cellular uptake of gold nanorods in vascular endothelium and smooth muscles of isolated rat blood vessel: Importance of surface modification // *Small.* 2012. V. 8. P. 1270–1278.
179. Arnida, Malugin A., Ghandehari H. Cellular uptake and toxicity of gold nanoparticles in prostate cancer cells: A comparative study of rods and spheres // *J. Appl. Toxicol.* 2010. V. 30. P. 212–217.
180. Arnida, Janát-Amsbury M.M., Ray A., Peterson C.M., Ghandehari H. Geometry and surface characteristics of gold nanoparticles influence their biodistribution and uptake by macrophages // *Eur. J. Pharm. Biopharm.* 2011. V. 77. P. 417–423.
181. Cho E.C., Liu Y., Xia Y. A simple spectroscopic method for differentiating cellular uptakes of gold nanospheres and nanorods from their mixtures // *Angew. Chem. Int. Ed.* 2010. V. 49. P. 1976–1980.
182. Pietuch A., Brückner B.R., Schneider D., Tarantola M., Rosman C., Sönnichsen C., Janshoff A. Mechanical properties of MDCK II cells exposed to gold nanorods // *Beilstein J. Nanotechnol.* 2015. V. 6. P. 223–231.
183. Puvanakrishnan P., Park J., Chatterjee D., Krishnan S., Tunnell J.W. *In vivo* tumor targeting of gold nanoparticles: Effect of particle type and dosing strategy // *Int. J. Nanomedicine.* 2012. V. 7. P. 1251–1258.
184. Larson T.A., Joshi P.P., Sokolov K. Preventing protein adsorption and macrophage uptake of gold nanoparticles *via* a hydrophobic shield // *ACS Nano.* 2012. V. 6. P. 9182–9190.
185. Takahashi H., Niidome T., Kawano T., Yamada S., Niidome Y. Surface modification of gold nanorods using layer-by-layer technique for cellular uptake // *J. Nanopart. Res.* 2008. V. 10. P. 221–228.
186. Thomas M., Klibanov A.M. Conjugation to gold nanoparticles enhances polyethylenimine's transfer of plasmid DNA into mammalian cells // *Proc. Natl. Acad. Sci. USA.* 2003. V. 100. P. 9138–9143.
187. Pyshnaya I.A., Razum K.V., Poletaeva J.E., Pyshnyi D.V., Zenkova M.A., Ryabchikova E.I. Comparison of behaviour in different liquids and in cells of gold nanorods and spherical nanoparticles modified by linear polyethyleneimine and bovine serum albumin // *BioMed Res. Int.* 2014. V. 2014. 908175.
188. Hauck T.S., Ghazani A.A., Chan W.C.W. Assessing the effect of surface chemistry on gold nanorod uptake, toxicity, and gene expression in mammalian cells // *Small.* 2008. V. 4. P. 153–159.
189. Alkilany A.M., Nagaria P.K., Hexel C.R., Shaw T.J., Murphy C.J., Wyatt M.D. Cellular uptake and cytotoxicity of gold nanorods: Molecular origin of cytotoxicity and surface effects // *Small.* 2009. V. 5. P. 701–708.
190. Parab H.J., Chen H.M., Lai T.-C., Huang J.H., Chen P.H., Liu R.-S., Hsiao M., Chen C.-H., Tsai D.-P., Hwu Y.-K. Biosensing, cytotoxicity, and cellular uptake studies of surface-modified gold nanorods // *J. Phys. Chem. C.* 2009. V. 113. P. 7574–7578.

191. Fan Z., Yang X., Li Y., Li S., Niu S., Wu X., Wei J., Nie G. Deciphering an underlying mechanism of differential cellular effects of nanoparticles: An example of Bach-1 dependent induction of HO-1 expression by gold nanorod // *Biointerphases*. 2012. V. 7. 10.

192. Goh D., Gong T., Dinish U.S., Maiti K.K., Fu C.Y., Yong K.-T., Olivo M. Pluronic triblock copolymer encapsulated gold nanorods as biocompatible localized plasmon resonance-enhanced scattering probes for dark-field imaging of cancer cells // *Plasmonics*. 2012. V. 7. P. 595–601.

193. Qiu Y., Liu Y., Wang L., Xu L., Bai R., Ji Y., Wu X., Zhao Y., Li Y., Chen C. Surface chemistry and aspect ratio mediated cellular uptake of Au nanorods // *Biomaterials*. 2010. V. 31. P. 7606–7619.

194. Wang L., Liu Y., Li W., Jiang X., Ji Y., Wu X., Xu L., Qiu Y., Zhao K., Wei T., Li Y., Zhao Y., Chen C. Selective targeting of gold nanorods at the mitochondria of cancer cells: Implications for cancer therapy // *Nano Lett*. 2011. V. 11. P. 772–780.

195. Vigderman L., Manna P., Zubarev E.R. Quantitative replacement of cetyl trimethylammonium bromide by cationic thiol ligands on the surface of gold nanorods and their extremely large uptake by cancer cells // *Angew. Chem. Int. Ed*. 2012. V. 51. P. 636–641.

196. Li J.-L., Wang L., Liu X.-Y., Zhang Z.-P., Guo H.-C., Liu W.-M., Tang S.-H. *In vitro* cancer cell imaging and therapy using transferrin-conjugated gold nanoparticles // *Cancer Lett*. 2009. V. 274. P. 319–326.

197. Salmaso S., Caliceti P., Amendola V., Meneghetti M., Magnusson J.P., Pasparakis G., Alexander C. Cell up-take control of gold nanoparticles functionalized with a thermoresponsive polymer // *J. Mater. Chem*. 2009. V. 19. P. 1608–1615.

198. Dragoni S., Franco G., Regoli M., Bracciali M., Morandi V., Sgaragli G., Bertelli E., Valoti M. Gold nanoparticles uptake and cytotoxicity assessed on rat liver precision-cut slices // *Toxicol. Sci*. 2012. V. 128. P. 186–197.

199. Liu X., Jin Q., Ji Y., Ji J. Minimizing nonspecific phagocytic uptake of biocompatible gold nanoparticles with mixed charged zwitterionic surface modification // *J. Mater. Chem*. 2012. V. 22. P. 1916–1927.

200. Cho E.C., Xie J., Wurm P.A., Xia Y. Understanding the role of surface charges in cellular adsorption versus internalization by selectively removing gold nanoparticles on the cell surface with a I2/KI etchant // *Nano Lett*. 2009. V. 9. P. 1080–1084.

201. Liang M., Lin I.-C., Whittaker M.R., Minchin R.F., Monteiro M.J., Toth I. Cellular uptake of densely packed polymer coatings on gold nanoparticles // *ACS Nano*. 2010. V. 4. P. 403–413.

202. Lin I.-C., Liang M., Liu T-Y., Ziora Z.M., Monteiro M.J., Toth I. Interaction of densely polymer-coated gold nanoparticles with epithelial Caco-2 monolayers // *Biomacromolecules*. 2011. V. 12. P. 1339–1348.

203. Freese C., Gibson M.I., Klok H.-A., Unger R.E., Kirkpatrick C.J. Size- and coating-dependent uptake of polymer-coated gold nanoparticles in primary human dermal microvascular endothelial cells // *Biomacromolecules*. 2012. V. 13. P. 1533–1543.

204. Ojea-Jiménez I., García-Fernández L., Lorenzo J., Puntes V.F. Facile preparation of cationic gold nanoparticle-bioconjugates for cell penetration and nuclear targeting // *ACS Nano*. 2012. V. 6. P. 7692–7702.

205. Choi S.Y., Jang S.H., Park J., Jeong S., Park J.H., Ock K.S., Lee K., Yang S.I., Joo S.-W., Ryu P.D., Lee S.Y. Cellular uptake and cytotoxicity of positively charged chitosan gold nanoparticles in human lung adenocarcinoma cells // *J. Nanopart. Res.* 2012. V. 14. 1234.

206. Hühn D., Kantner K., Geidel C., Brandholt S., De Cock I., Soenen S.J., Rivera Gil P., Montenegro J.-M., Braeckmans K., Müllen K., Nienhaus G.U., Klapper M., Parak W.J. Polymer-coated nanoparticles interacting with proteins and cells: Focusing on the sign of the net charge // *ACS Nano.* 2013. V. 7. P. 3253–3263.

207. Fytianos K., Rodriguez-Lorenzo L., Clift M.J.D., Blank F., Vanhecke D., von Garnier C., Petri-Fink A., Rothen-Rutishauser B. Uptake efficiency of surface modified gold nanoparticles does not correlate with functional changes and cytokine secretion in human dendritic cells *in vitro* // *Nanomedicine.* 2015. V. 11. P. 633–644.

208. Zhu X.-M., Fang C., Jia H., Huang Y., Cheng C.H.K., Ko C.-H., Chen Z., Wang J., Wang Y.-X.J. Cellular uptake behaviour, photothermal therapy performance, and cytotoxicity of gold nanorods with various coatings // *Nanoscale.* 2014. V. 6. P. 11462–11472.

209. Liang Z., Liu Y., Li X., Wu Q., Yu J., Luo S., Lai L., Liu S. Surface-modified gold nanoshells for enhanced cellular uptake // *J. Biomed. Mater. Res. A.* 2011. V. 98. P. 479–487.

210. Arvizo R.R., Miranda O.R., Thompson M.A., Pabelick C.M., Bhattacharya R., Robertson J.D., Rotello V.M., Prakash Y.S., Mukherjee P. Effect of nanoparticle surface charge at the plasma membrane and beyond // *Nano Lett.* 2010. V. 10. P. 2543–2548.

211. Lin I-C., Liang M., Liu T-Y., Monteiro M.J., Toth I. Cellular transport pathways of polymer coated gold nanoparticles // *Nanomedicine.* 2012. V. 8. P. 8–11.

212. Maiorano G., Sabella S., Sorce B., Brunetti V., Malvindi M.A., Cingolani R., Pompa P.P. Effects of cell culture media on the dynamic formation of protein–nanoparticle complexes and influence on the cellular response // *ACS Nano.* 2010. V. 4. P. 7481–7491.

213. Pal A., Shah S., Kulkarni V., Murthy R.S.R., Devi S. Template free synthesis of silver–gold alloy nanoparticles and cellular uptake of gold nanoparticles in Chinese Hamster Ovary cell // *Mater. Chem. Phys.* 2009. V. 113. P. 276–282.

214. Boca S.C., Potara M., Toderas F., Stephan O., Baldeck P.L., Astilean S. Uptake and biological effects of chitosan-capped gold nanoparticles on Chinese hamster ovary cells // *Mater. Sci. Eng. C.* 2011. V. 31. P. 184–189.

215. Chithrani D.B., Dunne M., Stewart J., Allen C., Jaffray D.A. Cellular uptake and transport of gold nanoparticles incorporated in a liposomal carrier // *Nanomedicine.* 2010. V. 6. P. 161–169.

216. Dhar S., Mali V., Bodhankar S., Shiras A., Prasad B.L.V., Pokharkar V. Biocompatible gellan gum-reduced gold nanoparticles: Cellular uptake and subacute oral toxicity studies // *J. Appl. Toxicol.* 2011. V. 31. P. 411–420.

217. Marisca O.T., Kantner K., Pfeiffer C., Zhang Q., Pelaz B., Leopold N., Parak W.J., Rejman J. Comparison of the *in vitro* uptake and toxicity of collagen- and synthetic polymer-coated gold nanoparticles // *Nanomaterials.* 2015. V. 5. P. 1418–1430.

218. Gao B., Xu J., He k.-w., Shen L., Chen H., Yang H.-j., Li A.-h., Xiao W.-h. Cellular uptake and intra-organ biodistribution of functionalized silica-coated gold nanorods // *Mol. Imaging Biol.* 2016. V. 18. P. 667–676.

219. Lee Y., Geckeler K.E. Cytotoxicity and cellular uptake of lysozyme-stabilized gold nanoparticles // *J. Biomed. Mater. Res. A*. 2012. V. 100. P. 848–855.

220. Hao Y., Yang X., Song S., Huang M., He C., Cui M., Chen J. Exploring the cell uptake mechanism of phospholipid and polyethylene glycol coated gold nanoparticles // *Nanotechnology*. 2012. V. 23. 045103.

221. Wang M., Petersen N.O. Lipid-coated gold nanoparticles promote lamellar body formation in A549 cells // *Biochim. Biophys. Acta*. 2013. V. 1831. P. 1089–1097.

222. Wang M., Petersen N.O. Characterization of phospholipid-encapsulated gold nanoparticles: A versatile platform to study drug delivery and cellular uptake mechanisms // *Can. J. Chem.* 2015. V. 93. P. 265–271.

223. Yang S., Damiano M.G., Zhang H., Tripathy S., Luthi A.J., Rink J.S., Ugolkov A.V., Singh A.T K., Dave S.S., Gordon L.I., Thaxton C.S. Biomimetic, synthetic HDL nanostructures for lymphoma // *Proc. Natl. Acad. Sci. USA*. 2013. V. 110. P. 2511–2516.

224. Amarnath K., Mathew N.L., Nellore J., Siddarth C.R.V., Kumar J. Facile synthesis of biocompatible gold nanoparticles from Vites vinefera and its cellular internalization against HBL-100 cells // *Cancer Nano*. 2011. V. 2. P. 121–132.

225. Faulk W., Taylor G. An immunocolloid method for the electron microscope // *Immunochemistry*. 1971. V. 8. P. 1081–1083.

226. Hu R., Yong K.-T., Roy I., Ding H., He S., Prasad P.N. Metallic nanostructures as localized plasmon resonance enhanced scattering probes for multiplex dark-field targeted imaging of cancer cells // *J. Phys. Chem. C*. 2009. V. 113. P. 2676–2684.

227. Rejiya C.S., Kumar J., Raji V., Vibin M., Abraham A. Laser immunotherapy with gold nanorods causes selective killing of tumour cells // *Pharmacol. Res*. 2012. V. 65. P. 261–269.

228. Pissuwan D., Valenzuela S.M., Killingsworth M.C., Xu X., Cortie M.B. Targeted destruction of murine macrophage cells with bioconjugated gold nanorods // *J. Nanoparticle Res*. 2007. V. 9. P. 1109–1124.

229. Lapotko D.O., Lukianova-Hleb E.Y., Oraevsky A.A. Clusterization of nanoparticles during their interaction with living cells // *Nanomedicine (Lond.)*. 2007. V. 2. P. 241–253.

230. Melancon M.P., Lu W., Yang Z., Zhang R., Cheng Z., Elliot A.M., Stafford J., Olson T., Zhang J.Z., Li C. *In vitro* and *in vivo* targeting of hollow gold nanoshells directed at epidermal growth factor receptors for photothermal ablation therapy // *Mol. Cancer Ther.* 2008. V. 7. P. 1730–1739.

231. Lu W., Xiong C., Zhang G., Huang Q., Zhang R., Zhang J.Z., Li C. Targeted photothermal ablation of murine melanomas with melanocyte-stimulating hormone analog-conjugated hollow gold nanospheres // *Clin. Cancer Res.* 2009. V. 15. P. 876–886.

232. Marega R., Karmani L., Flamant L., Nageswaran P.G., Valembois V., Masereel B., Feron O., Borght T.V., Lucas S., Michiels C., Gallez B., Bonifazi D. Antibody-functionalized polymer-coated gold nanoparticles targeting cancer cells: An *in vitro* and *in vivo* study // *J. Mater. Chem.* 2012. V. 22. P. 21305–21312.

233. Raoof M., Corr S.J., Kaluarachchi W.D., Massey K.L., Briggs K., Zhu C., Cheney M.A., Wilson L.J., Curley S.A. Stability of antibody-conjugated gold nanoparticles in the endolysosomal nanoenvironment: Implications for noninvasive radiofrequency-based cancer therapy // *Nanomedicine*. 2012. V. 8. P. 1096–1105.

234. Song J., Zhou J., Duan H. Self-assembled plasmonic vesicles of SERS-encoded amphiphilic gold nanoparticles for cancer cell targeting and traceable intracellular drug delivery // *J. Am. Chem. Soc.* 2012. V. 134. P. 13458–13469.

235. Liu T., Cousins A., Chien, C.-C. Kempson I., Thompson S., Hwu Y., Thierry B. Immunospecific targeting of CD45 expressing lymphoid cells: Towards improved detection agents of the sentinel lymph node // *Cancer Lett.* 2013. V. 328. P. 271–277.

236. Charan S., Sanjiv K., Singh N., Chien F.-C., Chen Y.-F., Nergui N.N., Huang S.-H., Kuo C.W., Lee T.-C., Chen P. Development of chitosan oligosaccharide-modified gold nanorods for *in vivo* targeted delivery and noninvasive imaging by NIR irradiation // *Bioconjug. Chem.* 2012. V. 23. P. 2173–2182.

237. Au L., Zhang Q., Cobley C.M., Gidding M., Schwartz A.G., Chen J., Xia Y. Quantifying the cellular uptake of antibody-conjugated Au nanocages by two-photon microscopy and inductively coupled plasma mass spectrometry // *ACS Nano*. 2010. V. 4. P. 35–42.

238. Cho E.C., Au L., Zhang Q., Xia Y. The effects of size, shape, and surface functional group of gold nanostructures on their adsorption and internalization by cells // *Small*. 2010. V. 6. P. 517–522.

239. Cho E.C., Zhang Y., Cai X., Moran C.M., Wang L.V., Xia Y. Quantitative analysis of the fate of gold nanocages *in vitro* and *in vivo* after uptake by U87-MG tumor cells // *Angew. Chem, Int. Ed.* 2013. V. 52. P. 1152–1155.

240. Wagstaff K.M., Jans D.A. Protein transduction: Cell penetrating peptides and their therapeutic applications // *Curr. Med. Chem.* 2006. V. 13. P. 1371–1387.

241. Okuyama M., Laman H., Kingsbury S.R., Visintin C., Leo E., Eward K.L., Stoeber K., Boshoff C., Williams G.H., Selwood D.L. Small-molecule mimics of an α-helix for efficient transport of proteins into cells // *Nat. Methods*. 2007. V. 4. P. 153–159.

242. Frankel A.D., Pabo C.O. Cellular uptake of the tat protein from human immunodeficiency virus // *Cell*. 1988. V. 55. P. 1189–1193.

243. Feldherr C.M., Lanford R.E., Akin D. Signal-mediated nuclear transport in simian virus 40-transformed cells is regulated by large tumor antigen // *Proc. Natl. Acad. Sci. USA*. 1992. V. 89. P. 11002–11005.

244. Vivès E., Schmidt J., Pèlegrin A. Cell-penetrating and cell-targeting peptides in drug delivery // *Biochim. Biophys. Acta*. 2008. V. 1786. P. 126–138.

245. Tkachenko A.G., Xie H., Coleman D., Glomm W., Ryan J., Anderson M.F., Franzen S., Feldheim D.L. Multifunctional gold nanoparticle–peptide complexes for nuclear targeting // *J. Am. Chem. Soc.* 2003. V. 125. P. 4700–4701.

246. Mandal D., Maran A., Yaszemski M.J., Bolander M.E., Sarkar G. Cellular uptake of gold nanoparticles directly cross-linked with carrier peptides by osteosarcoma cells // *J. Mater. Sci. Mater. Med.* 2009. V. 20. P. 347–350.

247. Tkachenko A.G., Xie H., Liu Y., Coleman D., Ryan J., Glomm W.R., Shipton M.K., Franzen S., Feldheim D.L. Cellular trajectories of peptide-modified gold particle complexes: Comparison of nuclear localization signals and peptide transduction domains // *Bioconjug. Chem.* 2004. V. 15. P. 482–490.

248. Ryan J., Overton K.W., Speight M.E., Oldenburg C.N., Loo L., Robarge W., Franzen S., Feldheim D.L. Cellular uptake of gold nanoparticles passivated with BSA-SV40 large T antigen conjugates // *Anal. Chem.* 2007. V. 79. P. 9150–9159.

249. Liu Y., Shipton M.K., Ryan J., Kaufman E.D., Franzen S., Feldheim D.L. Synthesis, stability, and cellular internalization of gold nanoparticles containing mixed peptide-poly(ethylene glycol) monolayers // *Anal. Chem.* 2007. V. 79. P. 2221–2229.

250. Liu Y., Franzen S. Factors determining the efficacy of nuclear delivery of antisense oligonucleotides by gold nanoparticles // *Bioconjug. Chem.* 2008. V. 19. P. 1009–1016.

251. Ghosh P., Yang X., Arvizo R., Zhu Z.-J., Agasti S.S., Mo Z., Rotello V.M. Intra cellular delivery of a membrane-impermeable enzyme in active form using func- tionalized gold nanoparticles // *J. Am. Chem. Soc.* 2010. V .132. P. 2642–2645.

252. de la Fuente J.M., Berry C.C. Tat peptide as an efficient molecule to translocate gold nanoparticles into the cell nucleus // *Bioconjug. Chem.* 2005. V. 16. P. 1176–1180.

253. Oh E., Delehanty J.B., Sapsford K.E., Susumu K., Goswami R., Blanco-Canosa J.B., Dawson P.E., Granek J., Shoff M., Zhang Q., Goering P.L., Huston A., Medintz I.L. Cellular uptake and fate of PEGylated gold nanoparticles is dependent on both cell-penetration peptides and particle size // *ACS Nano.* 2011. V. 5. P. 6434–6448.

254. Krpetić Ž., Saleemi S., Prior I.A., Sée V., Qureshi R., Brust M. Negotiation of intra- cellular membrane barriers by TAT-modified gold nanoparticles // *ACS Nano.* 2011. V. 5. 5195–5201.

255. Dekiwadia C.D., Lawrie A.C., Fecondo J.V. Peptide-mediated cell penetration and targeted delivery of gold nanoparticles into lysosomes // *J. Pept. Sci.* 2012. V. 18. P. 527–534.

256. Nativo P., Prior I.A., Brust M. Uptake and intracellular fate of surface-modified gold nanoparticles // *ACS Nano.* 2008. V. 2. P. 1639–1644.

257. Sun L., Liu D., Wang Z. Functional gold nanoparticle–peptide complexes as cell- targeting agents // *Langmuir.* 2008. V. 24. P. 10293–10297.

258. Morais T., Soares M.E., Duarte J.A., Soares L., Maia S., Gomes P., Pereira E., Fraga S., Carmo H., Bastos M.deL. Effect of surface coating on the biodistribution profile of gold nanoparticles in the rat // *Eur. J. Pharm. Biopharm.* 2012. V. 80. P. 185–193.

259. Oyelere A.K., Chen P.C., Huang X., El-Sayed I.H., El-Sayed M.A. Peptide- conjugated gold nanorods for nuclear targeting // *Bioconjug. Chem.* 2007. V. 18. P. 1490–1497.

260. Kang B., Mackey M.A., El-Sayed M.A. Nuclear targeting of gold nanoparticles in cancer cells induces DNA damage, causing cytokinesis arrest and apoptosis // *J. Am. Chem. Soc.* 2010. V. 132. P. 1517–1519.

261. Yuan H., Fales A.M., Vo-Dinh T. TAT peptide-functionalized gold nanostars: Enhanced intracellular delivery and efficient NIR photothermal therapy using ultra-low irradiance // *J. Am. Chem. Soc.* 2012. V. 134. P. 11358–11361.

262. Liu S.-Y., Liang Z.-S., Gao F., Luo S.-F., Lu G.-Q. *In vitro* photothermal study of gold nanoshells functionalized with small targeting peptides to liver cancer cells // *J. Mater. Sci. Mater. Med.* 2010. V. 21. P. 665–674.

263. Yang H., Fung S.-Y., Liu M. Programming the cellular uptake of physiologically stable peptide–gold nanoparticle hybrids with single amino acids // *Angew. Chem. Int. Ed.* 2011. V. 50. P. 9643–9646.

264. Wang G., Norton A.S., Pokharel D., Song Y., Hill R.A. KDEL peptide gold nano-constructs: A promising nanoplatform for drug delivery // *Nanomedicine.* 2013. V. 9. P. 366–374.

265. Shirazi A.N., Mandal D., Tiwari R.K., Guo L., Lu W., Parang K. Cyclic peptide-capped gold nanoparticles as drug delivery systems // *Mol. Pharm.* 2013. V. 10. P. 500–511.

266. Yao L., Danniels J., Moshnikova A., Kuznetsov S., Ahmed A., Engelman D.M., Reshetnyak Y.K., Andreev O.A. pHLIP peptide targets nanogold particles to tumors // *Proc. Natl. Acad. Sci. USA.* 2013. V. 110. P. 465–470.

267. Yang C., Uertz J., Yohan D., Chithrani B.D. Peptide modified gold nanoparticles for improved cellular uptake, nuclear transport, and intracellular retention // *Nanoscale.* 2014. V. 6. P. 12026–12033.

268. Todorova N., Chiappini C., Mager M., Simona B., Patel I.I., Stevens M.M., Yarovsky I. Surface presentation of functional peptides in solution determines cell internalization efficiency of TAT conjugated nanoparticles // *Nano Lett.* 2014. V. 14. P. 5229–5237.

269. Lin J., Alexander-Katz A. Cell membranes open "doors" for cationic nanoparticles/biomolecules: Insights into uptake kinetics // *ACS Nano.* 2013. V. 7. P. 10799–10808.

270. Nowinski A.K., White A.D., Keefe A.J., Jiang S. Biologically inspired stealth peptide-capped gold nanoparticles // *Langmuir.* 2014. V. 30. P. 1864–1870.

271. Park H., Tsutsumi H., Mihara H. Cell penetration and cell-selective drug delivery using α-helix peptides conjugated with gold nanoparticles // *Biomaterials.* 2013. V. 34. P. 4872–4879.

272. Wang G., Papasani M.R., Cheguru P., Hrdlicka P.J., Hill R.A. Gold-peptide nano-conjugate cellular uptake is modulated by serum proteins // *Nanomedicine.* 2012. V. 8. P. 822–832.

273. Rosi N.L., Giljohann D.A., Thaxton C.S., Lytton-Jean A.K.R., Han M.S., Mirkin C.A. Oligonucleotide-modified gold nanoparticles for intracellular gene regulation // *Science.* 2006. V. 312. P. 1027–1030.

274. Giljohann D.A., Seferos D.S., Patel P.C., Millstone J.E., Rosi N.L., Mirkin C.A. Oligonucleotide loading determines cellular uptake of DNA-modified gold nanoparticles // *Nano Lett.* 2007. V. 7. P. 3818–3821.

275. Patel P.C., Giljohann D.A., Seferos D.S., Mirkin C.A. Peptide antisense nanopar-ticles // *Proc. Natl. Acad. Sci. USA.* 2008. V. 105. P. 17222–17226.

276. Massich M.D., Giljohann D.A., Schmucker A.L., Patel P.C., Mirkin C.A. Cellular response of polyvalent oligonucleotide–gold nanoparticle conjugates // *ACS Nano.* 2010. V. 4. P. 5641–5646.

277. Cutler J.I., Auyeung E., Mirkin C.A. Spherical nucleic acids // *J. Am. Chem. Soc.* 2012. V. 134. P. 1376–1391.

278. Wu X.A., Choi C.H.J., Zhang C., Hao L., Mirkin C.A. Intracellular fate of spherical nucleic acid nanoparticle conjugates // *J. Am. Chem. Soc.* 2014. V. 136. P. 7726–7733.

279. Hurst S.J., Hill HD., Mirkin C.A. "Three-dimensional hybridization" with polyvalent DNA–gold nanoparticle conjugates // *J. Am. Chem. Soc.* 2008. V. 130. P. 12192–12200.

280. Giljohann D.A., Seferos D.S., Prigodich A.E., Patel P.C., Mirkin C.A. Gene regulation with polyvalent siRNA–nanoparticle conjugates // *J. Am. Chem. Soc.* 2009. V. 131. P. 2072–2073.

281. Jewell C.M., Jung J.-M., Atukorale P.U., Carney R.P., Stellacci F., Irvine D.J. Oligonucleotide delivery by cell-penetrating "striped" nanoparticles // *Angew. Chem. Int. Ed.* 2011. V. 50. P. 12312–12315.

282. Bonoiu A.C., Mahajan S.D., Ding H., Roy I., Yong K.-T., Kumar R., Hu R., Bergey E.J., Schwartz S.A., Prasad P.N. Nanotechnology approach for drug addiction therapy: Gene silencing using delivery of gold nanorod-siRNA nanoplex in dopaminergic neurons // *Proc. Natl. Acad. Sci. USA.* 2009. V. 106. P. 5546–5550.

283. Crew E., Rahman S., Razzak-Jaffar A., Mott D., Kamundi M., Yu G., Tchah N., Lee J., Bellavia M., Zhong C.-J. MicroRNA conjugated gold nanoparticles and cell transfection // *Anal. Chem.* 2012. V. 84. P. 26–29.

284. Guo S., Huang Y., Jiang Q., Sun Y., Deng L., Liang Z., Du Q., Xing J., Zhao Y., Wang P.C., Dong A., Liang X.-J. Enhanced gene delivery and siRNA silencing by gold nanoparticles coated with charge-reversal polyelectrolyte // *ACS Nano.* 2010. V. 4. P. 5505–5511.

285. Ghosh R., Singh L.C., Shohet J.M., Gunaratne P.H. A gold nanoparticle platform for the delivery of functional microRNAs into cancer cells // *Biomaterials.* 2013. V. 24. P. 807–816.

286. Braun G.B., Pallaoro A., Wu G., Missirlis D., Zasadzinski J.A., Tirrell M., Reich N.O. Laser-activated gene silencing using gold nanoshell-siRNA conjugates // *ACS Nano.* 2009. V. 3. P. 2007–2015.

287. Song W.J., Du J.Z., Sun T.M., Zhang P.Z., Wang J. Gold nanoparticles capped with polyethyleneimine for enhanced siRNA delivery // *Small.* 2009. V. 6. P. 239–246.

288. Tian H., Guo Z., Chen J., Lin L., Xia J., Dong X., Chen X. PEI conjugated gold nanoparticles: Efficient gene carriers with visible fluorescence // *Adv. Healthcare Mater.* 2012. V. 1. P. 337–341.

289. Sandhu K.K., McIntosh C.M., Simard J.M., Smith S.W., Rotello V.M. Gold nanoparticle-mediated transfection of mammalian cells // *Bioconjug. Chem.* 2002. V. 13. P. 3–6.

290. Sée V., Free P., Cesbron Y., Nativo P., Shaheen U., Rigden D.J., Spiller D.G., Fernig D.G., White M.R., Prior I.A., Brust M., Lounis B., Lévy R. Cathepsin L digestion of nanobioconjugates upon endocytosis // *ACS Nano.* 2009. V. 3. P. 2461–2468.

291. Yan X., Blacklock J., Li J., Möhwald H. One-pot synthesis of polypeptide–gold nanoconjugates for *in vitro* gene transfection // *ACS Nano.* 2012. V. 6. P. 111–117.

292. Lee M.-Y., Park S.-J., Park K., Kim K.S., Lee H., Hahn S.K. Target-specific gene silencing of layer-by-layer assembled gold–cysteamine/siRNA/PEI/HA nanocomplex // *ACS Nano.* 2011. V. 5. P. 6138–6147.

293. Durán M.C., Willenbrock S., Barchanski A., Müller J.-M.V., Maiolini A., Soller J.T., Barcikowski S., Nolte I., Feige K., Escobar H.M. Comparison of nanoparticle-mediated transfection methods for DNA expression plasmids: Efficiency and cytotoxicity // *J. Nanobiotechnology*. 2011. V. 9. 47.

294. Elbakry A., Wurster E.-C., Zaky A., Liebl R., Schindler E., Bauer-Kreisel P., Blunk T., Rachel R., Goepferich A., Breunig M. Layer-by-layer coated gold nanoparticles: Size-dependent delivery of DNA into cells // *Small*. 2012. V. 8. P. 3847–3856.

295. Lukianova-Hleb E.Y., Ren X., Constantinou P.E., Danysh B.P., Shenefelt D.L., Carson D.D., Farach-Carson M.C., Kulchitsky V.A., Wu X., Wagner D.S., Lapotko D.O. Improved cellular specificity of plasmonic nanobubbles versus nanoparticles in heterogeneous cell systems // *PLoS One*. 2012. V. 7. e34537.

296. Szlachcic A., Pala K., Zakrzewska M., Jakimowicz P., Wiedlocha A., Otlewski J. FGF1-gold nanoparticle conjugates targeting FGFR efficiently decrease cell viability upon NIR irradiation // *Int. J. Nanomedicine*. 2012. V. 7. P. 5915–5927.

297. Tsai S.-W., Liaw J.-W., Hsu F.-Y., Chen Y.-Y., Lyu M.-J., Yeh M.-H. Surface-modified gold nanoparticles with folic acid as optical probes for cellular imaging // *Sensors*. 2008. V. 8. P. 6660–6673.

298. Dixit V., Van den Bossche J., Sherman D.M., Thompson D.H., Andres R.P. Synthesis and grafting of thioctic acid-PEG-folate conjugates onto Au nanoparticles for selective targeting of folate receptor-positive tumor cells // *Bioconjug. Chem*. 2006. V. 17. P. 603–609.

299. Li G., Li D., Zhang L., Zhai J., Wang E. One-step synthesis of folic acid protected gold nanoparticles and their receptor-mediated intracellular uptake // *Chem. Eur. J*. 2009. V. 15. P. 9868–9873.

300. Tong L., Zhao Y., Huff T.B., Hansen M.N., Wei A., Cheng J.-X. Gold nanorods mediate tumor cell death by compromising membrane integrity // *Adv. Mater*. 2007. V. 19. P. 3136–3141.

301. Bhattacharya R., Patra C.R., Earl A., Wang S.F., Katarya A., Lu L., Kizhakkedathu J.N., Yaszemski M.J., Greipp P.R., Mukhopadhyay D., Mukherjee P. Attaching folic acid on gold nanoparticles using noncovalent interaction via different polyethylene glycol backbones and targeting of cancer cells // *Nanomed. Nanotechnol. Biol. Med*. 2007. V. 3. P. 224–238.

302. Park J., Jeon W.I., Lee S.Y., Ock K.-S., Seo J.H., Park J., Ganbold E.-O., Cho K., Song N.W., Joo S.-W. Confocal Raman microspectroscopic study of folate receptor-targeted delivery of 6-mercaptopurine-embedded gold nanoparticles in a single cell // *J. Biomed. Mater. Res*. 2012. V. 100A. P. 1221–1228.

303. Paciotti G.F., Kingston D.G.I., Tamarkin L. Colloidal gold nanoparticles: A novel nanoparticle platform for developing multifunctional tumor-targeted drug delivery vectors // *Drug Dev. Res*. 2006. V. 67. P. 47–54.

304. Shao J., Griffin R.J., Galanzha E.I., Kim J.-W., Koonce N., Webber J., Mustafa T., Biris A.S., Nedosekin D.A., Zharov V.P. // *Sci. Rep*. 2013. V. 3. 1293.

305. Dam D.H.M., Lee J.H., Sisco P.N., Co D.T., Zhang M., Wasielewski M.R., Odom T.W. Direct observation of nanoparticle–cancer cell nucleus interactions // *ACS Nano*. 2012. V. 6. P. 3318–3326.

306. Kumar S., Harrison N., Richards-Kortum R., Sokolov K. Plasmonic nanosensors for imaging intracellular biomarkers in live cells // *Nano Lett.* 2007. V. 7. P. 1338–1343.

307. Kumar S., Aaron J., Sokolov K. Directional conjugation of antibodies to nanoparticles for synthesis of multiplexed optical contrast agents with both delivery and targeting moieties // *Nat. Protocols.* 2008. V. 3. P. 314–320.

308. Lukianova-Hleb E.Y., Belyanin A., Kashinath S., Wu X., Lapotko D.O. Plasmonic nanobubble-enhanced endosomal escape processes for selective and guided intracellular delivery of chemotherapy to drug-resistant cancer cells // *Biomaterials.* 2012. V. 33. P. 1821–1826.

309. Mukherjee P., Bhattacharya R., Bone N., Lee Y.K., Patra C.R., Wang S., Lu L., Secreto C., Banerjee P.C., Yaszemski M.J., Kay N.E., Mukhopadhyay D. Potential therapeutic application of gold nanoparticles in B-chronic lymphocytic leukemia (BCLL): Enhancing apoptosis // *J. Nanobiotechnology.* 2007. V. 5. 4.

310. Kalishwaralal K., Sheikpranbabu S., BarathManiKanth S., Haribalaganesh R., Ramkumarpandian S., Gurunathan S. Gold nanoparticles inhibit vascular endothelial growth factor-induced angiogenesis and vascular permeability via Src dependent pathway in retinal endothelial cells // *Angiogenesis.* 2011. V. 14. P. 29–45.

311. Wang Y., Xu J., Xia X., Yang M., Vangveravong S., Chen J., Mach R.H., Xia Y. SV119-gold nanocage conjugates: A new platform for targeting cancer cells via sigma-2 receptors // *Nanoscale.* 2012. V. 4. P. 421–424.

312. Krpetic Z., Porta F., Scarì G. Selective entrance of gold nanoparticles into cancer cells // *Gold Bull.* 2006. V. 39. P. 66–68.

313. Kasten B.B., Liu T., Nedrow-Byers J.R., Benny P.D., Berkman C.E. Targeting prostate cancer cells with PSMA inhibitor-guided gold nanoparticles // *Bioorg. Med. Chem. Lett.* 2013. V. 23. P. 565–568.

314. Huang X., Peng X., Wang Y., Wang Y., Shin D.M., El-Sayed M.A., Nie S. A reexamination of active and passive tumor targeting by using rod-shaped gold nanocrystals and covalently conjugated peptide ligands // *ACS Nano.* 2010. V. 4. P. 5887–5896.

315. Ali M.R.K., Panikkanvalappil S.R., El-Sayed M.A. Enhancing the efficiency of gold nanoparticles treatment of cancer by increasing their rate of endocytosis and cell accumulation using rifampicin // *J. Am. Chem. Soc.* 2014. V. 136. P. 4464–4467.

316. Song K., Xu P., Meng Y., Geng F., Li J., Li Z., Xing J., Chen J., Kong B. Smart gold nanoparticles enhance killing effect on cancer cells // *Int. J. Oncol.* 2013. V. 42. P. 597–608.

317. Li X., Zhou H., Yang L., Du G., Pai-Panandiker A.S., Huang X., Yan B. Enhancement of cell recognition *in vitro* by dual-ligand cancer targeting gold nanoparticles // *Biomaterials.* 2011. V. 32. P. 2540–2545.

318. Bhattacharyya S., Khan J.A., Curran G.L., Robertson J.D., Bhattacharya R., Mukherjee P. Efficient delivery of gold nanoparticles by dual receptor targeting // *Adv. Mater.* 2011. V. 23. P. 5034–5038.

319. Bhattacharyya S., Bhattacharya R., Curley S., McNiven M.A., Mukherjee P. Nanoconjugation modulates the trafficking and mechanism of antibody induced receptor endocytosis // *Proc. Natl. Acad. Sci. USA.* 2010. V. 107. P. 14541–14546.

320. Bhattacharyya S., Singh R.D., Pagano R., Robertson J.D., Bhattacharya R., Mukherjee P. Switching the targeting pathways of a therapeutic antibody by nanodesign // *Angew. Chem. Int. Ed.* 2012. V. 51. P. 1563–1567.

321. Hosta-Rigau L., Olmedo I., Arbiol J., Cruz L.J., Kogan M.J., Albericio F. Multifunctionalized gold nanoparticles with peptides targeted to gastrin-releasing peptide receptor of a tumor cell line // *Bioconjug. Chem.* 2010. V. 21. P. 1070–1078.

322. Suresh D., Zambre A., Chanda N., Hoffman T.J., Smith C.J., Robertson J.D., Kannan R. Bombesin peptide conjugated gold nanocages internalize via clathrin mediated endocytosis // *Bioconjug. Chem.* 2014. V. 25. P. 1565–1579.

323. Kumar A., Ma H., Zhang X., Huang K., Jin S., Liu J., Wei T., Cao W., Zou G., Liang X.-J. Gold nanoparticles functionalized with therapeutic and targeted peptides for cancer treatment // *Biomaterials.* 2012. V. 33. P. 1180–1189.

324. Heo D.N., Yang D.H., Moon H.-J., Lee J.B., Bae M.S., Lee S.C., Lee W.J., Sun I.-C., Kwon I.K. Gold nanoparticles surface-functionalized with paclitaxel drug and biotin receptor as theranostic agents for cancer therapy // *Biomaterials.* 2012. V. 33. P. 856–866.

325. Song M., Wang X., Li J., Zhang R., Chen B., Fu D. Effect of surface chemistry modification of functional gold nanoparticles on the drug accumulation of cancer cells // *J. Biomed. Mater. Res. A.* 2008. V. 86. P. 942–946.

326. Eghtedari M., Liopo A.V., Copland J.A., Oraevsky A.A., Motamedi M. Engineering of hetero-functional gold nanorods for the *in vivo* molecular targeting of breast cancer cells // *Nano Lett.* 2009. V. 9. P. 287–291.

327. Dreaden E.C., Mwakwari S.C., Sodji Q.H., Oyelere A.K., El-Sayed M.A. Tamoxifen poly(ethylene glycol) thiol gold nanoparticle conjugates: Enhanced potency and selective delivery for breast cancer treatment // *Bioconjug. Chem.* 2009. V. 20. P. 2247–2253.

328. Kim B., Han G., Toley B., Kim C.-k., Rotello V.M., Forbes N.S. Tuning payload delivery in tumour cylindroids using gold nanoparticles // *Nat. Nanotechnol.* 2010. V. 5. P. 465–472.

329. Wang F., Wang Y.C., Dou S., Xiong M.H., Sun T.M., Wang J. Doxorubicin-tethered responsive gold nanoparticles facilitate intracellular drug delivery for overcoming multidrug resistance in cancer cells // *ACS Nano.* 2011. V. 5. P. 3679–3692.

330. Bibikova O.A., Staroverov S.A., Sokolov O.I., Dykman L.A., Bogatyrev V.A. Plasmon-resonance gold nanoparticles as drug carriers and optical labels for cytological investigations // *Proc. Saratov University. Ser. Physics.* 2011. V. 11. P. 58–61.

331. Joshi P., Chakraborti S., Ramirez-Vick J.E., Ansari Z.A., Shanker V., Chakrabarti P., Singh S.P. The anticancer activity of chloroquine-gold nanoparticles against MCF-7 breast cancer cells // *Colloids Surf. B.* 2012. V. 95. P. 195–200.

332. Kim M., Ock K., Cho K., Joo S.-W., Lee S.Y. Live-cell monitoring of the glutathione-triggered release of the anticancer drug topotecan on gold nanoparticles in serum-containing media // *Chem. Commun.* 2012. V. 48. P. 4205–4207.

333. Zhao P., Astruc D. Docetaxel nanotechnology in anticancer therapy // *ChemMedChem.* 2012. V. 7. P. 952–972.

334. Comenge J., Sotelo C., Romero F., Gallego O., Barnadas A., Parada T.G.-C., Dominguez F., Puntes V.F. Detoxifying antitumoral drugs via nanoconjugation: The case of gold nanoparticles and cisplatin // *PLoS One*. 2012. V. 7. e47562.

335. Zhang X., Chibli H., Mielke R., Nadeau J. Ultrasmall gold-doxorubicin conjugates rapidly kill apoptosis-resistant cancer cells // *Bioconjug. Chem.* 2011. V. 22. P. 235–243.

336. Jang H., Ryoo S.-R., Kostarelos K., Han S.W., Min D.-H. The effective nuclear delivery of doxorubicin from dextran-coated gold nanoparticles larger than nuclear pores // *Biomaterials*. 2013. V. 34. P. 3503–3510.

337. Soenen S.J., Manshian B., Montenegro J.M., Amin F., Meermann B., Thiron T., Cornelissen M., Vanhaecke F., Doak S., Parak W.J., De Smedt S., Braeckmans K. Cytotoxic effects of gold nanoparticles: A multiparametric study // *ACS Nano*. 2012. V. 6. P. 5767–5783.

338. Zhu Z.-J., Posati T., Moyano D.F., Tang R., Yan B., Vachet R.W., Rotello V.M. The interplay of monolayer structure and serum protein interactions on the cellular uptake of gold nanoparticles // *Small*. 2012. V. 8. P. 2659–2663.

339. Patel P.C., Giljohann D.A., Daniel W.L., Zheng D., Prigodich A.E., Mirkin C.A. Scavenger receptors mediate cellular uptake of polyvalent oligonucleotide-functionalized gold nanoparticles // *Bioconjug. Chem.* 2010. V. 21. P. 2250–2256.

Immunological Properties of Gold Nanoparticles

5.1 Interaction of Gold Nanoparticles with Immune Cells

The immune system cells constitute the first barrier to nanoparticle penetration of animal tissues and cells. Therefore, the study of gold nanoparticle (GNP) interactions with phagocytes, the mechanisms of intracellular uptake, and the responses of immune cells to GNPs is undoubtedly of major interest. Perhaps the first detailed consideration of these issues can be found in Shukla *et al.* [1], who, using three microscopic methods, examined the uptake of 3-nm GNPs into RAW264.7 macrophage cells. The conclusion from their study was that small GNPs enter macrophages through pinocytosis and get localized mostly in lysosomes and in the perinuclear space. According to the data of Shukla *et al.*, GNPs are biocompatible, noncytotoxic, and nonimmunogenic and they suppress the production of reactive oxygen species and do not cause elaboration of the proinflammatory cytokines tumor necrosis factor-α (TNF-α) and interleukin (IL)-1β. In contrast, Yen *et al.* [2] noted that upon the administration of GNPs, the number of macrophages decreases and their size increases, this being accompanied by elevated production of IL-1, IL-6, and TNF-α. Although the data of Shukla *et al.* [1] were obtained for very small (3-nm) particles, Lim *et al.* [3], using much larger (60-nm) hollow nanospheres (NSphs) capped with dextran, and Zhang *et al.* [4], using 60-nm GNPs, achieved results similar to the findings of Shukla *et al.* [1] for the same cell culture. Sumbayev *et al.* [5] showed that citrate-stabilized GNPs, in a size-dependent manner, specifically downregulate cellular responses induced by IL-1β both *in vitro* and *in vivo*. le Guevél *et al.* [6] demonstrated induction of cell-mediated responses by 12-nm GNPs accompanied by inflammatory natural killer (NK) cell stimulation, whereas 2-nm GNPs were more efficiently taken up without inducing dendritic cell maturation or lymphocyte proliferation.

With some inspiration from these data, Choi *et al.* [7] proposed a new method for the photothermal therapy of tumors that employs a "Trojan horse" in the form of monocytes and macrophages laden with phagocytosed GNSs. For these purposes, Dreaden *et al.* [8] suggested the use of GNPs conjugated with macrolide antibiotics, which can accumulate in tumor-specific macrophages and induce their cytotoxicity, causing tumor cells to die. Thus, particle size and structure in these studies were not critical to macrophage uptake.

The influence of colloidal gold on immunocompetent cells was examined *in vivo* also by Tian *et al.* [9] and by Lou *et al.* [10]. From those examinations, the injection of non-conjugated GNPs into mice enhances the proliferation of lymphocytes and normal killers, as well as an increase in IL-2 production.

Quite interesting data were acquired by Bastús *et al.* [11,12]. From their results, it follows that indeed, 10-nm nonconjugated GNPs, on entry into murine bone marrow macrophages, do not affect the production of proinflammatory cytokines. However, if the GNP surface is modified with the peptide amyloid growth inhibitory peptide (LPFFD) or sweet arrow peptide [(VRLPPP)$_3$], GNPs, on entry into the macrophages, involve the induction of NO synthase and proinflammatory cytokines such as TNF-α, IL-1β, and IL-6. In addition, they inhibit macrophage proliferation. The recognition of GNP–peptide conjugates was affected through toll-like receptors 4 (TLR-4) on the surface of the macrophages. Yet, Staroverov *et al.* [13,14] demonstrated that both 15-nm nonconjugated GNPs and their conjugates with high- and low-molecular-weight antigens, on entry into rat peritoneal macrophages, enhance their respiratory activity and the activity of macrophage mitochondrial enzymes (Figure 5.1). GNPs also greatly increased the production of IL-1, IL-6, and IFN-γ (Figure 5.2).

Lee *et al.* [15] reported that the penetration of gold nanorods (GNRs) and SiO$_2$-coated GNRs into macrophages induces the release of inflammatory mediators (cytokines, prostaglandins, *etc.*) and the activation of immune response genes. The activation of macrophages by GNPs, found by several authors [11–19], can serve as a basis for new vaccine adjuvants. As in the usual cellular uptake, immunoactivity depended strongly

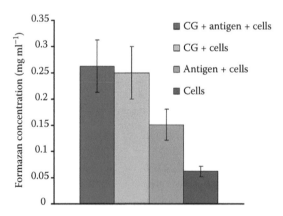

FIGURE 5.1 Changes in the concentration of reduced formazan depending on the cultivation conditions of antigen (AG) with peritoneal rat macrophages. (Adapted from Staroverov, S.A., Aksinenko, N.M., Gabalov, K.P., Vasilenko, O.A., Vidyasheva, I.V., Shchyogolev, S.Y., Dykman, L.A., *Gold Bull.*, 42, 153–156, 2009.)

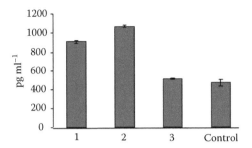

FIGURE 5.2 Changes in the serum IFN-γ concentrations in rats immunized with different antigens. 1—immunization with native antigen; 2—immunization with antigen conjugated with GNPs; 3—immunization with GNPs. (Adapted from Staroverov, S.A., Vidyasheva, I.V., Gabalov, K.P., Vasilenko, O.A., Laskavyi, V.N., Dykman, L.A., *Bull. Exp. Biol. Med.*, 151, 436–439, 2011. With permission.)

on the particle size: 5-nm particles conjugated with disaccharides performed far better than smaller, 2-nm ones [20].

Yet another means of activating macrophages with GNPs was proposed by Wei *et al.* [21]. Specifically, they used 15- and 30-nm GNPs conjugated to cytosine–phosphate–guanosine (CpG) oligodeoxynucleotides. As is known, these oligonucleotides are demethylated sites of microbial DNA that can activate macrophage immune response by interacting with the TLR-9 receptors and subsequently triggering a cascade of immune response signals. The immunostimulating activity of synthetic oligonucleotides containing CpG motifs may be analogous to that of oligonucleotides from bacterial DNA [22]. According to Wei *et al.* [21], GNP–CpG conjugates were effective in enhancing nanoparticle internalization in RAW264.7 macrophages, and they greatly increased the secretion of proinflammatory cytokines such as TNF-α and IL-6 (15-nm conjugates did so to a greater degree than 30-nm ones did). The immunostimulatory effect of GNP–CpG was much greater than that of native CpG at the same concentrations.

A recent study [23] examined the influence of the size of PEGylated GNPs on the activation of the TLR-9 receptors of RAW264.7 murine macrophages by CpG oligonucleotides. GNPs with diameters of 4, 11, 19, 35, and 45 nm inhibited CpG-induced elaboration of TNF-α and IL-6 and the activity of the TLR-9 receptors. This effect was markedly size dependent, with a peak for 4-nm GNPs, which penetrated the cells most intensively.

Massich *et al.* [24] reported on the immune response of macrophages after the phagocytosis of GNPs functionalized with polyvalent oligonucleotides. The effectiveness of uptake and the level of interferon production were found to depend on the density of DNA molecules on the GNP surface. According to Kim *et al.* [25], the uptake effectiveness of oligonucleotide-functionalized GNPs can differ for cells isolated from peripheral blood (mononuclear cells) and those introduced into a 293T culture. In addition, only in the first type of cell did the uptake of GNP conjugates activate the expression of immune response genes.

Walkey *et al.* [26] reported results of a thorough study of effects caused by coating of GNPs with serum proteins and poly(ethylene glycol) (PEG) on macrophage uptake. The authors studied the adsorption of 70 blood serum proteins to PEG-coated GNPs with different densities of PEG coating. Increasing the PEG coating density reduced serum protein adsorption and changed the composition of the adsorbed protein layer. Particle size also affected serum protein adsorption through a change in the steric interactions between the PEG molecules. Both the density of PEG molecules on the GNP surface and the size of GNPs determined the mechanism and effectiveness of macrophage uptake, possibly because of regulation of the composition of adsorbed blood serum proteins and their availability to cells. If the density of PEG coating was lower than ~0.16 PEG molecules/nm², the macrophage uptake of GNPs depended on the presence of adsorbed proteins (serum-dependent uptake). If the density was higher than ~0.64 PEG molecules/nm², serum-independent uptake was seen (Figure 5.3). Serum-dependent uptake was more effective than serum-independent uptake, apparently because of the difference in the energy of the GNP–cell interaction. Interestingly, serum-independent uptake was more effective for large GNPs (90 nm); serum-dependent uptake, for 50-nm GNPs.

Ma *et al.* [27] showed that GNPs attenuate lipopolysaccharide (LPS)-induced nitrogen oxide (NO) production through the inhibition of nuclear factor-κB (NF-κB) and interferon-β/signal transducer and activator of transcription 1 (IFN-β/STAT1) pathways in RAW264.7 cells. In contrast, Liu *et al.* [28] demonstrated that PEGylated GNPs were internalized more quickly by lipopolysaccharide-activated RAW264.7 cells than unstimulated cells, reaching saturation within 24 h. The PEGylated GNPs enhanced LPS-induced production of NO and IL-6 and inducible nitric oxide synthase expression in RAW264.7 cells, partly by activating p38 mitogen-activated protein kinases and NF-κB pathways. Goldstein *et al.* [29] showed that GNPs and their plasmonic excitation could activate the Nrf2-Keap1 pathway in macrophages.

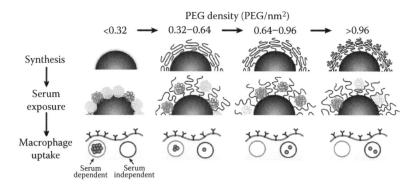

FIGURE 5.3 Scheme for the influence of the PEG coating density on the adsorption of serum proteins to GNPs and their subsequent uptake by macrophages. (Adapted from Walkey, C.D., Olsen, J.B., Guo, H., Emili, A., Chan, W.C.W., *J. Am. Chem. Soc.*, 134, 2139–2147, 2012. With permission.)

García *et al.* [30] studied the cellular uptake of GNPs with or without exposure of cells to Latrunculin-A, a phagocytosis inhibitor. The results indicate a size dependence of the internalization mechanisms for macrophage THP-1 cells. The internalization of larger GNPs (15 and 35 nm) was blocked in the presence of Latrunculin-A, although they could attach to the cell membrane. Smaller GNPs (5 nm), though, were not blocked by actin-dependent processes.

Of considerable interest are studies of GNP uptake not only by macrophages but also by other cells of the immune system, in particular dendritic cells. In the past decade, dendritic cells have attracted increased interest owing to the ease of their isolation from peripheral blood monocytes and to their ability to effectively present antigens to T cells. By now, a great deal of work has been done on the modulation of immune response in patients with chronic infections and oncological diseases by using antigen-primed dendritic cells [31]. GNPs have been selected among other carriers for application in antigen delivery to dendritic cells [32]. For example, Cheung *et al.* [33] described the use of 15-nm GNPs for presenting a peptide antigen associated with Epstein–Barr virus to dendritic cells. According to their transmission electron microscopy (TEM) data, peptide-functionalized GNPs penetrated into the dendritic cell cytoplasm but were not found in the nuclei. The uptake of GNPs by dendritic cells resulted in an increased content of γ-interferon, the presentation by major histocompatibility complex I (MHC-I) of the antigen to CD4+ T cells, and, correspondingly, activation of an epitope-specific immune response by cytotoxic T cells.

Cruz *et al.* [34] addressed dendritic cell uptake of and immune response activation by 13-nm GNPs conjugated to prostate cancer peptide antigens. By TEM, laser confocal microscopy, and flow cytometry, GNPs functionalized with the peptides and with Fc fragments of IgG were shown to interact with the Fcγ receptors of dendritic cells and were localized, upon uptake, in the cytoplasm in a diffuse way. Internalization of antigen-conjugated GNPs in dendritic cells brought about an increase in the immune response, as compared with the effect obtained from the use of the native antigen, which was manifested as enhanced lymphocyte proliferation. Such an approach, in the authors' opinion, opens up the way to the creation of an effective system for the development of antitumor and other vaccines.

Villiers *et al.* [35] described the effect of 10-nm non-antigen-functionalized GNPs on the immune functions of dendritic cells. From their findings, the GNPs that had entered cell endosomes were not cytotoxic and had no effect on the production of the proinflammatory cytokine IL-6. However, they did promote the secretion of interleukin IL-12p70, which is directly involved in the activation of T cells and, thus, in the regulation of an antigen-specific immune response. Villiers *et al.* also noted the development of long dendrites and an increase in the cell-surface amount of MHC-II molecules, which present antigens to T lymphocytes. Thus, even nonfunctionalized GNPs are immunostimulatory to both dendritic cells and macrophages [2].

Ye *et al.* [36] used TEM and flow fluorocytometry to quantify the uptake of GNRs by dendritic cells and the particle effect on their functions. Compared to spherical GNPs, GNRs entered dendritic cells more effectively and induced higher expression of CD86 immunocostimulatory molecules, which are characteristic of dendritic cells.

Lin *et al.* [37] reported that GNPs, in complex with peptides derived from tumor-associated antigens, are taken up effectively by dendritic cells. Moreover, dendritic cells take up GNPs with minimal toxicity and can process the vaccine peptides on the particles to stimulate cytotoxic T lymphocytes. A high peptide density on the GNP surface can stimulate cytotoxic T lymphocytes better than free peptides can. Thus, GNPs have great potential as carriers for various vaccine types. It is important that surface properties of GNPs, such as the chemical composition and surface charge, modulate uptake by dendritic cells and cytokines release [38].

By using mice model, Małaczewska [39,40] demonstrated an increased activity of phagocytes and some changes in the lymphocyte phenotypes, *i.e.*, increased percentage of B and CD4+/CD8+ double positive T cells, after oral administration of GNPs. The lowest dose had a proinflammatory or immunostimulating effect, enhancing the synthesis of proinflammatory cytokines (IL-1β, IL-2, IL-6, and TNF-α). The effect of the highest dose can be considered as a proinflammatory or immunotoxic one, because the stimulated cytokine synthesis was accompanied by a drastic decline in the proliferative activity of lymphocytes.

To estimate the functional impact of GNPs on B-lymphocytes, Sharma *et al.* [41] treated a murine B-lymphocyte cell line (CH12.LX) with 10-nm citrate-stabilized GNPs. This treatment activated an NF-κB-regulated luciferase reporter, and this activation correlated with the altered B lymphocyte function (*i.e.*, with increased antibody expression). According to TEM images, GNPs could penetrate the cellular membrane and therefore could interact with the intracellular components of the NF-κB signaling pathway.

From *in vitro*, *ex vivo*, and *in vivo* evidences, Lee *et al.* [42] suggested the GNP-mediated activation of B cells and enhancement of IgG secretion. GNP treatment upregulates blimp1, downregulates pax5, and enhances downstream IgG secretion. The enhancement was size dependent and time dependent. GNPs ranging from 2 to 12 nm had the maximum stimulatory activity for the production of antibody.

Moreover, GNPs augmented lymphocytes proliferation in response to phytohemagglutinin, and this effect was greater for as-synthesized than for capped GNPs. Release of IL-10 and IFN-γ from lymphocytes was increased and the effect was again more marked for as-synthesized than capped GNPs [43].

Bartneck *et al.* [44,45] reported on the interaction of GNPs with human neutrophil granulocytes, monocytes, and macrophages. On the basis of their study of the interaction of variously shaped and sized particles with human immune cells, the mechanism of nanoparticle trapping can be classified as macropinocytosis rather than phagocytosis. Particle shape was found to be able to affect strongly the particle trapping by immune cells; specifically, cetyltrimethylammonium bromide (CTAB)-coated GNRs (50 × 15 nm) could be trapped faster than CTAB-coated GNSphs (15 and 50 nm). Replacing CTAB by poly(ethylene oxide) greatly reduced uptake effectiveness for both types of GNPs. Nanoparticle uptake by the immune cells was accompanied by activation of the genes of proinflammatory cytokines and by a corresponding change in the cell phenotype. A characteristic fact is that the "professionally" phagocytic cells took up GNPs two orders of magnitude more effectively than, *e.g.*, HeLa cells did. In addition, the authors revealed an alternative elimination mechanism whereby GNPs can be cleared from peripheral blood *via* an extracellular network ("trap") produced by neutrophil granulocytes.

The same group presented data [46] on the uptake of GNPs into various cells of the reticuloendothelial system: monocytes, macrophages, immature and mature dendritic cells, and endothelial cells. The greatest uptake ability was demonstrated by macrophages, endothelial cells, and immature dendritic cells. Positively charged GNPs penetrated into cells of the reticuloendothelial system more effectively. Moreover, GNPs intensified the induction of several cytokines, including γ-interferon, IL-8 (both in dendritic cells and in macrophages), IL-1β, and IL-6 (only in dendritic cells). Interestingly, in mature dendritic cells, GNPs accumulate in the MHC-II compartment, and consequently, they may affect antigen processing.

Thus, GNPs can penetrate into various immune cells (Figure 5.4) and activate the production of proinflammatory cytokines (Table 5.1) [47,48].

Phagocytic cells of the immune system have a multitude of various receptors on their surface, through which they bind and take up foreign material. Six types of phagocytosis receptors are differentiated: (1) mannose receptors (or C-type lectin receptors); (2) integrins (the complement receptors); (3) Fc receptors (Ig receptors); (4) leucine-rich repeat receptors (or CD14, LPS receptors); (5) scavenger receptors (receptors of sialic acid derivatives); and tyrosine kinase receptors [49]. As distinct from these, TLRs are not directly involved in the uptake of foreign material; however, they do take part in the regulation of phagosome formation and in inflammatory reactions [50].

As shown in Chapter 4, interactions with various types of receptors and, consequently, various types of GNP endocytosis depend in many ways on nanoparticle size and shape but especially on surface functionalization (including opsonization by proteins from the culture medium or blood plasma [51]) and on the presence of mannose-containing polysaccharides on the GNP surface [52]. Some researchers are inclined to believe that the key role in macrophage uptake of GNPs is played by scavenger receptors [53,54]. These are mainly involved in the endocytosis of apoptotic cells. A characteristic peculiarity of their functioning, in contrast to the other macrophage receptors, is the absence of release of proinflammatory cytokines.

More specifically, Patel *et al.* [55] demonstrated that the uptake of GNPs functionalized with polyvalent oligonucleotides by mammalian cells is effected through scavenger receptors. Cell preincubation with fucoidan and polyinosinic acid, which are agonists for these receptors, decreased the level of uptake by 60% (Figure 5.5). However, bafilomycin A1 and methyl-β-cyclodextrin did not inhibit GNP uptake because these pharmacological agents are known to inhibit other modes of cellular entry. Coating of GNP conjugates with serum proteins also reduced uptake effectiveness.

An in-depth study on the involvement of scavenger receptors in macrophage uptake of GNPs was published by França *et al.* [56]. Their data show that macrophages take up opsonized GNPs through SR-mediated pathways (both 30- and 150-nm GNPs), as well as through clathrin- and caveolin-dependent pinocytosis (only 30-nm GNPs). Thus, the smaller (30-nm) particles use a broader range of internalization routes, in contrast to the larger (150-nm) GNPs. Noteworthy is the fact that as demonstrated by inhibition analysis, phagocytosis began with an interaction of GNPs with scavenger receptors and was not attended by induction of proinflammatory cytokines.

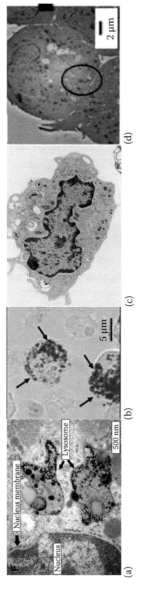

FIGURE 5.4 TEM images of (a) spleen macrophages, (b) dendritic cells, (c) monocytes, and (d) lymphocytes treated with GNPs. (Adapted from Dykman, L.A., Khlebtsov, N.G., *Chem. Sci.*, 8, 1719–1735, 2017. With permission.).

TABLE 5.1 Effect of GNPs on the Functions of Various Immune Cells

Macrophages	Dendritic cells	Lymphocytes
Induction of cytokines (IL-1β, IL-6, IL-8, IL-10, TNF-α) and prostaglandins.	Induction of IFN-γ, TNF-α, IL-1β, IL-6, IL-8, IL-12p70 cytokines.	Induction of IL-2 and IFN-γ cytokines.
Stimulation of CD8+ and CD4+ T cells.	Stimulated of CD8+ and CD4+ T cells.	Increasing proliferation of lymphocytes and NK cells.
Activation of immune response genes.	Induction of CD86 costimulatory molecules.	Activation of NF-κB signaling pathway.
Inhibition of macrophage proliferation, decreasing their amount and increasing their size.	Increasing in the cell-surface amount of MHC-II.	Regulation of blimp1/pax5 signaling pathway.
Activation of Keap1/Nrf2 signaling pathway.	Increasing the amount of dendritic cells.	Enhance antibody secretion in B cells.
	Activation of antigen processing.	

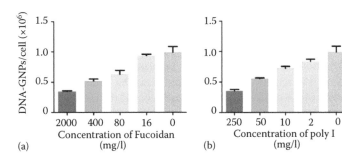

FIGURE 5.5 Cellular endocytosis of GNPs is mediated by scavenger receptors. Cell preincubation with fucoidan (a) and polyinosinic acid (b), which are agonists for these receptors, decreased the level of uptake by 60%. (Adapted from Patel, P.C., Giljohann, D.A., Daniel, W.L., Zheng, D., Prigodich, A.E., Mirkin, C.A., *Bioconjug. Chem.*, 21, 2250–2256, 2010. With permission.)

5.2 Production of Antibodies by Using GNPs

Since the 1920s, the immunological properties of colloidal metals (in particular, gold) have been attracting much research interest. This interest was mainly due to the physicochemical (nonspecific) theory of immunity proposed by J. Bordet, who had postulated that immunogenicity, along with antigenic specificity, depends predominantly on the physicochemical properties of antigens, first of all on their colloidal state. L.A. Zilber made successful attempts to obtain agglutinating sera to colloidal gold [57]. (Curiously, a repeated attempt to prepare antisera to colloidal gold was performed almost 80 years later, in 2006 [58].) Yet, several authors have shown that the introduction of a complete antigen together with colloidal metals promotes the production of antibodies [59]. Furthermore, some haptens may cause antibody production when adsorbed to colloidal particles [60]. Numerous data on the influence of colloidal gold on nonspecific immune reaction are given in one of the best early reviews [61]. In particular, it was noted that at 2 h after an intravenous injection of 5 ml of colloidal gold into rabbits, there was a sizable increase in total leucocytes in 1 ml of blood (from 9,999 to 19,800) against a slight decline in mononuclear cells (from 5,200 to 4,900) and a considerable increase in polynuclear cells (from 4,700 to 14,900) [62]. On injection of other colloidal metals, no such phenomena were observed. Unfortunately, with advances in immunology and denial of many postulates of Bordet's theory, interest in the immunological properties of colloids decreased. There is no doubt, though, that the data obtained on the enhancement of immune response to antigens adsorbed on colloidal particles were utilized for the development of various adjuvants [63,64].

It is known that antibody biosynthesis is induced by substances possessing sufficiently developed structures (immunogenicity). The substances include proteins, polysaccharides, and some synthetic polymers. However, many biologically active substances (vitamins, hormones, antibiotics, narcotics, *etc.*) have relatively small molecular masses and, as a rule, do not elicit a pronounced immune response. In standard methods of antibody preparation *in vivo*, this limitation is overcome by chemically attaching such substances (haptens) to high-molecular-weight carriers (most often proteins), which makes it possible to obtain specific antisera. However, such antisera usually contain attendant antibodies to the carrier's antigenic structures [65].

Let us take a brief look at two interrelated problems in current immunology that have attracted much research attention. These are the raising of antibodies to nonimmunogenic low-molecular-weight compounds (haptens) and the creation of next-generation vaccines based on natural (microbial) or synthetic peptides [66–71].

It is known that antibody biosynthesis is induced by substances possessing sufficiently developed structures (immunogenicity). These substances include proteins, polysaccharides, and some synthetic polymers [72]. However, many biologically active substances (neurotransmitters, hormones, vitamins, antibiotics, *etc.*) have relatively small molecular masses. Low-molecular-weight antigens are assigned to the category "weak antigens"; *i.e.*, they do not elicit a pronounced immune response.

Because haptens are weakly immunogenic, the choice of an optimal carrier (delivery system) providing a high immune response, in parallel with the obtainment of pure enough antibody preparations, is an important task when producing antibodies to low-molecular-weight compounds. Traditionally, this problem is solved by chemical attachment of a hapten to a protein matrix called a schlepper (from the German *schleppen* ["to drag"]) and by the use of adjuvants and intensive schemes of animal immunization with the obtained conjugate [65,73]. Bovine serum albumin (BSA), ovalbumin, thyreoglobulin, hemocyanin, and diphtheria or tetanus toxoids (in the case of synthetic peptides) are generally used as schleppers. However, this method yields antibodies to both the hapten and the immunodominant sites of the carrier. Note that when such a carrier is used, an expressed immune response to weak antigens far from always develops. Besides, the subsequent purification and screening of the obtained antibodies are laborious and expensive, and their titre and affinity are often low. Most currently used adjuvants based on oil emulsions and on suspensions of inorganic substances are, as a rule, liable to phase separation, are often reactogenic, and their immunogenic properties vary with time. Many of these adjuvants cause local and systemic toxicity [63].

In recent years, efforts have been made to develop what are called complex antigens, *i.e.*, artificial molecular complexes formed from both necessary antigens and carriers or/and adjuvants. In particular, synthetic polyelectrolytes (poly-L-lysine, polyacrylic acid, polyvinylpyridine, sulfonated polystyrene, ficoll, *etc.*) were proposed for use as adjuvants [74]. These polymer compounds are produced by chain-radical polymerization of the corresponding monomers. The simplicity of polyelectrolyte composition and synthesis, the possibility of obtainment of polymer chains with a wide range of molecular masses (*i.e.*, of various lengths), their solubility in water, and other properties (the capacity for conformational transitions, the formation of complexes with proteins, *etc.*) opened up possibilities for the use of polyelectrolytes in immunologic investigations. Such adjuvant carriers are capable of antigen deposition at the sites of injection, enhancement of antigen presentation to immunocompetent cells, and induction of production of necessary cytokines. However, the low immunogenicity of such complexes, due to their small epitope density, prompts researchers to look for new nontoxic and effective carriers, additionally possessing adjuvant properties.

In this respect, of special interest are nanoscale corpuscular carriers: polymer nanoparticles (*e.g.*, those made of polymethylmethacrylate, polyalkylcyanoacrylate, polylactide-co-glycolide, poly(γ-glutamic acid), polystyrene, *etc.*) [75–79]; liposomes, proteasomes, and microcapsules [80–82]; fullerenes [83,84]; carbon nanotubes [85–88]; graphene oxide [89]; dendrimers [90]; paramagnetic particles [91]; silica nanoparticles [92]; titanium dioxide nanoparticles [93]; aluminum [94] and aluminum oxide nanoparticles [95]; cobalt oxide nanoparticles [96]; silver nanoparticles [97,98]; selenium nanoparticles [99]; and others. When these are used, the forms of manifestation of immunogenicity of a given substance in the host's immune system vary. An antigen, once adsorbed or encapsulated by nanoparticles, may be used as an adjuvant for optimization of the immune response after vaccination [100–124].

In 1986, Japanese researchers [125] first reported success in generating antibodies against glutamate by using colloidal gold particles as a carrier. Subsequently, a number of papers were published whose authors applied and further developed this technique to obtain antibodies to the following haptens and complete antigens: amino acids [126,127]; platelet-activating factor [128,129]; quinolinic acid [130]; biotin [131]; recombinant peptides [132,133]; lysophosphatide acid [134]; endostatin [135]; the capsid peptide of the transmissible gastroenteritis [14,136], the hepatitis C [137], influenza [138], foot-and-mouth disease [139,140], and dengue [141] viruses; α-amidated peptides [142]; actin [143]; antibiotics [144]; ivermectin [145,146]; azobenzene [147]; Aβ-peptide [148]; clenbuterol [149]; α-methylacyl-CoA racemase [150]; *Yersinia* [97,151,152], *Listeria monocytogenes* [153], and *Escherichia coli* [154] surface antigens; *Neisseria meningitidis* [155], *Streptococcus pneumoniae* [156], and *Burkholderia mallei* [157,158] carbohydrate antigens; *Pseudomonas aeruginosa* flagellin [159]; tuberculin [160]; the peptides of the malaria plasmodium surface proteins [161,162]; opisthorchiasis excretory-secretory antigen [163]; tetanus toxoid [164], lycopene [165]. In all these studies, the haptens or complete antigens were directly conjugated to colloidal gold particles, mixed with complete Freund's adjuvant or alum, and used for animal immunization. As a result of this, high-titer antisera were obtained that did not need further purification from contaminant antibodies (Figure 5.6).

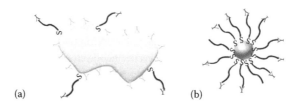

FIGURE 5.6   Schematic representation of immunogen localization on the surface of key-hole limpet hemocyanin (KLH) and GNPs, used as antigen carriers. (a) Antibodies toward the peptide–KLH conjugate are produced to the epitopes of both peptide and KLH. (b) Antibodies toward the peptide–GNP conjugate are produced only to the epitopes of the peptide. (Adapted from Dykman, L.A., Khlebtsov, N.G., *Chem. Sci.*, 8, 1719–1735, 2017. With permission.)

Staroverov and Dykman [160] described, for the first time, the use of antituberculin antibodies for immunoassay of mycobacteria. Figure 5.7 illustrates applications of the immunodot assay, TEM, and light microscopy imaging to mycobacteria, with the reaction products being visualized by using immunogold markers. In future work, the authors plan to use the GNP + tuberculin conjugates not only for obtainment of diagnostic antibodies but also for the development of tuberculin-based antituberculosis vaccines. This can be considered as a new variant of theranostics, which can be called "prophynostics" (prophylaxes + diagnostics).

In 1993, Pow and Crook [166] suggested attaching a hapten (γ-aminobutyric acid) to a carrier protein before conjugating this complex to colloidal gold. This suggestion was supported in papers devoted to the raising of antibodies to some peptides [167–171], amino acids [172–175], phenyl-β-D-thioglucoronide [176], and diminazene [177]. The antibodies obtained in this way possessed high specificities to the antigens under study and higher (as Pow and Crook [166] put it, "extremely high") titers—from 1:250,000 to 1:1,000,000, as compared with the antibodies produced routinely. At present, the Australian-based company ImmunoSolution offers antibodies, obtained according to the paper [166], to some neurotransmitters and amino acids.

In 1996, Demenev *et al.* [178] showed, for the first time, the possibility of using colloidal gold particles as part of an antiviral vaccine as carriers for the protein antigen of the tick-borne encephalitis virus capsid. According to the authors' data, the offered experimental vaccine had higher protective properties than its commercial analogs, despite the fact that the vaccine did not contain adjuvants.

Subsequently, GNPs have been used to generate antibodies and design experimental vaccines (both peptide and carbohydrate) against influenza A virus [179,180], West Nile virus [181], the respiratory syncytial virus [182], hepatitis E virus [183], infectious bronchitis virus [184], as well as against tuberculosis [160] and listeriosis [185]. In addition, GNPs are being used in the development of experimental vaccines against tumors [186–192] and HIV/AIDS [193–196]. In 2011, Wang *et al.* [197] suggested a new therapeutic vaccine based on the combination of myelin-associated inhibitors and GNPs for the treatment of rat medullispinal traumas. Also, for GNP-assisted antigens, several groups

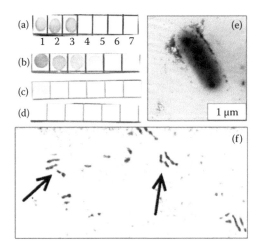

FIGURE 5.7 (a) Specificity of antituberculin antibodies as determined by dot analysis using primary labeling with rabbit antituberculin antibodies and secondary labeling with conjugates of antirabbit antibodies with 160/20 nm (SiO$_2$ core/Au shell) nanoshells. Sampled antigens: 1, rabbit antituberculin antibodies; 2, tuberculin; 3, *Mycobacterium bovis* BCG; 4, *Escherichia coli* XL-1 blue; 5, *Staphylococcus aureus* 209-R; 6, *Brucella abortus* vaccine strain 82; 7, brucellin. For samples 1, 2, and 7, the concentrations were 1 mg/ml. (b–d) Dot immunoassay of the mycobacteria *M. bovis* (b), *Mycobacterium smegmatis* (c), and *Mycobacterium phlei* (d) by using polyclonal antibodies to tuberculin (primary antibodies) and conjugates of antirabbit antibodies with 15-nm GNPs (secondary antibodies). Note the weak nonspecific coloration of *M. smegmatis* bacteria. (e) TEM image of an *M. bovis* cell treated with antituberculin antibodies and labeled with conjugates of antirabbit antibodies with 15-nm GNPs. The GNP accumulation on the bacterial surface may reflect the localization of the tuberculin antigen. (f) Light microscopy of *M. bovis* BCG treated with rabbit antituberculin antibodies and labeled with conjugates of antirabbit antibodies with 15-nm GNPs. The arrows point to mycobacteria. (Adapted from Staroverov, S.A., Dykman, L.A., *Nanotechnol. Russia*, 8, 816–822, 2013.)

reported new administration ways: oral, pulmonary, transcutaneous, and transmucosal immunization [198–203].

Table 5.2 summarizes the literature data on antigens and haptens that have been conjugated with GNP carriers and then used for immunization of animals. The titers of the antibodies have been increased owing to GNPs.

A considerable number of papers devoted to the use of GNPs for creating DNA vaccines have emerged as well. The principle of DNA immunization is as follows: gene constructions coding for the proteins to which one needs to obtain antibodies are introduced into an organism. If the gene expression is effective, these proteins serve as antigens for the development of an immune response [204,205]. In the early papers, immunization was conducted by a subcutaneous or intramuscular injection of a "naked" DNA. However, for this purpose, a "biolistic" transfection, using GNPs, began to be applied almost simultaneously. It was found to be very effective, apparently because of the multiplicity of sites of transgene interaction with tissues and because of transgene penetration

TABLE 5.2　Conjugate GNPs with Antigens and Haptens Used for Immunization and Vaccination of Animals

Amino Acids	Neurotransmitters and Hormones	Antibiotics and Other Drugs	Bacterial, Protozoan, and Viral Antigens	Others Substances
Glutamate	Acetylcholine	Chloramphenicol	*Yersinia pseudotuberculosis*	Platelet-activating factor
Aspartate	Serotonin	Gentamicin	*Yersinia pestis*	Quinolinic acid
Glycine	Norepinephrine	Neomycin	*Salmonella typhimurium*	Biotin
Serine	Histamine	Lincomycin	*Brucella abortus*	Lysophosphatide acid
Cysteine	Testosterone	Kanamycin	*Mycobacterium tuberculosis*	Immunophilin
Taurine	γ-Aminobutyric acid	Clindamycin	*Streptococcus pneumoniae*	Endostatin
Citrulline	Nortestosterone	Ofloxacinum	*Neisseria meningitides*	Azobenzene
	Estradiol	Tilmicosin	*Burkholderia mallei*	Bacteriorhodopsin
		Ivermectin	*Escherichia coli*	Phenyl-β-D-thioglucoronide
		Diminazene	*Listeria monocytogenes*	Indole-3-acetic acid
		Clenbuterol	*Clostridium tetani*	Actin
		Xylazine	*Pseudomonas aeruginosa*	Bovine serum albumin
			Francisella tularensis	Ferritin
			Plasmodium malariae	Tuberculin
			Plasmodium falciparum	Tetanus toxoid
			Opisthorchis felineus	α-Methylacyl-CoA racemase
			Hepatitis B virus	Protein kinase
			Hepatitis C virus	Carbonic anhydrase
			Hepatitis E virus	Lycopene
			Influenza virus	Tumor antigens
			Foot-and-mouth disease virus	Recombinant and natural peptides
			Transmissible gastroenteritis virus	Oligosaccharides
			Tick-borne encephalitis virus	
			West Nile virus	
			Respiratory syncytial virus	
			Rabies virus	
			Dengue virus	
			Infectious bronchitis virus	
			HIV-1	

directly into cells and nuclei [206,207]. The method of gene immunization, often called DNA vaccination, which was well developed in experiments with animals, has shown high efficiency especially in respect of viral infections: tick-borne encephalitis, HIV infection, hepatitis B, and some others [208].

DNA immunization has some advantages over routine vaccination. A single recombinant vector can govern the synthesis of several antigens simultaneously, reducing the number of separate immunizations. This results in erasing problems connected with difficulties of protein penetration into the organism and in reducing significantly the risk of side effects, which depend on the toxicity of the contaminant proteins introduced during a routine immunization or on the virulence of the bacteria and viruses used. One can expect that DNA immunization will be among the most effective gene-therapy methods in the coming years [209–211].

Recently, intramuscular injection of a "naked" DNA was abandoned in DNA vaccination. Investigators have come to use nanoparticles as a carrier for genetic material and to introduce the injection substance subcutaneously, intracutaneously, epicutaneously, and intranasally [212–214]. Among the nanoparticles used as DNA carriers, GNPs, both spherical and cylindrical (multivalent Au–Ni nanorods), are especially popular with researchers [215–221]. Besides DNA, polysaccharides, peptides, and glycopeptides are used as vectors in such vaccines [33,222–228]. Moreover, whereas gold was earlier used only as a carrier, Zhao *et al.* [229] noted: "Although the mechanism behind this is not well understood, it appears that gold cartridges might enhance immune responses *in vivo.*"

5.3 Adjuvant Properties of GNPs

Dykman *et al.* [144,230–232] proposed a technology for the preparation of antibodies to various antigens, which uses colloidal gold as a carrier and as an adjuvant. In their method, antigens are adsorbed directly on the GNP surface, with no cross-linking reagents. It was found that animal immunization with colloidal gold–antigen conjugates (with or without the use of Freund's complete adjuvant) yielded specific, high-titer antibodies to a variety of antigens, with no concomitant antibodies. GNPs can stimulate antibody synthesis in rabbits, rats, and mice, and the amount of antigen required is reduced, as compared with that needed when using some conventional adjuvant (Table 5.3).

TABLE 5.3 Antibody Titers Obtained during Immunization of Rabbits with *Yersinia* Antigen

Preparation	First Immunization	Second Immunization	Boosting
Colloidal gold + antigen (1 mg)	1:32	1:256	1:10240
Complete Freund's adjuvant + antigen (100 mg)	1:32	1:256	1:10240
Physiological saline + antigen (100 mg)	1:2	1:16	1:512

Source: Adapted from Dykman, L.A., Khlebtsov, N.G., *Chem. Sci.*, 8, 1719–1735, 2017.

In summary, experimental results give grounds to state that:

1. The method of "gold immunization" can be used for obtaining antibodies to those haptens to which it is very difficult to obtain antibodies conventionally (in particular, antibiotics, vitamins, and nonimmunogenic peptides);
2. The amount of antigen used for immunization in this case is much smaller than that used in conventional methods, even when the latter allow one to obtain an immune response;
3. For several antigens conjugated with GNPs, an effective immune response was obtained without the use of other adjuvants;
4. GNPs used as an antigen carrier stimulate the phagocytic activity of lymphoid cells and induce the release of inflammatory mediators.

We believe that all the earlier facts show decisively that GNPs possess adjuvant properties. With use of GNPs as an antigen carrier, they activated the phagocytic activity of macrophages and influenced the functioning of lymphocytes, which apparently may be responsible for their immunomodulating effect. It also was found that GNPs and their conjugates with low- and high-molecular weight antigens stimulate the respiratory activity of cells of the reticuloendothelial system and the activity of macrophage mitochondrial enzymes [13], which is possibly one of the causative factors determining the adjuvant properties of colloidal gold. That GNPs act as both an adjuvant and a carrier (*i.e.*, they present haptens to T cells) seems the most interesting aspect of manifestation of immunogenic properties by colloidal gold. In particular, GNPs conjugated to antigens were found to influence the activation of T cells: a 10-fold increase in proliferation, as compared with that observed on the addition of the native antigen, was found. This fact shows that there is a fundamental possibility of targeted activation of T cells followed by macrophage activation and pathogen killing.

As already mentioned in Chapter 2, there have been reports of successful therapy of rheumatoid arthritis with a colloidal gold solution. According to the data of Graham [233], the effect of GNPs in this case is an inhibition of monocyte-induced lymphocyte proliferation. The transformation of Au(0) to Au(I) in the immune-system cells under the action of several amino acids is discussed by Merchant [234]. It was noted by Eisler [235] that injection of GNPs into laboratory animals could result in an inflammatory response, accumulation of gold in the reticular cells of lymphoid tissue, and activation of cellular and humoral immunity.

However, not a single paper available to us has reported data on the mechanism of such properties of gold particles. In our opinion, the reasoning given by Pow and Crook [166] on the preferable macrophage response to corpuscular antigens, as opposed to soluble ones, is certainly valid. This fact has also been confirmed by researchers studying the mechanism of action of DNA vaccines and using gold particles to deliver genetic material to cells [229]. The role of Kupffer and Langerhans cells in the development of immune response was shown in those investigations. The influence of dendritic cells on the development of immune response upon injection of a GNP-conjugated antigen is discussed by Vallhov *et al.* [236]. In addition, those authors note that when using nanoparticles in medical practice, one has to ensure that there are no lipopolysaccharides on their surface. Similar results for interaction GNPs with macrophage were reported by Kingston *et al.* [237]. The interaction of cells of the immune system with GNPs has been very actively examined by Dobrovolskaia's group [53,54,56,123,238–240].

Modern trends in the use of GNPs for vaccination are the application of multivalent glycopolymers [225] and peptides [37]; combined use of GNPs with other immunostymulators, in particular, CpG (including conjugated with GNPs) [241–246], polyvalent nucleic acid [24,247], and plant adjuvants, *e.g.*, extracts from *Quillaja saponaria* [248], *Asparagus racemosus* [249], or *Tamarindus indica* [250]; and application of GNPs of various sizes and shapes (including nanorods, nanocubics, nanocages, nanoclusters) [181,183,251–253].

When animals are injected with a model antigen (BSA) coupled to gold NSphs (diameters, 15 and 50 nm), nanorods, nanoshells, and nanostars, the titers of the resultant antibodies differ substantially. The antibody titers decrease in the sequence GNPs-50 nm > GNPs-15 nm > nanoshells > nanostars > nanorods > native BSA. Thus, 50- and 15-nm gold NSphs are the optimal antigen carrier and adjuvant for immunization. Although the highest titer was obtained with GNPs-50, we recommend using GNPs-15 because conjugating antigens to large particles can be difficult. Large nanoparticles tend to aggregate when conjugated to proteins. The highest titer of anti-BSA antibodies is observed in the blood serum of mice immunized simultaneously with BSA–GNP and CpG–GNP conjugates (unpublished data).

However, those data do not answer the question about the further mechanisms of antigen presentation to T helpers. According to the current view [72], the presentation of an antigen to T cells is preceded by its processing, *i.e.*, cleavage into peptide fragments followed by the formation of bonds with molecules of the MHC, which transport the antigen fragment to the surface of the antigen-presenting cell. It remains unclear, then, how this process can proceed with a hapten. The hypothesis of the multivalent antigen, *i.e.*, the antigen formed because of the high local concentration of univalent antigens on the surface of a gold particle, does not answer this question either. Hypothetical mechanisms of immunomodulatory effects of nanoparticles are shown in Figures 5.8 and 5.9 [113,254].

Recently, many papers have been published (see Chapter 2) in which the problems of GNP use for targeted drug delivery were discussed. In our opinion, one should deal with this question very carefully, taking into account the possibility of production in animals or humans of antibodies specific to the administered drug adsorbed on gold particles. We believe that the discovery of adjuvant properties of GNPs creates favorable conditions for designing next-generation vaccines.

Alongside GNPs, other nonmetallic nanoparticles also can serve as antigen carriers. The published examples include liposomes, proteosomes, microcapsules, fullerenes, carbon nanotubes, dendrimers, and paramagnetic particles [231]. In our view, especially promising carriers are synthetic and natural polymeric biodegradable nanomaterials [polymethyl methacrylate, poly(lactid-co-glycolid acid), chitosan, gelatin]. With the use of such nanoparticles, the immunogenicity of a loaded substance and its representation

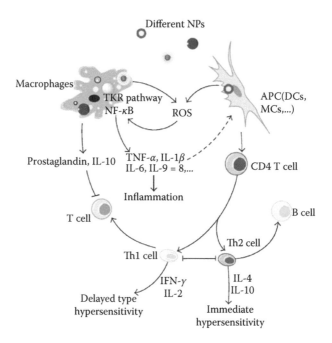

FIGURE 5.8 Mechanisms involved in NP-induced immunomodulation. The stimulation/ suppression to immune system depends on the nature of NPs and results in different outcomes. NPs, nanoparticles; NF-κB, nuclear factor kappa B; TLR pathway: toll-like receptor pathway; APC, antigen-presenting cell; DCs, dendritic cells; MCs, mast cells; GM-CSF, granulocyte-macrophage colonystimulating factor; Th0, type 0 T-helper lymphocyte; Th1, type 1 T-helper lymphocyte; Th2, type 2 T-helper lymphocyte; solid line with arrow, activate/release/induce; solid line with vertical dashes at ends, inhibit; dotted line, possible influence; broken line, polarization/ differentiation. (Adapted from Jiao, Q., Li, L., Mu, Q., Zhang, Q., *BioMed Res. Int.*, 2014, 426028, 2014.)

in a host immune system will be changed. A nanoparticle conjugate with an absorbed or a capsulated antigen can serve as an adjuvant for the optimization of immune response after vaccination.

The evident advantages of biodegradable nanoparticles is their complete utilization in the vaccinated organism, high loading efficiency for the target substance, enhanced ability to cross various physiological barriers, and low systemic side effects. In all likelihood, the immune action of biodegradable nanoparticles and GNPs as corpuscular carriers are similar. Keeping in mind the recent data for the low toxicity of GNPs and their efficient excretion by the hepatobiliary system, we expect that both nanoparticle classes—GNPs and biodegradable nanoparticles—will compete on equal footing for the development of next-generation vaccines.

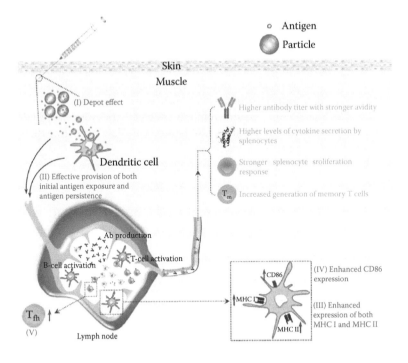

FIGURE 5.9 Schematic illustration of the proposed mode of action of the combined vaccine formulation composed of nanoparticles-encapsulated antigen and soluble antigen mixed with blank nanoparticles. (Adapted from Zhang, W., Wang, L., Liu, Y., Chen, X., Liu, Q., Jia, J., Yang, T., Qiu, S., Ma, G., *Biomaterials*, 35, 6086–6097, 2014. With permission.)

5.4 Concluding Remarks

Thus, GNP uptake into cells of the immune system activates the production of proinflammatory cytokines, a finding that indicates directly that GNPs are immunostimulatory. The activation of immune cells by GNPs, shown by several authors, may serve as a basis to develop new vaccine adjuvants [255,256]. As in the case of the usual cells, interactions with various types of receptors on the surface of immune cells and, correspondingly, various types of GNP endocytosis depend largely on the surface functionalization of GNPs. Many researchers believe that the key role in macrophage uptake of GNPs is played by scavenger receptors. However, the interaction of functionalized GNPs with cells of the immune system is still far from being understood in more or less detail and requires further study [257–259].

In conclusion, it may be said that the time is probably right to talk of not only the biochemistry but also the biophysics of immune response, because it is the unique biophysical properties of metallic particles—in particular, the surface charge and the electrostatic field of the particle (influencing, in a certain manner, the charge, orientation, and polarization of the antigen molecules adsorbed on the particles)—that have to significantly affect the immune-response process. Thus, the GNPs can serve as adjuvants to improve the effectiveness of vaccines, stimulate antigen presenting cells, and provide controlled release of antigens. In addition, the immunogenicity of CNPs is determined by the physicochemical properties of particles such as size, shape, charge, and surface functionalization. Study of the immune response characteristics when using GNPs as a carrier and adjuvant for the production of antibodies will allow evaluating their potential for the development of effective vaccines.

References

1. Shukla R., Bansal V., Chaudhary M., Basu A., Bhonde R.R., Sastry M. Biocompatibility of gold nanoparticles and their endocytotic fate inside the cellular compartment: A microscopic overview // *Langmuir*. 2005. V. 21. P. 10644–10654.
2. Yen H.-J., Hsu S.-h., Tsai C.-L. Cytotoxicity and immunological response of gold and silver nanoparticles of different sizes // *Small*. 2009. V. 5. P. 1553–1561.
3. Lim Y.T., Cho M.Y., Choi B.S., Noh Y.-W., Chung B.H. Diagnosis and therapy of macrophage cells using dextran-coated near-infrared responsive hollow-type gold nanoparticles // *Nanotechnology*. 2008. V. 19. 375105.
4. Zhang Q., Hitchins V.M., Schrand A.M., Hussain S.M., Goering P.L. Uptake of gold nanoparticles in murine macrophage cells without cytotoxicity or production of pro-inflammatory mediators // *Nanotoxicology*. 2011. V. 5. P. 284–295.
5. Sumbayev V.V., Yasinska I.M., Garcia C.P., Gilliland D., Lall G.S., Gibbs B.F., Bonsall D.R., Varani L., Rossi F., Calzolai L. Gold nanoparticles downregulate interleukin-1β-induced pro-inflammatory responses // *Small*. 2013. V. 9. P. 472–477.
6. le Guevél X., Palomares F., Torres M.J., Blanca M., Fernandez T.D., Mayorga C. Nanoparticle size influences the proliferative responses of lymphocyte subpopulations // *RSC Adv*. 2015. V. 5. P. 85305–85309.
7. Choi M.-R., Stanton-Maxey K.J., Stanley J.K., Levin C.S., Bardhan R., Akin D., Badve S., Sturgis J., Robinson J.P., Bashir R., Halas N.J., Clare S.E. A cellular Trojan horse for delivery of therapeutic nanoparticles into tumors // *Nano Lett*. 2007. V. 7. P. 3759–3765.
8. Dreaden E.C., Mwakwari S.C., Austin L.A., Kieffer M.J., Oyelere A.K., El-Sayed M.A. Small molecule–gold nanorod conjugates selectively target and induce macrophage cytotoxicity towards breast cancer cells // *Small*. 2012. V. 8. P. 2819–2822.
9. Tian Y., Cui Y., Lou H., Li J., Yan P. Effect of colloidal gold on immunological function in mice // *Chin. Agric. Sci. Bull*. 2007. V. 23. P. 7–12 (in Chinese).
10. Lou H., Tian Y., Gao J.-Q., Deng S.-Y., Li J.-L. Effect of colloidal gold on the cytoactive of NK and the function of macrophage in mice // *J. Foshan Univ*. 2007. V. 25. P. 24–27 (in Chinese).
11. Bastús N.G., Sánchez-Tilló E., Pujals S., Farrera C., Kogan M.J., Giralt E., Celada A., Lloberas J., Puntes V. Peptides conjugated to gold nanoparticles induce macrophage activation // *Mol. Immunol*. 2009. V. 46. P. 743–748.

12. Bastús N.G., Sánchez-Tilló E., Pujals S., Farrera C., López C., Kogan M.J., Giralt E., Celada A., Lloberas J., Puntes V. Homogeneous conjugation of peptides onto gold nanoparticles enhances macrophage response // *ACS Nano*. 2009. V. 3. P. 1335–1344.

13. Staroverov S.A., Aksinenko N.M., Gabalov K.P., Vasilenko O.A., Vidyasheva I.V., Shchyogolev S.Y., Dykman L.A. Effect of gold nanoparticles on the respiratory activity of peritoneal macrophages // *Gold Bull*. 2009. V. 42. P. 153–156.

14. Staroverov S.A., Vidyasheva I.V., Gabalov K.P., Vasilenko O.A., Laskavyi V.N., Dykman L.A. Immunostimulatory effect of gold nanoparticles conjugated with transmissible gastroenteritis virus // *Bull. Exp. Biol. Med*. 2011. V. 151. P. 436–439.

15. Lee J.Y., Park W., Yi D.K. Immunostimulatory effects of gold nanorod and silica-coated gold nanorod on RAW 264.7 mouse macrophages // *Toxicol. Lett*. 2012. V. 209. P. 51–57.

16. Xu L., Liu Y., Chen Z., Li W., Liu Y., Wang L., Liu Y., Wu X., Ji Y., Zhao Y., Ma L., Shao Y., Chen C. Surface-engineered gold nanorods: Promising DNA vaccine adjuvant for HIV-1 treatment // *Nano Lett*. 2012. V. 12. P. 2003–2012.

17. Zlobina O.V., Bugaeva I.O., Maslyakova G.N., Firsova S.S., Bucharskaya A.B., Khlebtsov N.G., Khlebtsov B.N., Dykman L.A. Morphokinetics of mesenterial lymphatic node cell populations at exposure of gold nanoparticles in experiment // *Russ. Open Med. J*. 2012. V. 1. 0302.

18. Brown D.M., Johnston H., Gubbins E., Stone V. Cytotoxicity and cytokine release in rat hepatocytes, C3A cells and macrophages exposed to gold nanoparticles— Effect of biological dispersion media or corona // *J. Biomed. Nanotechnol*. 2014. V. 10. P. 3416–3429.

19. Bancos S., Stevens D.L., Tyner K.M. Effect of silica and gold nanoparticles on macrophage proliferation, activation markers, cytokine production, and phagocytosis *in vitro* // *Int. J. Nanomed*. 2015. V. 10. P. 183–206.

20. Fallarini S., Paoletti T., Battaglini C.O., Ronchi P., Lay L., Bonomi R., Jha S., Mancin F., Scrimin P., Lombardi G. Factors affecting T cell responses induced by fully synthetic glyco-gold-nanoparticles // *Nanoscale*. 2013. V. 5. P. 390–400.

21. Wei M., Chen N., Li J., Yin M., Liang L., He Y., Song H., Fan C., Huang Q. Polyvalent immunostimulatory nanoagents with self-assembled CpG oligonucleotide-conjugated gold nanoparticles // *Angew. Chem. Int. Ed*. 2012. V. 51. P. 1202–1206.

22. Rothenfusser S., Tuma E., Wagner M., Endres S., Hartmann G. Recent advances in immunostimulatory CpG oligonucleotides // *Curr. Opin. Mol. Ther*. 2003. V. 5. P. 98–106.

23. Tsai C.-Y., Lu S.-L., Hu C.-W., Yeh C.-S., Lee G.-B., Lei H.-Y. Size-dependent attenuation of TLR9 signaling by gold nanoparticles in macrophages // *J. Immunol*. 2012. V. 188. P. 68–76.

24. Massich M.D., Giljohann D.A., Seferos D.S., Ludlow L.E., Horvath C.M., Mirkin C.A. Regulating immune response using polyvalent nucleic acid–gold nanoparticle conjugates // *Mol. Pharm*. 2009. V. 6. P. 1934–1940.

25. Kim E.-Y., Schulz R., Swantek P., Kunstman K., Malim M.H., Wolinsky S.M. Gold nanoparticle-mediated gene delivery induces widespread changes in the expression of innate immunity genes // *Gene Ther*. 2012. V. 19. P. 347–353.

26. Walkey C.D., Olsen J.B., Guo H., Emili A., Chan W.C.W. Nanoparticle size and surface chemistry determine serum protein adsorption and macrophage uptake // *J. Am. Chem. Soc.* 2012. V. 134. P. 2139–2147.

27. Ma J.S., Kim W.J., Kim J.J., Kim T.J., Ye S.K., Song M.D., Kang H., Kim D.W., Moon W.K., Lee K.H. Gold nanoparticles attenuate LPS-induced NO production through the inhibition of NF-κB and IFN-β/STAT1 pathways in RAW264.7 cells // *Nitric Oxide.* 2010. V. 23. P. 214–219.

28. Liu Z., Li W., Wang F., Sun C., Wang L., Wang J., Sun F. Enhancement of lipopolysaccharide-induced nitric oxide and interleukin-6 production by PEGylated gold nanoparticles in RAW264.7 cells // *Nanoscale.* 2012. V. 4. P. 7135–7142.

29. Goldstein A., Soroka Y., Frusic-Zlotkin M., Lewis A., Kohen R. The bright side of plasmonic gold nanoparticles; activation of Nrf2, the cellular protective pathway // *Nanoscale.* 2016. V. 8. P. 11748–11759.

30. García C.P., Sumbayev V., Gilliland D., Yasinska I.M., Gibbs B.F., Mehn D., Calzolai L., Rossi F. Microscopic analysis of the interaction of gold nanoparticles with cells of the innate immune system // *Sci. Rep.* 2013. V. 3. 1326.

31. Ueno H., Klechevsky E., Morita R., Aspord C., Cao T., Matsui T., Di Pucchio T., Connolly J., Fay J.W., Pascual V., Palucka A.K., Banchereau J. Dendritic cell subsets in health and disease // *Immunol. Rev.* 2007. V. 219. P. 118–142.

32. Kang S., Ahn S., Lee J., Kim J.Y., Choi M., Gujrati V., Kim H., Kim J., Shin E.C., Jon S. Effects of gold nanoparticle-based vaccine size on lymph node delivery and cytotoxic T-lymphocyte responses // *J. Control. Release.* 2017. V. 256. P. 56–67.

33. Cheung W.-H., Chan V.S.-F., Pang H.-W., Wong M.-K., Guo Z.-H., Tam P.K.-H., Che C.-M., Lin C.-L., Yu W.-Y. Conjugation of latent membrane protein (LMP)-2 epitope to gold nanoparticles as highly immunogenic multiple antigenic peptides for induction of Epstein–Barr virus-specific cytotoxic T-lymphocyte responses *in vitro* // *Bioconjug. Chem.* 2009. V. 20. P. 24–31.

34. Cruz L.J., Rueda F., Cordobilla B., Simón L., Hosta L., Albericio F., Domingo J.C. Targeting nanosystems to human DCs via Fc receptor as an effective strategy to deliver antigen for immunotherapy // *Mol. Pharm.* 2011. V. 8. P. 104–116.

35. Villiers C.L., Freitas H., Couderc R., Villiers M.-B., Marche P.N. Analysis of the toxicity of gold nano particles on the immune system: Effect on dendritic cell functions // *J. Nanopart. Res.* 2010. V. 12. P. 55–60.

36. Ye F., Vallhov H., Qin J., Daskalaki E., Sugunan A., Toprak M.S., Fornara A., Gabrielsson S., Scheynius A., Muhammed M. Synthesis of high aspect ratio gold nanorods and their effects on human antigen presenting dendritic cells // *Int. J. Nanotechnology.* 2011. V. 8. P. 631–652.

37. Lin A.Y., Lunsford J., Bear A.S., Young J.K., Eckels P., Luo L., Foster A.E., Drezek R.A. High-density sub-100-nm peptide-gold nanoparticle complexes improve vaccine presentation by dendritic cells *in vitro* // *Nanoscale Res. Lett.* 2013. V. 8. 72.

38. Fytianos K., Rodriguez-Lorenzo L., Clift M.J., Blank F., Vanhecke D., von Garnier C., Petri-Fink A., Rothen-Rutishauser B. Uptake efficiency of surface modified gold nanoparticles does not correlate with functional changes and cytokine secretion in human dendritic cells *in vitro* // *Nanomedicine.* 2015. V. 11. P. 633–644.

39. Małaczewska J. The splenocyte proliferative response and cytokine secretion in mice after oral administration of commercial gold nanocolloid // *Pol. J. Vet. Sci.* 2015. V. 18. P. 181–189.

40. Małaczewska J. Effect of oral administration of commercial gold nanocolloid on peripheral blood leukocytes in mice // *Pol. J. Vet. Sci.* 2015. V. 18. P. 273–282.

41. Sharma M., Salisbury R.L., Maurer E.I., Hussain S.M., Sulentic C.E.W. Gold nanoparticles induce transcriptional activity of NF-κB in a B-lymphocyte cell line // *Nanoscale.* 2013. V. 5. P. 3747–3756.

42. Lee C.-H., Syu S.-H., Chen Y.-S., Hussain S.M, Onischuk A.A., Chen W.L., Huang G.S. Gold nanoparticles regulate the blimp1/pax5 pathway and enhance antibody secretion in B cells // *Nanotechnology.* 2014. V. 25. 125103.

43. Liptrott N.J., Kendall E., Nieves D.J., Farrell J., Rannard S., Fernig D.G., Owen A. Partial mitigation of gold nanoparticle interactions with human lymphocytes by surface functionalization with a 'mixed matrix' // *Nanomedicine (Lond).* 2014. V. 9. P. 2467–2479.

44. Bartneck M., Keul H.A., Singh S., Czaja K., Bornemann J., Bockstaller M., Möller M., Zwadlo-Klarwasser G., Groll J. Rapid uptake of gold nanorods by primary human blood phagocytes and immunomodulatory effects of surface chemistry // *ACS Nano.* 2010. V. 4. P. 3073–3086.

45. Bartneck M., Keul H.A., Zwadlo-Klarwasser G., Groll J. Phagocytosis independent extracellular nanoparticle clearance by human immune cells // *Nano Lett.* 2010. V. 10. P. 59–63.

46. Bartneck M., Keul H.A., Wambach M., Bornemann J., Gbureck U., Chatain N., Neuss S., Tacke F., Groll J., Zwadlo-Klarwasser G. Effects of nanoparticle surface-coupled peptides, functional endgroups, and charge on intracellular distribution and functionality of human primary reticuloendothelial cells // *Nanomedicine.* 2012. V. 8. P. 1282–1292.

47. Cho W.-S., Cho M., Jeong J., Choi M., Cho H.-Y., Han B.S., Kim S.H., Kim H.O., Lim Y.T., Chung B.H., Jeong J. Acute toxicity and pharmacokinetics of 13 nm-sized PEG-coated gold nanoparticles // *Toxicol. Appl. Pharmacol.* 2009. V. 236. P. 16–24.

48. Dykman L.A., Khlebtsov N.G. Immunological properties of gold nanoparticles // *Chem. Sci.* 2017. V. 8. P. 1719–1735.

49. DeFranco A.L., Locksley R.M., Robertson M. *Immunity: The Immune Response to Infection.* Oxford: Oxford University Press. 2007.

50. Blander J.M., Medzhitov R. On regulation of phagosome maturation and antigen presentation // *Nat. Immunol.* 2006. V. 7. P. 1029–1035.

51. Deng Z.J., Liang M., Monteiro M., Toth I., Minchin R.F. Nanoparticle-induced unfolding of fibrinogen promotes Mac-1 receptor activation and inflammation // *Nat. Nanotechnol.* 2011. V. 6. P. 39–44.

52. Arnáiz B., Martinez-Ávila O., Falcon-Perez J.M., Penadés S. Cellular uptake of gold nanoparticles bearing HIV gp120 oligomannosides // *Bioconjugate Chem.* 2012. V. 23. P. 814–825.

53. Dobrovolskaia M.A., McNeil S.E. Immunological properties of engineered nanomaterials // *Nat. Nanotechnol.* 2007. V. 2. P. 469–478.

54. Dobrovolskaia M.A., Aggarwal P., Hall J.B., McNeil S.E. Preclinical studies to understand nanoparticle interaction with the immune system and its potential effects on nanoparticle biodistribution // *Mol. Pharm.* 2008. V. 5. P. 487–495.

55. Patel P.C., Giljohann D.A., Daniel W.L., Zheng D., Prigodich A.E., Mirkin C.A. Scavenger receptors mediate cellular uptake of polyvalent oligonucleotide-functionalized gold nanoparticles // *Bioconjug. Chem.* 2010. V. 21. P. 2250–2256.

56. França A., Aggarwal P., Barsov E.V., Kozlov S.V., Dobrovolskaia M.A., González-Fernández Á. Macrophage scavenger receptor A mediates the uptake of gold colloids by macrophages *in vitro* // *Nanomedicine (Lond).* 2011. V. 6. P. 1175–1188.

57. Zilber L.A., Friese W.W. Über die antigene Eigenschaften kolloidaler Metalle // *Zh. Eksp. Biol.* 1929. V. 11. P. 128–135.

58. Huang G.S., Chen Y.-S., Yeh H.-W. Measuring the flexibility of immunoglobulin by gold nanoparticles // *Nano Lett.* 2006. V. 6. P. 2467–2471.

59. Steabben D.B. Studies on the physiological action of colloids. The action of colloidal substances on blob-elements and antibody content // *Br. J. Exp. Pathol.* 1925. V. 6. P. 1–13.

60. Zozaya J., Clark J. Active immunization of mice with the polysaccharides of pneumococci types I, II and III // *J. Exp. Med.* 1933. V. 57. P. 21–40.

61. Pacheco G. Studies on the action of metallic colloids on immunization // *Mem. Inst. Oswaldo Cruz.* 1925. V. 18. P. 119–149.

62. Gros O., O'Connor J.M. Einige Beobachtungen bei colloidalen Metallen mit Rücksicht auf ihre physikalisch-chemischen Eigenschaften und deren pharmakologische Wirkungen // *Archiv f. Exp. Path. u. Pharmakol.* 1911. Bd. 64. S. 456–467.

63. Stills H.F., Jr. Adjuvants and antibody production: Dispelling the myths associated with Freund's complete and other adjuvants // *ILAR J.* 2005. P. 46. P. 280–293.

64. Reed S.G., Orr M.T., Fox C.B. Key roles of adjuvants in modern vaccines // *Nat. Med.* 2013. V. 19. P. 1597–1608.

65. Kovalev I.E., Polevaya O.Y. *Biochemical Foundations of Immunity to Low-Molecular Chemical Compounds.* Moscow: Nauka. 1985 (in Russian).

66. Arnon R., Horwitz R.J. Synthetic peptides as vaccines // *Curr. Opin. Immunol.* 1992. V. 4. P. 449–453.

67. Ben-Yedidia T., Arnon R. Design of peptide and polypeptide vaccines // *Curr. Opin. Biotechnol.* 1997. V. 8. P. 442–448.

68. Bloom B.R., Lambert P.-H. *The Vaccine Book.* San Diego: Academic Press. 2003.

69. Moisa A.A., Kolesanova E.F. Synthetic peptide vaccines // *Biochemistry (Moscow) Suppl. Ser. B: Biomed. Chem.* 2010. V. 4. P. 321–332.

70. Li W., Joshi M.D., Singhania S., Ramsey K.H., Murthy A.K. Peptide vaccine: Progress and challenges // *Vaccines.* 2014. V. 2. P. 515–536.

71. Vartak A., Sucheck S.J. Recent advances in subunit vaccine carriers // *Vaccines.* 2016. V. 4. 12.

72. Male D., Brostoff J., Roth D., Roitt I. *Immunology.* Philadelphia: Saunders. 2012.

73. Kumar B.S., Ashok V., Kalyani P., Nair G.R. Conjugation of ampicillin and enrofloxacin residues with bovine serum albumin and raising of polyclonal antibodies against them // *Vet. World.* 2016. V. 9. P. 410–416.

74. Petrov R.V., Khaitov R.M. *Immunogenes and Vaccines of New Generation.* Moscow: GEOTAR-Media. 2011 (in Russian).

75. Kreuter J. Nanoparticles as adjuvants for vaccines // *Pharm. Biotechnol.* 1995. V. 6. P. 463–472.

76. De Koker S., Lambrecht B.N., Willart M.A., van Kooyk Y., Grooten J., Vervaet C., Remona J.P., De Geest B.G. Designing polymeric particles for antigen delivery // *Chem. Soc. Rev.* 2011. V. 40. P. 320–339.

77. George S.E., Elliott C.T., McLaughlin D.P., Delahaut P., Akagi T., Akashi M., Fodey T.L. An investigation into the potential use of nanoparticles as adjuvants for the production of polyclonal antibodies to low molecular weight compounds // *Vet. Immunol. Immunopathol.* 2012. V. 149. P. 46–53.

78. Gamvrellis A., Gloster S., Jefferies M., Mottram P.L., Smooker P., Plebanski M., Scheerlinck J.-P.Y. Characterisation of local immune responses induced by a novel nano-particle based carrier-adjuvant in sheep // *Vet. Immunol. Immunopathol.* 2013. V. 155. P. 21–29.

79. Maldonado R.A., LaMothe R.A., Ferrari J.D., Zhang A.-H., Rossi R.J., Kolte P.N., Griset A.P., O'Neil C., Altreuter D.H., Browning E., Johnston L., Farokhzad O.C., Langer R., Scott D.W., von Andrian U.H., Kishimoto T.K. Polymeric synthetic nanoparticles for the induction of antigen-specific immunological tolerance // *Proc. Natl. Acad. Sci. USA.* 2015. V. 112. P. 156–165.

80. Gregoriadis G. Immunological adjuvants: A role for liposomes // *Immunol. Today.* 1990. V. 11. P. 89–97.

81. Fukasawa M. Liposome oligomannan-coated with neoglycolipid, a new candidate for a safe adjuvant for induction of CD8+ cytotoxic T-lymphocytes // *FEBS Lett.* 1998. V. 441. P. 353–356.

82. Morris W. Potential of polymer microencapsulation technology for vaccine innovation // *Vaccine.* 1994. V. 12. P. 5–11.

83. Masalova O.V., Shepelev A.V., Atanadze S.N., Parnes Z.N., Romanova V.S., Volpina O.M., Semiletov Y.A., Kushch A.A. Immunostimulation effect of water-soluble fullerene derivatives – potential adjuvants for the new generation of vaccines // *Dokl. Biochem.* 1999. V. 369. P. 180–183.

84. Andreev S.M., Babakhin A.A., Petrukhina A.O., Romanova V.S., Parnes Z.N., Petrov R.V. Immunogenic and allergenic properties of fulleren conjugates with aminoacids and proteins // *Dokl. Biochem.* 2000. V. 370. P. 4–7.

85. Pantarotto D., Partidos C.D., Hoebeke J., Brown F., Kramer E., Briand J.-P., Muller S., Prato M., Bianco A. Immunization with peptide-functionalized carbon nanotubes enhances virus-specific neutralizing antibody responses // *Chem. Biol.* 2003. V. 10. P. 961–966.

86. Parra J., Abad-Somovilla A., Mercader J.V., Taton T.A., Abad-Fuentes A. Carbon nanotube-protein carriers enhance size-dependent self-adjuvant antibody response to haptens // *J. Control. Release.* 2013. V. 170. P. 242–251.

87. Silvestre B.T., Rabelo É.M.L., Versiani A.F., da Fonseca F.G., Silveira J.A.G., Bueno L.L., Fujiwara R.T., Ribeiro M.F.B. Evaluation of humoral and cellular immune response of BALB/c mice immunized with a recombinant fragment of MSP1a from *Anaplasma marginale* using carbon nanotubes as a carrier molecule // *Vaccine*. 2014. V. 32. P. 2160–2166.

88. Ceballos-Alcantarilla E., Abad-Somovilla A., Agulló C., Abad-Fuentes A., Mercader J.V. Protein-free hapten-carbon nanotube constructs induce the secondary immune response // *Bioconjug. Chem.* 2017. V. 28. P. 1630–1638.

89. Cao Y., Ma Y., Zhang M., Wang H., Tu X., Shen H., Dai J., Guo H., Zhang Z. Ultrasmall graphene oxide supported gold nanoparticles as adjuvants improve humoral and cellular immunity in mice // *Adv. Funct. Mater.* 2014. V. 24. P. 6963–6971.

90. Balenga N.A., Zahedifard F., Weiss R., Sarbolouki M.N., Thalhamer J., Rafati S. Protective efficiency of dendrosomes as novel nano-sized adjuvants for DNA vaccination against birch pollen allergy // *J. Biotechnol.* 2006. V. 124. P. 602–614.

91. Müller K., Skepper J.N., Posfai M., Trivedi R., Howarth S., Corot C., Lancelot E., Thompson P.W., Brown A.P., Gillard J.H. Effect of ultrasmall superparamagnetic iron oxide nanoparticles (Ferumoxtran-10) on human monocyte-macrophages *in vitro* // *Biomaterials*. 2007. V. 28. P. 1629–1642.

92. Zhao L., Mahony D., Cavallaro A.S., Zhang B., Zhang J., Deringer J.R., Zhao C.-X., Brown W.C., Yu C., Mitter N., Middelberg A.P.J. Immunogenicity of outer membrane proteins VirB9-1 and VirB9-2, a novel nanovaccine against *Anaplasma marginale* // *PLoS One*. 2016. e0154295.

93. Larsen S.T., Roursgaard M., Jensen K.A., Nielsen G.D. Nano titanium dioxide particles promote allergic sensitization and lung inflammation in mice // *Basic Clin. Pharmacol. Toxicol.* 2010. V. 106. P. 114–117.

94. Wang T., Zhen Y., Ma X., Wei B., Wang N. Phospholipid bilayer-coated aluminum nanoparticles as an effective vaccine adjuvant-delivery system // *ACS Appl. Mater. Interfaces*. 2015. V. 7. P. 6391–6396.

95. Maquieira Á., Brun E.M., Garcés-García M., Puchades R. Aluminum oxide nanoparticles as carriers and adjuvants for eliciting antibodies from non-immunogenic haptens // *Anal. Chem.* 2012. V. 84. P. 9340–9348.

96. Cho W.-S., Dart K., Nowakowska D.J., Zheng X., Donaldson K., Howie S.E.M. Adjuvanticity and toxicity of cobalt oxide nanoparticles as an alternative vaccine adjuvant // *Nanomedicine*. 2012. V. 7. P. 1495–1505.

97. Kireev M.N., Polunina T.A., Guseva N.P., Podboronova N.A., Krasnov Y.M., Taranenko T.M. Studying the immunogenic properties of plague microbe capsule antigen F1 conjugated with nanoparticles of colloid gold and silver // *Problems of Particularly Dangerous Infections*. 2008. V. 96. P. 43–46 (in Russian).

98. Xu Y., Tang H., Liu J.-h., Wang H., Liu Y. Evaluation of the adjuvant effect of silver nanoparticles both *in vitro* and *in vivo* // *Toxicol. Lett.* 2013. V. 219. P. 42–48.

99. Staroverov S.A., Volkov A.A., Larionov S.V., Mezhennyy P.V., Kozlov S.V., Fomin A.S., Dykman L.A. Study of transmissible-gastroenteritis-virus-antigen-conjugated immunogenic properties of selenium nanoparticles and gold // *Life Sci. J.* 2014. V. 11. P. 456–460.

100. Xiang S.D., Scholzen A., Minigo G., David C., Apostolopoulos V., Mottram P.L., Plebanski M. Pathogen recognition and development of particulate vaccines: Does size matter? // *Methods*. 2006. V. 40. P. 1–9.

101. Peek L.J., Middaugh C.R., Berkland C. Nanotechnology in vaccine delivery // *Adv. Drug Deliv. Rev.* 2008. V. 60. P. 915–928.

102. Hubbell J.A., Thomas S.N., Swartz M.A. Materials engineering for immunomodulation // *Nature*. 2009. V. 462. P. 449–460.

103. Bachmann M.F., Jennings G.T. Vaccine delivery: A matter of size, geometry, kinetics and molecular patterns // *Nat. Rev. Immunol.* 2010. V. 10. P. 787–796.

104. Fujita Y., Taguchi H. Current status of multiple antigen-presenting peptide vaccine systems: Application of organic and inorganic nanoparticles // *Chem. Cent. J.* 2011. V. 5. 48.

105. Krishnamachari Y., Geary S.M., Lemke C.D., Salem A.K. Nanoparticle delivery systems in cancer vaccines // *Pharmaceut. Res.* 2011. V. 28. P. 215–236.

106. Gregory A.E., Titball R., Williamson D. Vaccine delivery using nanoparticles // *Front. Cell Infect. Microbiol.* 2013. V. 3. 13.

107. Zaman M., Good M.F., Toth I. Nanovaccines and their mode of action // *Methods*. 2013. V. 60. P. 226–231.

108. Leleux J., Roy K. Micro and nanoparticle-based delivery systems for vaccine immunotherapy: An immunological and materials perspective // *Adv. Healthcare Mater.* 2013. V. 2. P. 72–94.

109. Hajizade A., Ebrahimi F., Salmanian A.-H., Arpanaei A., Amani J. Nanoparticles in vaccine development // *J. Appl. Biotechnol. Rep.* 2014. V. 1. P. 125–134.

110. Zhao L., Seth A., Wibowo N., Zhao C.-X., Mitter N., Yu C., Middelberg A.P.J. Nanoparticle vaccines // *Vaccine*. 2014. V. 32. P. 327–337.

111. Liu Y., Xu Y., Tian Y., Chen C., Wang C., Jiang X. Functional nanomaterials can optimize the efficacy of vaccines // *Small*. 2014. V. 10. P. 4505–4520.

112. Prashant C.K., Kumar M., Dinda A.K. Nanoparticle based tailoring of adjuvant function: The role in vaccine development // *J. Biomed. Nanotechnol.* 2014. V. 10. P. 2317–2331.

113. Jiao Q., Li L., Mu Q., Zhang Q. Immunomodulation of nanoparticles in nanomedicine applications // *BioMed Res. Int.* 2014. V. 2014. 426028.

114. Zhu M., Wang R., Nie G. Applications of nanomaterials as vaccine adjuvants // *Hum. Vaccin. Immunother.* 2014. V. 10. P. 2761–2774.

115. Aklakur M., Rather M.A., Kumar N. Nano delivery: An emerging avenue for nutraceuticals and drug delivery // *Crit. Rev. Food Sci. Nutr.* 2016. V. 56. P. 2352–2361.

116. Farrera C., Fadeel B. It takes two to tango: Understanding the interactions between engineered nanomaterials and the immune system // *Eur. J. Pharm. Biopharm.* 2015. V. 95. P. 3–12.

117. Irvine D.J., Hanson M.C., Rakhra K., Tokatlian T. Synthetic nanoparticles for vaccines and immunotherapy // *Chem. Rev.* 2015. V. 115. P. 11109–11146.

118. Maughan C.N., Preston S.G., Williams G.R. Particulate inorganic adjuvants: Recent developments and future outlook // *J. Pharm. Pharmacol.* 2015. V. 67. P. 426–449.

119. Seth A., Oh D.-B., Lim Y.T. Nanomaterials for enhanced immunity as an innovative paradigm in nanomedicine // *Nanomedicine (Lond.)*. 2015. V. 10. P. 959–975.

120. Salazar-González J.A., González-Ortega O., Rosales-Mendoza S. Gold nanoparticles and vaccine development // *Expert Rev. Vaccines*. 2015. V. 14. P. 1197–1211.

121. Kononenko V., Narat M., Drobne D. Nanoparticle interaction with the immune system // *Arh. Hig. Rada Toksikol*. 2015. V. 66. P. 97–108.

122. Gupta A., Das S., Schanen B., Seal S. Adjuvants in micro- to nanoscale: Current state and future direction // *Wiley Interdiscip. Rev. Nanomed. Nanobiotechnol.* 2016. V. 8. P. 61–84.

123. Ilinskaya A.N., Dobrovolskaia M.A. Understanding the immunogenicity and antigenicity of nanomaterials: Past, present and future // *Toxicol. Appl. Pharmacol.* 2016. V. 299. P. 70–77.

124. Moyano D.F., Liu Y., Peer D., Rotello V.M. Modulation of immune response using engineered nanoparticle surfaces // *Small*. 2016. V. 12. P. 76–82.

125. Shiosaka S., Kiyama H., Wanaka A., Tohyama M. A new method for producing a specific and high titer antibody against glutamate using colloidal gold as a carrier // *Brain Res*. 1986. V. 382. P. 399–403.

126. Wanaka A., Shiotani Y., Kiyama H., Matsuyama T., Kamada T., Shiosaka S., Tohyama M. Glutamate-like immunoreactive structures in primary sensory neurons in the rat detected by a specific antiserum against glutamate // *Exp. Brain Res.* 1987. V. 65. P. 691–694.

127. Ottersen O.P., Storm-Mathisen J. Localization of amino acid neurotransmitters by immunocytochemistry // *Trends Neurosci*. 1987. V. 10. P. 250–255.

128. Tomii A., Masugi F. Production of anti-platelet-activating factor antibodies by the use of colloidal gold as carrier // *Jpn. J. Med. Sci. Biol*. V. 1991. V. 44. P. 75–80.

129. Tatsumi N., Terano Y., Hashimoto K., Hiyoshi M., Matsuura S. An anti-platelet activating factor antibody and its effects on platelet aggregation // *Osaka City Med. J.* 1993. V. 39. P. 167–174.

130. Moffett J.R., Espey M.G., Namboodiri M.A.A. Antibodies to quinolinic acid and the determination of its cellular-distribution within the rat immune-system // *Cell Tissue Res*. 1994. V. 278. P. 461–469.

131. Dykman L.A., Matora L.Y., Bogatyrev V.A. Use of colloidal gold to obtain antibiotin antibodies // *J. Microbiol. Meth*. 1996. V. 24. P. 247–248.

132. Walensky L.D., Gascard P., Fields M.E., Blackshaw S., Conboy J.G., Mohandas N., Snyder S.H. The 13-kD FK506 binding protein, FKBP13, interacts with a novel homologue of the erythrocyte membrane cytoskeletal protein 4.1. // *J. Cell Biol.* 1998. V. 141. P. 143–153.

133. Walensky L.D., Dawson T.M., Steiner J.P., Sabatini D.M., Suarez J.D., Klinefelter G.R., Snyder S.H. The 12 kD FK506 binding protein FKBP12 is released in the male reproductive tract and stimulates sperm motility // *Mol. Med*. 1998. V. 4. P. 502–514.

134. Chen J., Zou F., Wang N., Xie S., Zhang X. Production and application of LPA polyclonal antibody // *Bioorg. Med. Chem. Lett*. 2000. V. 10. P. 1691–1693.

135. Feldman A.L., Tamarkin L., Paciotti G.F., Simpson B.W., Linehan W.M., Yang J.C., Fogler W.E., Turner E.M., Alexander H.R., Libutti S.K. Serum endostatin levels are elevated and correlate with serum vascular endothelial growth factor levels in patients with stage IV clear cell renal cancer // *Clin. Cancer Res.* 2000. V. 6. P. 4628–4634.

136. Mezhenny P.V., Staroverov S.A., Fomin A.S., Volkov A.A., Domnitsky I.Y., Kozlov S.V., Laskavy V.N., Dykman L.A. Study of immunogenic properties of transmissible gastroenteritis virus antigen conjugated to gold nanoparticles // *J. Biomed. Photonics Eng.* 2016. V. 2. 040308.

137. Olenina L.V., Kolesanova E.F., Gervaziev Y.V., Zaitseva I.S., Kuraeva T.E., Sobolev B.N., Archakov A.I. Preparation of anti-peptide antibodies to the protein binding sites E2 of hepatitis C virus with CD81 // *Med. Immunol.* 2001. V. 3. P. 231 (in Russian).

138. Chen Y.-S., Hung Y.-C., Liau I., Huang G.S. Assessment of the *in vivo* toxicity of gold nanoparticles // *Nanoscale Res. Lett.* 2009. V. 4. P. 858–864.

139. Chen Y.-S., Hung Y.-C., Liau I., Huang G.S. Assessment of gold nanoparticles as a size-dependent vaccine carrier for enhancing the antibody response against synthetic foot-and-mouth disease virus peptide // *Nanotechnology.* 2010. V. 21. 195101.

140. Dykman L.A., Staroverov S.A., Mezhenny P.V., Fomin A.S., Kozlov S.V., Volkov A.A., Laskavy V.N., Shchyogolev S.Y. Use of a synthetic foot-and-mouth disease virus peptide conjugated to gold nanoparticles for enhancing immunological response // *Gold Bull.* 2015. V. 48. P. 93–101.

141. Versiani A.F., Andrade L.M., Martins E.M.N., Scalzo S., Geraldo J.M., Chaves C.R., Ferreira D.C., Ladeira M., Guatimosim S., Ladeira L.O., da Fonseca F.G. Gold nanoparticles and their applications in biomedicine // *Future Virol.* 2016. V. 11. P. 293–309.

142. Mueller G.P., Driscoll W.J. α-Amidated peptides: Approaches for analysis // In: *Posttranslational Modification of Proteins: Tools for Functional Proteomics* / Ed. Kannicht C. Totowa: Humana Press. 2002. P. 241–257.

143. Dykman L.A., Bogatyrev V.A., Zaitseva I.S., Sokolova M.K., Ivanov V.V., Sokolov O.I. Use of colloid gold conjugates for identification of actins of various origin // *Biophysics.* 2002. V. 47. P. 587–594.

144. Dykman L.A., Sumaroka M.V., Staroverov S.A., Zaitseva I.S., Bogatyrev V.A. Immunogenic properties of colloidal gold // *Biol. Bull. Russ. Acad. Sci.* 2004. V. 31. P. 75–79.

145. Staroverov S.A., Pristensky D.V., Yermilov D.N., Semenov S.V., Aksinenko N.M., Shchyogolev S.Y., Dykman L.A. Obtaining of polyclonal antibodies to ivermectin and their detection in animal biological liquids // *Biotechnol. Russia.* 2007. No. 6. P. 100–109.

146. Pristensky D.V., Staroverov S.A., Ermilov D.N., Shchyogolev S.Y., Dykman L.A. Analysis of effectiveness of intracellular penetration of ivermectin immobilized onto corpuscular carriers // *Biochemistry (Moscow) Suppl. Ser. B: Biomed. Chem.* 2007. V. 1. P. 249–253.

147. Ishii N., Fitrilawati F., Manna A., Akiyama H., Tamada Y., Tamada K. Gold nanoparticles used as a carrier enhance production of anti-hapten IgG in rabbit: A study with azobenzene-dye as a hapten presented on the entire surface of gold nanoparticles // *Biosci. Biotechnol. Biochem.* 2008. V. 72. P. 124–131.

148. Kayed R., Head E., Thompson J.L., McIntire T.M., Milton S.C., Cotman C.W., Glabe C.G. Common structure of soluble amyloid oligomers implies common mechanism of pathogenesis // *Science.* 2003. V. 300. P. 486–489.

149. Vasilenko O.A., Staroverov S.A., Yermilov D.N., Pristensky D.V., Shchyogolev S.Y., Dykman L.A. Obtainment of polyclonal antibodies to clenbuterol with the use of colloidal gold // *Immunopharmacol. Immunotoxicol.* 2007. V. 29. P. 563–568.

150. Dykman L.A., Staroverov S.A., Fomin A.S., Panfilova E.V., Shirokov A.A., Bucharskaya A.B., Maslyakova G.N., Khlebtsov N.G. Gold nanoparticle-aided preparation of antibodies to α-methylacyl-CoA racemase and its immunochemical detection // *Gold Bull.* 2016. V. 49. P. 87–94.

151. Staroverov S.A., Ermilov D.N., Shcherbakov A.A., Semenov S.V., Shchyegolev S.Y., Dykman L.A. Generation of antibodies to *Yersinia pseudotuberculosis* antigens using the colloid gold particles as an adjuvant // *Zh. Mikrobiol. Epidemiol. Immunobiol.* 2003. No. 3. P. 54–57 (in Russian).

152. Gregory A.E., Williamson E.D., Prior J.L., Butcher W.A., Thompson I.J., Shaw A.M., Titball R.W. Conjugation of *Y. pestis* F1-antigen to gold nanoparticles improves immunogenicity // *Vaccine.* 2012. V. 30. P. 6777–6782.

153. Rodriguez-Del Rio E., Marradi M., Calderon-Gonzalez R., Frande-Cabanes E., Penadés S., Petrovsky N., Alvarez-Dominguez C. A gold glyco-nanoparticle carrying a listeriolysin O peptide and formulated with Advax™ delta inulin adjuvant induces robust T-cell protection against listeria infection // *Vaccine.* 2015. V. 33. P. 1465–1473.

154. Gao W., Fang R.H., Thamphiwatana S., Luk B.T., Li J., Angsantikul P., Zhang Q., Hu C.-M.J., Zhang L. Modulating antibacterial immunity via bacterial membrane-coated nanoparticles // *Nano Lett.* 2015. V. 15. P. 1403–1409.

155. Manea F., Bindoli C., Fallarini S., Lombardi G., Polito L., Lay L., Bonomi R., Mancin F., Scrimin P. Multivalent, saccharide-functionalized gold nanoparticles as fully synthetic analogs of type A *Neisseria meningitidis* antigens // *Adv. Mater.* 2008. V. 20. P. 4348–4352.

156. Safari D., Marradi M., Chiodo F., Dekker H.A.T., Shan Y., Adamo R., Oscarson S., Rijkers G.T., Lahmann M., Kamerling J.P., Penadés S., Snippe H. Gold nanoparticles as carriers for a synthetic *Streptococcus pneumoniae* type 14 conjugate vaccine // *Nanomedicine (Lond).* 2012. V. 5. P. 651–662.

157. Gregory A.E., Judy B.M., Qazi O., Blumentritt C.A., Brown K.A., Shaw A.M., Torres A.G., Titball R.W. A gold nanoparticle-linked glycoconjugate vaccine against *Burkholderia mallei* // *Nanomedicine.* 2015. V. 11. P. 447–456.

158. Torres A.G., Gregory A.E., Hatcher C.L., Vinet-Oliphant H., Morici L.A., Titball R.W., Roy C.J. Protection of non-human primates against glanders with a gold nanoparticle glycoconjugate vaccine // *Vaccine.* 2015. V. 33. P. 686–692.

159. Dakterzada F., Mohabati Mobarez A., Habibi Roudkenar M., Mohsenifar A. Induction of humoral immune response against *Pseudomonas aeruginosa* flagellin$_{(1-161)}$ using gold nanoparticles as an adjuvant // *Vaccine*. 2016. V. 34. P. 1472–1479.

160. Staroverov S.A., Dykman L.A. Use of gold nanoparticles for the preparation of antibodies to tuberculin, the immunoassay of mycobacteria, and animal vaccination // *Nanotechnol. Russia*. 2013. V. 8. P. 816–822.

161. Parween S., Gupta P.K., Chauhan V.S. Induction of humoral immune response against PfMSP-119 and PvMSP-119 using gold nanoparticles along with alum // *Vaccine*. 2011. V. 29. P. 2451–2460.

162. Kumar R., Ray P.C., Datta D., Bansal G.P., Angov E., Kumar N. Nanovaccines for malaria using *Plasmodium falciparum* antigen Pfs25 attached gold nanoparticles // *Vaccine*. 2015. V. 33. P. 5064–5071.

163. Bulashev A.K., Serikova S.S., Eskendirova S.Z. Anti-idiotypic antibodies in diagnosis of opisthorchiasis // *Biotechnol. Theory Pract.* 2014. No. 1. P. 36–42 (in Russian).

164. Barhate G.A., Gaikwad S.M., Jadhav S.S., Pokharkar V.B. Structure function attributes of gold nanoparticle vaccine association: Effect of particle size and association temperature // *Int. J. Pharm.* 2014. V. 471. P. 439–448.

165. Tsibezov V.V., Bashmakov Y.K., Pristenskiy D.V., Zigangirova N.A., Kostina L.V., Chalyk N.E., Kozlov A.Y., Morgunova E.Y., Chernyshova M.P., Lozbiakova M.V., Kyle N.H., Petyaev I.M. Generation and application of monoclonal antibody against lycopene // *Monoclon. Antib. Immunodiagn. Immunother.* 2017. V. 36. P. 62–67.

166. Pow D.V., Crook D.K. Extremely high titre polyclonal antisera against small neurotransmitter molecules: Rapid production, characterisation and use in light and electron-microscopic immunocytochemistry // *J. Neurosci. Meth.* 1993. V. 48. P. 51–63.

167. Baude A., Nusser Z., Molnár E., McIlhinney R.A.J., Somogyi P. High-resolution immunogold localization of AMPA type glutamate receptor subunits at synaptic and non-synaptic sites in rat hippocampus // *Neuroscience*. 1995. V. 69. P. 031–1055.

168. Harris D.P., Vordermeier H.-M., Arya A., Bogdan K., Moreno C., Ivanyi J. Immunogenicity of peptides for B cells is not impaired by overlapping T-cell epitope topology // *Immunology*. 1996. V. 88. P. 348–354.

169. Pickard L., Noël J., Henley J.M., Collingridge G.L., Molnar E. Developmental changes in synaptic AMPA and NMDA receptor distribution and AMPA receptor subunit composition in living hippocampal neurons // *J. Neurosci.* 2000. V. 20. P. 7922–7931.

170. Schäfer M.K.-H., Varoqui H., Defamie N., Weihe E., Erickson J.D. Molecular cloning and functional identification of mouse vesicular glutamate transporter 3 and its expression in subsets of novel excitatory neurons // *J. Biol. Chem.* 2002. V. 277. P. 50734–50748.

171. Holmseth S., Dehnes Y., Bjørnsen L.P., Boulland J.-L., Furness D.N., Bergles D., Danbolt N.C. Specificity of antibodies: Unexpected cross-reactivity of antibodies directed against the excitatory amino acid transporter 3 (EAAT3) // *Neuroscience*. 2005. V. 136. P. 649–660.

172. Schell M.J., Molliver M.E., Snyder S.H. D-serine, an endogenous synaptic modulator: Localization to astrocytes and glutamate-stimulated release // *Proc. Natl. Acad. Sci. USA*. 1995. V. 92. P. 3948–3952.

173. Schell M.J., Cooper O.B., Snyder S.H. D-aspartate localizations imply neuronal and neuroendocrine roles // *Proc. Natl. Acad. Sci. USA*. 1997. V. 94. P. 2013–2018.

174. Eliasson M.J.L., Blackshaw S., Schell M.J., Snyder S.H. Neuronal nitric oxide synthase alternatively spliced forms: Prominent functional localizations in the brain // *Proc. Natl. Acad. Sci. USA*. 1997. V. 94. P. 3396–3401.

175. Huster D., Hjelle O.P., Haug F.-M., Nagelhus E.A., Reichelt W., Ottersen O.P. Subcellular compartmentation of glutathione and glutathione precursors. A high resolution immunogold analysis of the outer retina of guinea pig // *Anat. Embryol*. 1998. V. 198. P. 277–287.

176. Staimer N., Gee S.J., Hammock B.D. Development of a sensitive enzyme immunoassay for the detection of phenyl-β-d-thioglucoronide in human urine // *Fresenius J. Anal. Chem*. 2001. V. 369. P. 273–279.

177. Staroverov S.A., Vasilenko O.A., Gabalov K.P., Pristensky D.V., Yermilov D.N., Aksinenko N.M., Shchyogolev S.Y., Dykman L.A. Preparation of polyclonal antibodies to diminazene and its detection in animal blood plasma // *Int. Immunopharmacol*. 2008. V. 8. P. 1418–1422.

178. Demenev V.A., Shchinova M.A., Ivanov L.I., Vorobeva R.N., Zdanovskaia N.I., Nebaikina N.V. Perfection of methodical approaches to designing vaccines against tick-borne encephalitis // *Vopr. Virusol*. 1996. V. 41. P. 107–110 (in Russian).

179. Mezhenny P.V., Staroverov S.A., Volkov A.A., Kozlov S.V., Laskavy V.N., Dykman L.A., Isayeva A.Y. Construction of conjugates of colloidal selenium and colloidal gold with the protein of influenza virus and the study of their immunogenic properties // *Bull. Saratov State Agrarian University*. 2013. No. 2. P. 29–32 (in Russian).

180. Tao W., Ziemer K.S, Gill H.S. Gold nanoparticle–M2e conjugate coformulated with CpG induces protective immunity against influenza A virus // *Nanomedicine (Lond.)*. 2014. V. 9. P. 237–251.

181. Niikura K., Matsunaga T., Suzuki T., Kobayashi S., Yamaguchi H., Orba Y., Kawaguchi A., Hasegawa H., Kajino K., Ninomiya T., Ijiro K., Sawa H. Gold nanoparticles as a vaccine platform: Influence of size and shape on immunological responses *in vitro* and *in vivo* // *ACS Nano*. 2013. V. 7. P. 3926–3938.

182. Stone J.W., Thornburg N.J., Blum D.L., Kuhn S.J., Wright D.W., Crowe J.E., Jr. Gold nanorod vaccine for respiratory syncytial virus // *Nanotechnology*. 2013. V. 24. 295102.

183. Wang H., Ding Y., Su S., Meng D., Mujeeb A., Wu Y., Nie G. Assembly of hepatitis E vaccine by '*in situ*' growth of gold clusters as nano-adjuvants: An efficient way to enhance the immune responses of vaccination // *Nanoscale Horiz*. 2016. V. 1. P. 394–398.

184. Chen H.-W., Huang C.-Y., Lin S.-Y., Fang Z.-S., Hsu C.-H., Lin J.-C., Chen Y.I., Yao B.-Y., Hu C.-M.J. Synthetic virus-like particles prepared via protein corona formation enable effective vaccination in an avian model of coronavirus infection // *Biomaterials*. 2016. V. 106. P. 111–118.

185. Calderón-Gonzalez R., Terán-Navarro H., Frande-Cabanes E., Ferrández-Fernández E., Freire J., Penadés S., Marradi M., García I., Gomez-Román J., Yañez-Díaz S., Álvarez-Domínguez C. Pregnancy vaccination with gold glyco-nanoparticles carrying *Listeria monocytogenes* peptides protects against listeriosis and brain- and cutaneous-associated morbidities // *Nanomaterials*. 2016. V. 6. 151.

186. Almeida J.P.M., Figueroa E.R., Drezek R.A. Gold nanoparticle mediated cancer immunotherapy // *Nanomedicine*. 2013. V. 10. P. 503–514.

187. Cao-Milán R., Liz-Marzán L.M. Gold nanoparticle conjugates: Recent advances toward clinical applications // *Expert Opin. Drug Deliv*. 2014. V. 11. P. 741–752.

188. Lee I.-H., Kwon H.-K., An S., Kim D., Kim S., Yu M.K., Lee J.-H., Lee T.-S., Im S.-H., Jon S. Imageable antigen-presenting gold nanoparticle vaccines for effective cancer immunotherapy *in vivo* // *Angew. Chem., Int. Ed*. 2012. V. 51. P. 8800–8805.

189. Park Y.-M., Lee S.J., Kim Y.S., Lee M.H., Cha G.S., Jung I.D., Kang T.H., Han H.D. Nanoparticle-based vaccine delivery for cancer immunotherapy // *Immune Netw*. 2013. V. 13. P. 177–183.

190. Ahn S., Lee I.-H., Kang S., Kim D., Choi M., Saw P.E., Shin E.-C., Jon S. Gold nanoparticles displaying tumor-associated self-antigens as a potential vaccine for cancer immunotherapy // *Adv. Healthc. Mater*. 2014. V. 3. P. 1194–1199.

191. Almeida J.P.M., Lin A.Y., Figueroa E.R., Foster A.E., Drezek R.A. *In vivo* gold nanoparticle delivery of peptide vaccine induces anti-tumor immune response in prophylactic and therapeutic tumor models // *Small*. 2015. V. 11. P. 1453–1459.

192. Biswas S., Medina S.H., Barchi J.J., Jr. Synthesis and cell-selective antitumor properties of amino acid conjugated tumor-associated carbohydrate antigen-coated gold nanoparticles // *Carbohydr. Res*. 2015. V. 405. P. 93–101.

193. Chiodo F., Enríquez-Navas P.M., Angulo J., Marradi M., Penadés S. Assembling different antennas of the gp120 high mannose-type glycans on gold nanoparticles provides superior binding to the anti-HIV antibody 2G12 than the individual antennas // *Carbohydr. Res*. 2015. V. 405. P. 102–109.

194. Gianvincenzo P.D., Calvo J., Perez S., Álvarez A., Bedoya L.M., Alcamí J., Penadés S. Negatively charged glyconanoparticles modulate and stabilize the secondary structures of a gp120 V3 loop peptide: Toward fully synthetic HIV vaccine candidates // *Bioconjug. Chem*. 2015. V. 26. P. 755–765.

195. Liu Y., Chen C. Role of nanotechnology in HIV/AIDS vaccine development // *Adv. Drug Deliv. Rev*. 2016. V. 103. P. 76–89.

196. Lin F. Development of gold nanoparticle-based antigen delivery platform for vaccines against HIV-1. Iowa State University. Graduate Theses and Dissertations. 2015. Paper 14546.

197. Wang Y.-T., Lu X.-M., Zhu F., Huang P., Yu Y., Zeng L., Long Z.-Y., Wu Y.-M. The use of a gold nanoparticle-based adjuvant to improve the therapeutic efficacy of hNgR-Fc protein immunization in spinal cord-injured rats // *Biomaterials*. 2011. V. 32. P. 7988–7998.

198. Marasini N., Skwarczynski M., Toth I. Oral delivery of nanoparticle-based vaccines // *Expert Rev. Vaccines*. 2014. V. 13. P. 1361–1376.

199. Ballester M., Nembrini C., Dhar N., de Titta A., de Piano C., Pasquier M., Simeoni E., van der Vlies A.J., McKinney J.D., Hubbell J.A., Swartz M.A. Nanoparticle conjugation and pulmonary delivery enhance the protective efficacy of Ag85B and CpG against tuberculosis // *Vaccine*. 2011. V. 29. P. 6959–6966.

200. Gupta P.N., Vyas S.P. Colloidal carrier systems for transcutaneous immunization // *Curr. Drug Targets*. 2011. V. 12. P. 579–597.

201. Chadwick S., Kriegel C., Amiji M. Nanotechnology solutions for mucosal immunization // *Adv. Drug Deliv. Rev.* 2010. V. 62. P. 394–407.

202. Pokharkar V., Bhumkar D., Suresh K., Shinde Y., Gairola S., Jadhav S.S. Gold nanoparticles as a potential carrier for transmucosal vaccine delivery // *J. Biomed. Nanotechnol.* 2011. V. 7. P. 57–59.

203. Pissuwan D., Nose K., Kurihara R., Kaneko K., Tahara Y., Kamiya N., Goto M., Katayama Y., Niidome T. A solid-in-oil dispersion of gold nanorods can enhance transdermal protein delivery and skin vaccination // *Small*. 2011. V. 7. P. 215–220.

204. Kowalczyk D.W., Ertl H.C.J. Immune responses to DNA vaccines // *Cell. Mol. Life Sci.* 1999. V. 55. P. 751–770.

205. Hasan U.A., Abai A.M., Harper D.R., Wren B.W., Morrow W.J.W. Nucleic acid immunization: Concepts and techniques associated with third generation vaccines // *J. Immunol. Meth.* 1999. V. 229. P. 1–22.

206. Yang N.S., Christou P. *Particle Bombardment Technology for Gene Transfer*. Oxford: Oxford University Press. 1994.

207. O'Brien J.A., Lummis S.C.R. Nano-biolistics: A method of biolistic transfection of cells and tissues using a gene gun with novel nanometer-sized projectiles // *BMC Biotechnology*. 2011. V. 11. 66.

208. Donnelly J.J., Wahren B., Liu M.A. DNA vaccines: Progress and challenges // *J. mmunol.* 2005. V. 175. P. 633–639.

209. *DNA Vaccines: A New Era in Vaccinology* / Eds. Liu M.A., Hillerman M.R., Kurth R. New York: New York Academy of Sciences. 1995.

210. Gurunathan S., Klinman D.M., Seder R.A. DNA vaccines: Immunology, application, and optimization // *Ann. Rev. Immunol.* 2000. V. 18. P. 927–974.

211. Yang J., Li Y., Jin S., Xu J., Wang P.C., Liang X.-J., Zhang X. Engineered biomaterials for development of nucleic acid vaccines // *Biomater. Res.* 2015. V. 19. 5.

212. Sundaram P., Xiao W., Brandsma J.L. Particle-mediated delivery of recombinant expression vectors to rabbit skin induces high-titered polyclonal antisera (and circumvents purification of a protein immunogen) // *Nucleic Acids Res.* 1996. V. 24. P. 1375–1377.

213. Cui Z., Mumper R.J. The effect of co-administration of adjuvants with a nanoparticle-based genetic vaccine delivery system on the resulting immune responses // *Eur. J. Pharm. Biopharm.* 2003. V. 5. P. 11–18.

214. Zhang L., Widera G., Rabussay D. Enhancement of the effectiveness of electroporation-augmented cutaneous DNA vaccination by a particulate adjuvant // *Bioelectrochemistry*. 2004. V. 63. P. 369–373.

215. Roy M.J., Wu M.S., Barr L.J., Fuller J.T., Tussey L.G., Speller S., Culp J., Burkholder J.K., Swain W.F., Dixon R.M., Widera G., Vessey R., King A., Ogg G., Gallimore A., Haynes J.R., Heydenburg Fuller D. Induction of antigen-specific CD8+ T cells, T helper cells, and protective levels of antibody in humans by particle-mediated administration of a hepatitis B virus DNA vaccine // *Vaccine.* 2000. V. 19. P. 764–778.

216. Leutenegger C.M., Boretti F., Mislin C.N., Flynn J.N., Schroff M., Habel A., Junghans C., Koenig-Merediz S.A., Sigrist B., Aubert A., Pedersen N.C., Wittig B., Lutz H. Immunization of cats against feline immunodeficiency virus (FIV) infection by using minimalistic immunogenic defined gene expression vector vaccines expressing FIV gp140 alone or with feline interleukin-12 (IL-12), IL-16, or a CpG motif // *J. Virol.* 2000. V. 74. P. 10447–10457.

217. Chen D., Payne L.G. Targeting epidermal Langerhans cells by epidermal powder immunization // *Cell Res.* 2002. V. 12. P. 97–104.

218. Dean H.J., Fuller D., Osorio J.E. Powder and particle-mediated approaches for delivery of DNA and protein vaccines into the epidermis // *Comp. Immun. Microbiol. Infect. Dis.* 2003. V. 26. P. 373–388.

219. Thomas M., Klibanov A.M. Conjugation to gold nanoparticles enhances polyethylenimine's transfer of plasmid DNA into mammalian cells // *Proc. Natl. Acad. Sci. USA.* 2003. V. 100. P. 9138–9143.

220. Salem A.K., Hung C.F., Kim T.W., Wu T.C., Searson P.C., Leong K.W. Multicomponent nanorods for vaccination applications // *Nanotechnology.* 2005. V. 16. P. 484–487.

221. Xu L., Liu Y., Chen Z., Li W., Liu Y., Wang L., Liu Y., Wu X., Ji Y., Zhao Y., Ma L., Shao Y., Chen C. Surface-engineered gold nanorods: Promising DNA vaccine adjuvant for HIV-1 treatment // *Nano Lett.* 2012. V. 12. P. 2003–2012.

222. Ojeda R., de Paz J.L., Barrientos A.G., Martín-Lomas M., Penadés S. Preparation of multifunctional glyconanoparticles as a platform for potential carbohydrate-based anticancer vaccines // *Carbohydr. Res.* 2007. V. 342. P. 448–459.

223. Marradi M., Di Gianvincenzo P., Enríquez-Navas P.M., Martínez-Ávila O.M., Chiodo F., Yuste E., Angulo J., Penadés S. Gold nanoparticles coated with oligomannosides of HIV-1 glycoprotein gp120 mimic the carbohydrate epitope of antibody 2G12 // *J. Mol. Biol.* 2011. V. 410. P. 798–810.

224. Brinãs R.P., Sundgren A., Sahoo P., Morey S., Rittenhouse-Olson K., Wilding G.E., Deng W., Barchi J.J., Jr. Design and synthesis of multifunctional gold nanoparticles bearing tumor-associated glycopeptide antigens as potential cancer vaccines // *Bioconjug. Chem.* 2012. V. 23. P. 1513–1523.

225. Parry A.L., Clemson N.A., Ellis J., Bernhard S.S.R., Davis B.G., Cameron N.R. 'Multicopy multivalent' glycopolymer-stabilized gold nanoparticles as potential synthetic cancer vaccines // *J. Am. Chem. Soc.* 2013. V. 135. P. 9362–9365.

226. Mocan T., Matea C., Tabaran F., Iancu C., Orasan R., Mocan L. *In vitro* administration of gold nanoparticles functionalized with MUC-1 protein fragment generates anticancer vaccine response via macrophage activation and polarization mechanism // *J. Cancer.* 2015. V. 6. P. 583–592.

227. Tavernaro I., Hartmann S., Sommer L., Hausmann H., Rohner C., Ruehl M., Hoffmann-Roeder A., Schlecht S. Synthesis of tumor-associated MUC1-glycopeptides and their multivalent presentation by functionalized gold colloids // *Org. Biomol. Chem.* 2015. V. 13. P. 81–97.

228. Cai H., Degliangeli F., Palitzsch B., Gerlitzki B., Kunz H., Schmitt E., Fiammengo R., Westerlind U. Glycopeptide-functionalized gold nanoparticles for antibody induction against the tumor associated mucin-1 glycoprotein // *Bioorg. Med. Chem.* 2016. V. 24. P. 1132–1135.

229. Zhao Z., Wakita T., Yasui K. Inoculation of plasmids encoding Japanese encephalitis virus PrM-E proteins with colloidal gold elicits a protective immune response in BALB/c mice // *J. Virol.* 2003. V. 77. P. 4248–4260.

230. Dykman L.A., Bogatyrev V.A., Staroverov S.A., Pristensky D.V., Shchyogolev S.Y., Khlebtsov N.G. The adjuvanticity of gold nanoparticles // *Proc. SPIE.* 2006. V. 6164. P. 616401-1–616401-10.

231. Dykman L.A., Staroverov S.A., Bogatyrev V.A., Shchyogolev S.Y. *Gold Nanoparticles as an Antigen Carrier and an Adjuvant.* New York: Nova Science Publishers. 2010.

232. Dykman L.A., Staroverov S.A., Bogatyrev V.A., Shchyogolev S.Y. Adjuvant properties of gold nanoparticles // *Nanotechnol. Russia.* 2010. V. 5. P. 748–761.

233. Graham G. Medicinal chemistry of gold // *Agents Actions Suppl.* 1993. V. 44. P. 209–217.

234. Merchant B. Gold, the Noble metal and the paradoxes of its toxicology // *Biologicals.* 1998. V. 26. P. 49–59.

235. Eisler R. Mammalian sensitivity to elemental gold (Au0) // *Biol. Trace Element Res.* 2004. V. 100. P. 1–18.

236. Vallhov H., Qin J., Johansson S.M., Ahlborg N., Muhammed M.A., Scheynius A., Gabrielsson S. The importance of an endotoxin-free environment during the production of nanoparticles used in medical applications // *Nano Lett.* 2006. V. 6. P. 1682–1686.

237. Kingston M., Pfau J.C., Gilmer J., Brey R. Selective inhibitory effects of 50-nm gold nanoparticles on mouse macrophage and spleen cells // *J. Immunotoxicol.* 2016. V. 13. P. 198–208.

238. Dobrovolskaia M.A., Germolec D.R., Weaver J.L. Evaluation of nanoparticle immunotoxicity // *Nat. Nanotechnol.* 2009. V. 4. P. 411–414.

239. Zolnik B.S., González-Fernández A., Sadrieh N., Dobrovolskaia M.A. Nanoparticles and the immune system // *Endocrinology.* 2010. V. 151. P. 458–465.

240. *Handbook of Immunological Properties of Engineered Nanomaterials* / Eds. Dobrovolskaia M.A., McNeil S.E. Singapore: World Scientific Publ. 2013.

241. Lin A.Y., Almeida J.P.M., Bear A., Liu N., Luo L., Foster A.E., Drezek R.A. Gold nanoparticle delivery of modified CpG stimulates macrophages and inhibits tumor growth for enhanced immunotherapy // *PLoS One.* 2013. V. 8. e63550.

242. Tao Y., Zhang Y., Ju E., Ren H., Ren J. Gold nanoclusters–based vaccines for dual-delivery of antigens and immunostimulatory oligonucleotides // *Nanoscale.* 2015. V. 7. P. 12419–12426.

243. Zhou Q., Zhang Y., Du J., Li Y., Zhou Y., Fu Q., Zhang J., Wang X., Zhan L. Different-sized gold nanoparticle activator/antigen increases dendritic cells accumulation in liver-draining lymph nodes and CD8+ T cell responses // *ACS Nano.* 2016. V. 10. P. 2678–2692.

244. Zhang H., Gao X.-D. Nanodelivery systems for enhancing the immunostimulatory effect of CpG oligodeoxynucleotides // *Mater. Sci. Eng. C.* 2017. V. 70. P. 935–946.

245. Wang Y., Wang Y., Kang N., Liu Y., Shan W., Bi S., Ren L., Zhuang G. Construction and immunological evaluation of CpG-Au@HBc virus-like nanoparticles as a potential vaccine // *Nanoscale Res. Lett.* 2016. V. 11. 338.

246. Klinman D.M., Sato T., Shimosato T. Use of nanoparticles to deliver immunomodulatory oligonucleotides // *WIREs Nanomed. Nanobiotechnol.* 2016. V. 8. P. 631–637.

247. Zhang P., Chiu Y.-C., Tostanoski L.H., Jewell C.M. Polyelectrolyte multilayers assembled entirely from immune signals on gold nanoparticle templates promote antigen-specific T cell response // *ACS Nano.* 2015. V. 9. P. 6465–6477.

248. Barhate G., Gautam M., Gairola S., Jadhav S., Pokharkar V. *Quillaja saponaria* extract as mucosal adjuvant with chitosan functionalized gold nanoparticles for mucosal vaccine delivery: Stability and immunoefficiency studies // *Intern. J. Pharm.* 2013. V. 441. P. 636–642.

249. Barhate G., Gautam M., Gairola S., Jadhav S., Pokharkar V. Enhanced mucosal immune responses against tetanus toxoid using novel delivery system comprised of chitosan-functionalized gold nanoparticles and botanical adjuvant: Characterization, immunogenicity, and stability assessment // *J. Pharm. Sci.* 2014. V. 103. P. 3448–3456.

250. Joseph M.M., Aravind S.R., Varghese S., Mini S., Sreelekha T.T. PST-Gold nanoparticle as an effective anticancer agent with immunomodulatory properties // *Colloids Surf. B.* 2013. V. 104. P. 32–39.

251. Ye F., Vallhov H., Qin J., Daskalaki E., Sugunan A., Toprak M.S., Fornara A., Gabrielsson S., Scheynius A., Muhammed M. Synthesis of high aspect ratio gold nanorods and their effects on human antigen presenting cells // *Int. J. Nanotechnology.* 2011. V. 8. P. 631–652.

252. Wang Y.-T., Lu X.-M., Zhu F., Zhao M. The preparation of gold nanoparticles and evaluation of their immunological function effects on rats // *Biomed. Mater. Eng.* 2014. V. 24. P. 885–892.

253. Yavuz E., Sakalak H., Cavusoglu H., Uyar P., Yavuz M.S., Bagriacik E.U. Evaluation of the adjuvant effect of gold nanocages *in vitro* // *Eur. J. Immunol., Suppl. 1.* 2016. V. 46. P. 1223–1224.

254. Zhang W., Wang L., Liu Y., Chen X., Liu Q., Jia J., Yang T., Qiu S., Ma G. Immune responses to vaccines involving a combined antigen–nanoparticle mixture and nanoparticle-encapsulated antigen formulation // *Biomaterials.* 2014. V. 35. P. 6086–6097.

255. Marques Neto L.M., Kipnis A., Junqueira-Kipnis A.P. Role of metallic nanoparticles in vaccinology: Implications for infectious disease vaccine development // *Front. Immunol.* 2017. V. 8. 239.

256. Carabineiro S.A.C. Applications of gold nanoparticles in nanomedicine: Recent advances in vaccines // *Molecules*. 2017. V. 22. 857.

257. Comber J.D., Bamezai A. Gold nanoparticles (AuNPs): A new frontier in vaccine delivery // *J. Nanomedine Biotherapeutic Discov*. 2015. V. 5. 4.

258. David C.A.W., Owen A., Liptrott N.J. Determining the relationship between nanoparticle characteristics and immunotoxicity: Key challenges and approaches // *Nanomedicine (Lond.)*. 2016. V. 11. P. 1447–1464.

259. Chattopadhyay S., Chen J.-Y., Chen H.-W., Hu C.-M.J. Nanoparticle vaccines adopting virus-like features for enhanced immune potentiation // *Nanotheranostics*. 2017. V. 1. P. 244–260.

6

Multifunctional Gold-Based Composites for Theranostics

6.1 Preliminary Remarks

Although colloidal gold nanoparticles (GNPs) have been employed in pioneering medical treatments since the Middle Ages, the past two decades have witnessed an extraordinary renascence of studies and trials (biomedical and, in part, preclinical) of these particles (see Chapter 2). This renewed interest in GNPs and their applications is due at least to two main reasons. First, owing to their size-, shape-, and structure-dependent localized plasmon resonances, GNPs have unique physical properties, which differ appreciably from those of both bulk gold and other nanoparticles. In particular, they can effectively scatter light, convert it into heat, and enhance the local electromagnetic fields near their surfaces, thus producing enormous enhancement of linear and nonlinear phenomena. These plasmon-related properties form a physical basis for the diagnostic and therapeutic applications of GNPs. Second, owing to their facile surface chemistry, GNPs can be functionalized with various molecules, ensuring preferential accumulation in their biological targets. This target-specific accumulation can be accomplished by the size-dependent enhanced permeability and retention mechanism. The nanoscale size of GNPs permits them to act as local heaters and as drugs or adjuvants through the modification of cellular responses. Finally, owing to their multivalent functionalization, they can deliver unstable or poorly soluble drugs and contrast agents to target cells, tissues, and organs. This beneficial combination of physical and chemical properties have given rise to important applications of GNPs in a range of areas, including biological sensing, clinical analytics, genomics, immunology, optical bioimaging, photodynamic and photothermal (PD and PT) therapy of tumors and bacterial infections, and targeted delivery of drugs, peptides, DNA, small-interfering RNA (siRNA), and antigens.

In the past few years, hybrid nanoparticle systems have attracted significant interest, as they combine different nanomaterials in a single multifunctional nanostructure that exhibits the modalities of its component modules [1]. Theranostics, an emerging trend in nanomedicine, is capable of combining all the previously mentioned advanced properties into a single nanostructure with simultaneous diagnostic and therapeutic functions, which can be physically and chemically tailored for a particular organ, disease, or patient [2,3]. The first time that the term "theranostics" appeared in the literature was in

2002 [4,5]. In 2010, Lukianova-Hleb *et al.* [6] presented a complete concept of theranostics. Theranostics is closely related to the fabrication and application of multifunctional nanoparticles, which combine therapeutic and diagnostic possibilities within a single structure [7,8].

In particular, the emerging theranostics nanoconstructs promise to integrate various functionalities by incorporating different nanomaterials into a single and efficient anticancer agent [9,10]. In principle, by changing only the size, shape, or structure of gold or composite nanoparticles, it is possible to tune their plasmonic properties so that their radiative and nonradiative properties can be utilized in theranostics with minimal surface functionalization with PEG molecules. Presumably, the earliest examples of such particles were gold nanoshells (GNSs) deposited onto silica cores, which combined the properties of contrast agents in optical coherence tomography or in dark-field light microscopy with those of heating agents in plasmonic PT therapy [11,12]. The material constituting composite metallic particles may be a regular alloy of different metals or an inhomogeneous nanostructure expressing several plasmonic modalities. As the theranostic applications of usual spherical GNPs, gold nanorods (GNRs), gold nanostars (GNSts), GNSs, gold nanocages (GNCs), and gold nanoclusters (GNCls) have been described in many research and review articles (see, *e.g.*, [12]), they will not be dealt with here.

The focus of this chapter is on multifunctional theranostic nanocomposites (NCs), which can be fabricated by three main routes. The first route is to create *composite* (or *hybrid*) *nanoparticles*, whose components enable diagnostic and therapeutic functions. Most often, such nanoparticles are made up of one or many plasmonic particles (which themselves may have a composite structure, *e.g.*, silica/GNSs or gold/silver nanocages) that are surrounded by a biopolymeric, silica, or another dielectric shell. The shell may be doped with diverse reporter molecules (fluorescent, Raman, PD, *etc.*) and/or molecules of the substances being delivered (drugs, peptides, siRNA, *etc.*) (Figure 6.1). Composites of the first type also include an important group of gold and magnetic nanoparticles, which combine plasmonic and magnetic features supplemented with other modalities at the cost of encapsulation of various cargos [13].

Second, by use of smart bioconjugation techniques [14], GNPs can be functionalized with a set of different molecules, enabling them to perform targeting, diagnostic, and therapeutic functions in a single treatment procedure. This second class of *multifunctionalized nanoparticles* has found exciting applications in proof-of-concept theranostic experiments [15], and recent accomplishments in this area will also be discussed here.

Finally, the third route for multifunctional nanoparticles is a combination of the first two and involves additional functionalization of a hybrid (composite) nanoparticle with several molecules possessing different properties (*multifunctionalized composite nanoparticles*).

The number of articles dealing with multifunctional and hybrid nanomaterials has been increasing steeply in recent years. Although reviews have addressed certain problems in the use of multifunctional nanomaterials [12,16–26], there is a strong need to systematize the constantly renewed data in this area so as to help researchers evaluate the already published results and plan new studies.

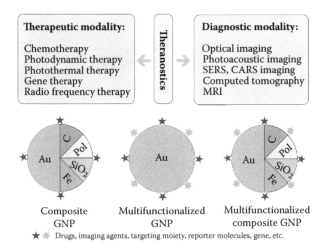

FIGURE 6.1 Schematic representation of three types of composites and their theranostic applications discussed in this chapter. The symbol Pol stands for a polymer incorporated in composite structure.

6.2 Composite GNPs

The composite GNPs used in biomedicine most often include polymeric nanoparticles or nanoparticles of other metals and of semiconductors [27–29]. An earliest example of this type of nanoparticle is a hybrid of GNPs and poly(amidoamine) (PAMAM) dendrimer [30]. Owing to improved cellular penetration, such composites increase the contrast in optical imaging and x-ray computed tomography (CT) [31], as well as the efficacy of tumor radiotherapy with PAMAM–GNP–^{198}Au [32]. By conjugating doxorubicin (DOX) to the dendrimer layer of PEG–PAMAM–GNR, Li *et al.* [33] demonstrated synergistic PT and chemotherapy treatment of HeLa cells and colon carcinoma tumors in mice (Figure 6.2).

Other NCs combine plasmonic heating of GNPs with thermally responsible polymers and encapsulated drugs [34,35]. At low temperature, these polymers typically exhibit swelling conformation stabilized by hydrogen bonds between water and polar chain groups. Under light-mediated plasmonic heating, the water molecules move into the external solution, causing the polymers to shrink and the encapsulated drugs to be released. More sophisticated composites may also contain reporter molecules as bioimaging and diagnostic agents.

Still other commonly used NCs are GNPs complexed with nanoparticles of the cationic biodegradable polyaminosaccharide chitosan (ChT). The antioxidative potential of these constructs is greater than that of each component alone [36]. In particular, such NCs are suited for the electrochemical study of cancer cells and for the electrochemical detection of myoglobin [37,38]. GNPs coated with ChT and polyacrylic acid and doped with the anticancer drug cisplatin were effective as a means of drug delivery into the cell

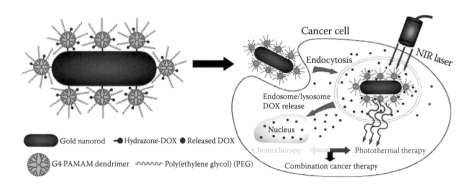

FIGURE 6.2 Schematic illustration of pH-sensitive PEGylated PAMAM dendrimer-doxorubicin conjugate-hybridized GNRs (PEG@DOX@PAMAM@GNR) for combined PT therapy and chemotherapy. (Adapted from Li, X., Takashima, M., Yuba, E., Harada, A., Kono, K., *Biomaterials*, 35, 6576–6584, 2014. With permission.)

and its nucleus and as a contrast agent for tumor cell imaging [39]. ChT-coated GNRs doped with cisplatin enhanced the drug's antitumor effect if PT therapy was included as an additional option [40]. ChT-modified GNRs with encapsulated indocyanine green, a PD dye, were suitable for combined PT and PD therapy [41]. Biodegradable polymers other than ChT are also utilized; *e.g.*, GNPs encapsulated within insulin-doped chondroitin sulfate were employed for oral treatment of diabetes in laboratory animals [42].

A composite built up from small (2–3 nm) GNPs encased in a lipid or polymer matrix, which was termed "nanobeacon," was developed for the optoacoustic detection of tumor cells [43]. Interesting results were generated by near-infrared (NIR) imaging of tumor cells with an NC consisting of 5-nm GNPs incorporated into a biodegradable poly(ethylene glycol) (PEG)/polylactone capsule [44]. Similar NCs that contained GNRs were reported for surface-enhanced Raman spectroscopy (SERS) detection and PT therapy of cancer cells [45]. GNPs deposited onto polylactide cores [46] enabled ultrasound imaging and PT therapy to be combined. GNPs encapsulated in PEG/polycaprolactone were used for photoacoustic (PA) imaging complemented with PT therapy [47]. Analogous structures containing the photosensitizer Ce6 were employed for NIR fluorescence (FL) detection and combined PT and PD therapy [48]. Polymeric biocompatible DOX–GNP micelles were utilized for PT therapy in combination with chemotherapy, as well as an effective contrast agent for computed and PA tomography *in vivo* [49]. For the same purpose, it has been proposed to enclose GNPs in lipid capsules [50,51], GNRs in polymerosomes [52], and GNSs in liposomes [53]. Conversely, liposomes can be loaded with drug and FL cargos (*e.g.*, DOX and carboxyfluorescein [54]) and covered with GNSs. Such nanocontainers [54] are leakage-free throughout the storage and can be efficiently released by NIR laser irradiation, at least in principle. However, strong heating of melted GNSs may interfere with the native drug efficacy.

Liu *et al.* directly fabricated multifunctional conducting polymeric materials on the GNR surfaces *via* facile oxidative polymerization. These dye-free NIR SERS nanoprobes served as both biocompatible surface coatings and NIR-active reporters and exhibited good structural stability, good biocompatibility, intriguing NIR SERS activity, and an extraordinary NIR PT transduction efficiency, indicating their potential applicability in cancer therapy [55]. Multifunctional biodegradable poly(lactide-co-glycolide) nanoparticles with encapsulated rhodamine (a drug model) that were coated successively with a magnetic shell and a gold shell [56] enabled PT-controlled targeted drug delivery and enhanced contrast in magnetic resonance tomography. Targeted drug delivery in conjunction with PT therapy was realized with DOX-encapsulated poly(lactide-co-glycolide) nanoparticles coated with a gold half-shell [57].

Biocompatible GNRs covered with a bovine serum albumin (BSA) capsule, with antibodies to vascular endothelial growth factor incorporated through the avidin–biotin system, increased the efficacy of PT therapy [58]. Peralta *et al.* [59] employed desolvation and cross-linking to embed GNPs, GNRs, or hollow nanoshells within 100–300-nm human serum albumin spherical particles. Simon *et al.* [60] designed theranostic agents based on Pluronic-stabilized gold nanoaggregates loaded with methylene blue for multimodal cell imaging and enhanced PT therapy. Costa Lima and Reis [61] incorporated methotrexate and GNPs into PEG–poly(D,L-lactic-co-glycolic acid) nanospheres as a novel platform for PA imaging and NIR PT application in rheumatoid arthritis [61]. All these examples clearly demonstrate improved functionality and, in some cases, a certain synergism of NCs through a rational choice of polymeric components.

Huang and Liu [62] developed a new polymer self-assembly strategy for preparing octahedral, triangular, and multicore Au@poly(styrene-*alt*-maleic acid) nanoparticles. The Au–polymer particles with Pt drugs tethered to the polymer structure on the surface performed combined chemo–PT cancer therapy together with NIR-excited emission. The drug delivery system was directly visualized by nonlinear optical microscopy.

For high-efficiency capturing of breast cancer cells, Park *et al.* [63] fabricated silicon nanowires covered uniformly with GNCls and with antibodies. The nanoclusters (NCls) demonstrated high PT efficiency when captured cancer cells were irradiated with a NIR laser. For combined chemotherapy and PT therapy in *in vitro* testing, Botella *et al.* [64] assembled 15-nm GNPs into clusters of well-defined size, covered them with a porous silica shell, and doped the shell with the antitumor drug 20(S)-camptothecin. Under femtosecond pulse 790-nm laser irradiation of the developed NCs, 42-MG-BA human glioma cells were killed by three mechanisms: PT, vapor bubble generation, and cytotoxic drug release. Similar NCs were made with GNRs as a plasmonic heater and DOX as an antitumor drug for *in vitro* and *in vivo* dual PT and chemotherapy [65–67].

In a separate research direction, NCs are built up from a plasmonic core and a mesoporous silica shell doped with PD dyes. Such constructs are suitable for PD therapy, FL microscopy, and SERS [68]. GNRs and GNCs coated with a mesoporous silica shell doped with the PD dye hematoporphyrin have been used successfully for combined PD and PT inactivation of antibiotic-resistant *Staphylococcus aureus* [69] (Figure 6.3a), cancer cells [70] (Figure 6.4), and large-volume (about 3 cm³) implanted tumors in rats [71,72] (Figure 6.3b). GNRs enclosed in a folate-doped silica shell were employed both for CT and for PT *in vivo* [73].

Halas's group [74] compared the PT performance of PEG-coated sub-100-nm gold–silica–gold "nanomatryoshkas" (NMs) with that of conventional 150-nm SiO₂/Au shells when both particle types were administered at equal doses of Au mass. Treatment of mice bearing highly aggressive triple negative breast tumors with the NMs led to

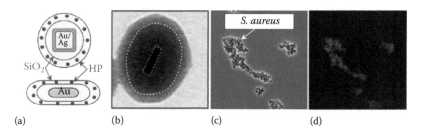

(a) (b) (c) (d)

FIGURE 6.3 Schematic illustration of multifunctional NCs containing a plasmonic core, a primary silica shell, and a secondary mesoporous silica shell doped with HP molecules (a), a transmission electron microscopy image of a GNR covered by a primary silica shell and a secondary mesoporous silica shell doped with HP (b). Panels (c) and (d) show transmission and FL imaging of *S. aureus* bacteria incubated with NCs. (Adapted from Khlebtsov, B.N., Tuchina, E.S., Khanadeev, V.A., Panfilova, E.V., Petrov, P.O., Tuchin, V.V., Khlebtsov, N.G., *J. Biophotonics.*, 6, 338–351, 2013. With permission.)

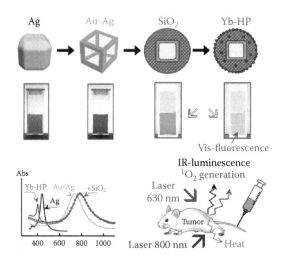

FIGURE 6.4 Schematic illustration summarizing how FL composite nanoparticles can be fabricated starting with Ag nanocubes and ending with silica-coated Au–Ag nanocages functionalized with Yb–hematoporphyrin (Yb–HP) molecules. The right, center photo shows visible FL of the particles under UV excitation. The left, bottom plots show the absorbance spectra of Ag nanocubes (black), Au–Ag nanocages (yellow-gold), silica-coated nanocages (blue circles), and final Au–Ag/SiO₂/Yb–HP NCs (red). The right-bottom image illustrates potential applications for *in vivo* imaging and PD therapy (using IR-luminescence and singlet oxygen generated by Yb–HP) and PT therapy (using the heat generated by plasmonic nanocages). (Adapted from Khlebtsov, B., Panfilova, E., Khanadeev, V., Bibikova, O., Terentyuk, G., Ivanov A., Rumyantseva V., Shilov I., Ryabova A., Loshchenov V., Khlebtsov, N., *ACS Nano*, 5, 7077–7089, 2011. With permission.)

enhanced accumulation of these composites in large tumors and doubled the animal survival time owing to their smaller size and higher adsorption cross-section.

Alongside the "polymeric" and "silica" NCs, composites made up of GNPs and iron oxide magnetic nanoparticles are also being used widely. The combination of the magnetic properties of a ferromagnetic material and the optical properties of plasmon resonant particles increases the capabilities of these NCs in biomedical research [75]. Both GNPs with a magnetic shell and nanoparticles with a magnetic core and gold shell have been reported. Perhaps the first example of such composites came from Mirkin's group [76], who synthesized hybrid DNA-functionalized magnetic–GNSs (SiO_2–Fe_3O_4–GNSs) with a gold surface, silica core, and magnetic inner layer. These magnetic–gold composites had the optochemical properties of DNA-functionalized GNPs and the superparamagnetic properties of iron oxide nanoparticles. In a typical protocol [77], 100-nm silica cores were combined with 7-nm magnetite (Fe_3O_4) nanoparticles, and 1–3-nm gold seeds were immobilized on their surfaces (Figure 6.5) [78].

A continuous gold shell was then formed that was stabilized with PEG thiol (PEG-SH), and SiO_2–Fe_3O_4–GNSs conjugated with anti-HER2/neu antibodies were applied to magnetic resonance imaging (MRI) and PT therapy of cancer cells [77]. For the same purposes, the spherical magnetic Fe_2O_3 [79] or Fe_3O_4 [80] was decorated with a gold shell and was functionalized with anti-EGFR Nemoarker antibodies [79] or with fluorescein-labeled integrin $\alpha_\nu\beta_3$ monoclonal antibodies [80], respectively.

Similar GNSs, modified with PEG [81] or amphiphilic polymers [82], were employed as contrast agents in computer-guided magnetic resonance therapy and magneto-acoustic imaging of cancer cells [83]. Fe_2O_3/Au–PEG core/shell NCs were used for magnetic resonance/PA multimodal imaging and PT therapy [84]. In other work, PT efficiency was enhanced with the GNR/Fe_3O_4 [85] and GNS/Fe_3O_4 [86] gold–magnetic composites. The utility of herceptin-tagged GNRs/Fe_3O_4 was demonstrated in dual-mode imaging and PT ablation of SK-BR-3 breast cancer cells [85].

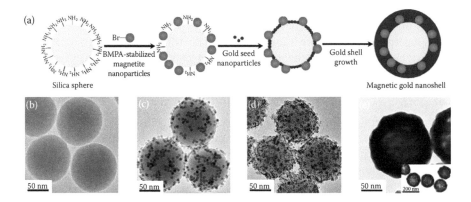

FIGURE 6.5 (a) Synthesis of magnetic GNSs. Transmission electron microscopy images of (b) amine-modified silica spheres, (c) silica spheres with Fe_3O_4 immobilized on their surfaces, (d) silica spheres with Fe_3O_4 and gold seeds, (e) magnetic GNSs. (Adapted from Dykman, L.A., Khlebtsov, N.G., *Biomaterials*, 108, 13–34, 2016. With permission.)

Dumbbell-like Au/Fe_3O_4 nanoparticles were functionalized with a single-chain antibody, scFv, that binds to the A33 antigen present on colorectal cancer cells [87]. This new class of hybrid nanoparticles can potentially serve as an effective antigen-targeted PT therapeutic agent for cancer treatment, as well as a probe for magnetic resonance-based imaging.

For simultaneous use in magnetic resonance and CT and in PT of tumor cells, Feng *et al.* [88] developed an NC consisting of a GNR core coated with polypyrrole and a Fe_3O_4 nanoparticle shell. To the same end, Li *et al.* [89] fabricated composite nanostars formed from a magnetic core with a gold shell functionalized with hyaluronic acid, which interacts with CD44 receptors, overexpressed on the surface of cancer cells. The NC proposed by Ji *et al.* [90] was composed of a magnetic core coated sequentially with a silica and a gold shell. The use of $Fe_2O_3/SiO_2/Au$ nanoshells, with their combination of unique magnetic and optical properties, promises to enhance the efficacy of nanoshell-mediated PT therapy by making it possible to navigate more nanoparticles to tumors by applying an external magnetic field and by permitting real-time *in vivo* MRI of the particle distribution before, during, and after PT therapy.

We have given above some examples of NCs with additional possibilities arising from functionalization with targeting molecules. In a study by Hu *et al.* [91], GNRs were capped with a silica layer containing magnetic nanoparticles and were conjugated to folic acid (FA). As folate receptors are overexpressed on the surface of most cancer cells, these NCs were used to make a "magnetic trap" and to perform optoacoustic imaging of circulating tumor cells [91]. Joint effects of chemotherapy and magnetic fields on tumors were examined with DOX-coated Fe_3O_4/GNP NCs [92,93].

More complex NCs have also been synthesized. Ma *et al.* [94] obtained a composite consisting of a magnetic core with a GNR-doped silica shell and demonstrated its capabilities in combined PT therapy and chemotherapy of tumors, as coupled with magnetic resonance tomography and NIR-thermal imaging. In another example [95], a magnetic core ($MnFe_2O_4$) was coated sequentially with layers of silica and gold and was stabilized with PEG-SH and functionalized with Erbitux therapeutic antibodies. This nanostructure served for targeted cancer detection both by MRI and by localized synchronous therapy through Erbitux antibodies.

If a superparamagnetic core is covered sequentially with a silica and a gold shell, such a particle becomes suitable both for PT therapy and for MRI of tumors [96]. Cheng *et al.* [97] performed effective FL imaging and magnetically targeted PT therapy of tumors with an NC based on small iron oxide nanoparticles and lanthanide nanocrystals wrapped in a thin layer of gold.

For a successful PT therapy of malignant cells, it is desirable to exert magnetic resonance control at all stages of treatment. NCs made up of GNPs and silica-coated magnetite nanoparticles [98], as well as iron–GNPs enclosed in a polymeric capsule [99,100], are some of the examples to illustrate this approach.

Carril *et al.* [101] fabricated gold-decorated iron oxide glyconanoparticles and demonstrated their trimodal functions as contrast agents in three clinically applicable imaging techniques—x-ray CT, ultrasound, and T_2-weighted MRI.

Interesting data emerged from a study by Huang *et al.* [102], who embedded a magnetic core within a gold shell surrounded by the antibiotic vancomycin. The resultant composite structure was used for magnetic separation of vancomycin-bound pathogenic bacteria followed by their PT treatment. A similar nanostructure, conjugated to S6 aptamer labeled with the FL dye Cy3, was applied to the targeted diagnosis, isolation, and PT therapy of cancer cells [103].

In Chen *et al.* [104], magnetic nanoparticles encapsulated in silica shells with a layer of PEGylated GNPs were functionalized with the antitumor agent curcumin for use in targeted drug delivery and MRI of tumor cells. In Cheng *et al.* [105], a polymeric capsule containing GNRs, quantum dots, and magnetic nanoparticles was coated with the anti-cancer agent paclitaxel, making it possible to perform chemotherapy and simultaneous PT therapy under the control of FL microscopy and magnetic resonance tomography.

Gold-nanoshelled liquid perfluorocarbon nanocapsules for combined dual-modal ultrasound/CT imaging and PT therapy of cancer were described by Ke *et al.* [106]. In their next study [107], multifunctional nanocapsules were fabricated through loading perfluorocarbon and superparamagnetic iron oxide nanoparticles into poly(lactic acid) nanocapsules, followed by the formation of PEGylated GNSs on the surface. The resulting multicomponent NCs proved to be able to act as a nanotheranostic agent to achieve successful bimodal ultrasound/MRI-guided PT ablation in human tumor xenograft models noninvasively.

Baek *et al.* [108] described GNRs that were coated successively with a mesoporous silica shell, DOX, and a thermoresponsive polymer. By combining different modes of treatment (*e.g.*, chemo and hyperthermia) and the temperature and pH responsivity of the nanoparticles, it is possible (1) to enhance the drug uptake, (2) to trigger the drug release locally by using internal and external stimuli, and (3) to combine synergistically drug and temperature effects to induce cell death in a desired location. Furthermore, the attenuation of the X-ray signal achieved with the nanohybrids suggests their suitability as a contrast agent for CT.

Other metals (palladium, cobalt, gadolinium, manganese, platinum) or their oxides have been used to make GNP NCs much more rarely and have only been employed in various diagnostic methods, including SERS, magnetic resonance tomography, opto-acoustic imaging, and positron emission tomography. We believe, however, that similarly to gold and carbon (graphene, fullerene, and nanotube) nanoparticle composites [109,110], they may well find their place in theranostics also.

Specific applications so far include (1) hybrid nanostructures based on carbon nano-tubes [111] or graphene [112] for effective DOX delivery and tumor cell imaging, (2) a PEG–folate-coated GNP–carbon nanotube composite for visualization and PT therapy of cancer cells *in vitro* [113], (3) an anti-GD2 antibody-coated GNP-conjugated carbon nanotube for two-photon imaging and PT killing of targeted melanoma cells [114], (4) a PEG-coated complex consisting of GNPs, magnetic nanoparticles, and graphene for magnetic resonance tomography and PT therapy of tumors *in vivo* [115], (5) a gold core–graphene oxide nanocolloid shell construct coated with the PD dye zinc phthalocya-nine for simultaneous Raman bioimaging, PT therapy, and PD therapy [116], and (6) a graphene-oxide-based NC coated with a GNP- and folate-doped mesoporous silica shell for the detection and selective killing of tumor cells [117].

Chen *et al.* [118] presented the synthesis and application of new graphene-based magnetic and plasmonic NCs for magnetic-field-assisted drug delivery and chemo/PT synergistic therapy. The NCs were prepared *via* conjugation of DOX-loaded PEGylated Fe_2O_3/Au core/shell nanoparticles with reduced graphene oxide.

In the past decade, a novel class of "SERS tag" nanoprobes has been developed to address unmet needs in chemical and biological sensing, biomedical diagnostics, and therapy [119], including just emerging point-of-care capabilities [120]. Owing to electromagnetic near field (NF) enhancement by metal nanoparticles and spectral fingerprints of Raman-active molecules, SERS tags demonstrate extraordinary sensitivity, multiplexing and quantitative abilities together with optical stability and biosafety as compared to common fluorescent dyes and quantum dots. However, SERS tags with outer Raman reporters have several serious drawbacks, *e.g.*, strong dependence of SERS response on the environmental conditions and the particle aggregation. To overcome these difficulties, Lim *et al.* [121] suggested a new highly efficient SERS tags in which Raman molecules are embedded in a nanometer-sized interior gap between the metallic core and the shell. Since that pioneering publication, several research groups have reported on layered plasmonic probes with embedded Raman molecules [122–132]. Such multilayered structures, also called nanomatryoshkas [74] (or nanorattles in the case of porous outer shell [133,134]) have great potential for biomedical applications owing to several advantages: (1) Raman molecules are protected from desorption, subjected to a strongly enhanced electromagnetic field in the gap [135] and their SERS response does not depend on the environmental conditions and NM aggregation; (2) owing to bright and uniform spectral pattern, NMs provide a linear correlation between probe concentration and SERS intensity and allow for a real-time *in vivo* imaging and high throughput sensing with short integration times; (3) NM size (50–100 nm) and NIR spectral properties can be designed for effective cellular uptake and tissue imaging with negligible background from autofluorescence of biological samples; and (4) NM probes can be multiplexed by incorporating different Raman molecules into two-layered or multilayered NMs [130].

The data on the theranostic use of composite GNPs are summarized in Table 6.1.

6.3 Multifunctionalized GNPs

Initially, multifunctionalized GNPs began to be used to improve the efficacy of targeted delivery of drugs to tumor cells and tissues. Most often, they were developed by double-functionalizing GNPs with drugs and targeting molecules; *e.g.*, Mukherjee *et al.* [142] double-conjugated GNPs to the anticancer drug gemcitabine and to antibodies against vascular endothelial growth factor, receptors which are overexpressed on the cancer cell surface. GNPs functionalized simultaneously with paclitaxel and tumor necrosis factor were highly effective both *in vitro* and *in vivo* [143]. Good efficacy was also achieved by concurrent attachment of paclitaxel and biotin to GNPs [144] and of DOX and biotin to GNRs [145]. GNPs were also combined simultaneously with the antitumor agent cetuximab and antibodies against the folate receptors of tumor cells [146] and simultaneously with folate and cisplatin [147]. Another promising option is the double functionalization of GNPs with antitumor and cell-penetrating peptides, which ensure greater efficacy of payload delivery to the target [148–150].

As tyrosine kinase receptors are overexpressed at the surface of tumor cells, hollow GNPs were modified concurrently with tyrosine kinase and DOX and were found to be highly effective in PT and chemotherapy [151]. GNPs coupled to DOX and folate were

TABLE 6.1 Composite GNPs

Composite	Functionalization	Therapeutic Modality	Diagnostic Modality	References
		Au/Polymer NCs		
GNRs/PAMAM	DOX	Drug delivery PT therapy	FL imaging	Li [33]
GNPs/ChT/poly(acrylic acid)	Cisplatin	Drug delivery	Dark-field imaging Transmission electron microscopy (TEM) imaging	Hu [39]
GNRs/ChT	Cisplatin	Drug delivery PT therapy	FL imaging	Chen [40]
GNRs/ChT	Indocyanine green	PT therapy PD therapy	FL imaging	Chen [41]
GNRs/PEG/polylactide	Raman reporter	PT therapy	SERS spectroscopy	Song [45]
Polylactide/polyvinyl alcohol/poly(allyl-amine hydrochloride)/Au shell	–	PT therapy	Ultrasound imaging	Ke [46]
GNRs/PEG/b-poly(ε-caprolactone)	–	PT therapy	PA imaging	Huang [47]
GNPs/polyethylene oxide-b-polystyrene	Ce6	PT therapy PD therapy	PA imaging	Lin [48]
Poly(ε-caprolactone)/poly(2-hydroxyethyl methacrylate)/poly[2-(2-methoxyethoxy) ethyl methacrylate]/GNPs	DOX	Drug delivery PT therapy	PA imaging CT imaging	Deng [49]
GNCs/phosphatidylcholine	Hypocrellin B	PT therapy PD therapy	Two-photon luminescence imaging	Gao [50]
GNRs/lipid GNPs/lipid	–	PT therapy PD therapy	NIR FL imaging	Vankayala [51]
GNRs/polyaniline GNRs/polypyrrole	–	PT therapy	SERS spectroscopy	Liu [55]

(*Continued*)

TABLE 6.1 (CONTINUED)   Composite GNPs

Composite	Functionalization	Therapeutic Modality	Diagnostic Modality	References
Polylactide/Au half shell	DOX	Drug delivery PT therapy	FL imaging	Lee [57]
GNRs/albumin	Anti-VEGFR2 antibodies	PT therapy	Two-photon luminescence imaging	Wang [58]
GNPs/Pluronic F127	Methylene blue	PD therapy	FL imaging	Simon [60]
GNPs/PEG-poly(D,L-lactic-co-glycolic acid)	Methotrexate	PT therapy	PA imaging	Costa Lima [61]
GNPs/poly(styrene-alt-maleic acid)	Pt(II)/Pt(IV)-based drugs	Drug delivery PT therapy	Multiphoton FL imaging	Huang [62]
GNPs/gelatin	DOX	Drug delivery	FL imaging Dark-field imaging TEM imaging SERS spectroscopy	Suarasan [136]
Au/SiO_2 NCs				
$GNRs/SiO_2$	DOX	Drug delivery PT therapy	Two-photon luminescence imaging	Zhang [65]
$GNRs/SiO_2$	FA	PT therapy	CT imaging	Huang [73]
$GNRs/SiO_2$	Trastuzumab	Drug delivery PT therapy	PA imaging	Cao [137]
Au/Fe NCs				
Fe_2O_3/Au shell	Anti-EGFR antibodies	PT therapy	MRI	Larson [79]
Fe_2O_3/Au/PEG	—	PT therapy	MRI PA imaging	Li [84]
$GNRs/Fe_3O_4$	Herceptin	Drug delivery PT therapy	MRI FL imaging	Wang [85]

(Continued)

TABLE 6.1 (CONTINUED) Composite GNPs

Composite	Functionalization	Therapeutic Modality	Diagnostic Modality	References
Dumbbell-like Au/Fe$_3$O$_4$ nanoparticles	scFv	PT therapy	MRI	Kirui [87]
Fe$_3$O$_4$/Au shell	Hyaluronic acid	PT therapy	MRI	Li [89]
			CT imaging	
Fe$_3$O$_4$/Au shell	Vancomycin	Drug delivery	Magnetic separation	Huang [102]
		PT therapy		
Fe$_3$O$_4$/Au shell	Aptamer	PT therapy	Magnetic separation	Fan [103]
GNRs/Fe$_3$O$_4$ nanoparticles/PEG	–	PT therapy	MRI	Khafaji [138]
		Au/C NCs		
Carbon nanotube/Au shell	DOX	Drug delivery	FL imaging	Minati [111]
Graphene oxide/Au shell	DOX	Drug delivery	SERS imaging	Ma [112]
Carbon nanotube/Au shell	FA	PT therapy	SERS imaging	Wang [113]
Carbon nanotube/GNPs	Anti-GD2 antibodies	PT therapy	Two-photon luminescence imaging	Tchounwou [114]
GNPs/graphene oxide	Zinc phthalocyanine	PT therapy	SERS imaging	Kim [116]
		PD therapy		
Carbon nanotubes ring/GNPs	–	PT therapy	SERS spectroscopy	Song [139]
			PA imaging	
		Complex NCs		
Polylactide/magnetic layer/Au shell	Drug model	Drug delivery	MRI	Park [56]
GNRs/polypyrrole/Fe^{3}O$_4$	–	PT therapy	MRI	Feng [88]
			CT imaging	
Fe$_2$O$_3$/SiO$_2$/Au shell	–	PT therapy	MRI	Ji [90]
Fe$_2$O$_3$/SiO$_2$/GNRs	DOX	Drug delivery	MRI	Ma [94]
		PT therapy	Infrared thermal imaging	

(Continued)

TABLE 6.1 (CONTINUED) Composite GNPs

Composite	Functionalization	Therapeutic Modality	Diagnostic Modality	References
$MnFe_2O_4/SiO_2/Au$ shell	Erbitux	Drug delivery PT therapy	MRI FL imaging	Lee [95]
$Fe_2O_3/SiO_2/Au$ shell	–	PT therapy	MRI	Melancon [96]
Lanthanide + Fe_3O_4/Au shell	–	PT therapy	Luminescence imaging MRI	Cheng [97]
$GNPs/Fe_2O_3/SiO_2$ shell	–	PT therapy	MRI	Sotiriou [98]
Fe+GNPs/poly(N-isopropylacrylamide-co-acrylamide)-block-poly(ε-caprolactone)	–	Magnetic hyperthermia	MRI	Kim [99]
GNRs/poly(N-isopropylacrylamide-comethacrylic acid)/Fe_3O_4	–	PT therapy	PA imaging MRI	Yang [100]
$Fe_2O_3/SiO_2/GNPs$	Curcumin	Drug delivery	MRI	Chen [104]
Poly(lactic-co-glycolic acid) NPs/GNRs + QDs + Fe_3O_4	Paclitaxel	Drug delivery PT therapy	FL imaging MRI	Cheng [105]
Perfluorocarbon/Au shell/PEG	–	PT therapy	Ultrasound imaging CT imaging	Ke [106]
Perfluorocarbon/Fe_3O_4/poly(lactic acid)/Au shell/PEG	–	PT therapy	Ultrasound imaging MRI	Ke [107]
GNRs/SiO_2/poly(NIPAAm-co-BVIm)	DOX	Drug delivery PT therapy	CT imaging	Baek [108]
Graphene oxide/$Fe_3O_4/GNPs$	–	PT therapy	X-ray/MRI	Shi [115]
Graphene oxide/$SiO_2/GNPs$	FA	Selective killing	Colorimetric detection	Maji [117]
Graphene oxide/Fe_2O_3/Au shell	DOX	Drug delivery PT therapy	MRI	Chen [118]
$GNRs/SiO_2$/carbon dots	–	PT therapy PD therapy	FL imaging PA imaging	Jia [140]
GNPs + Fe_3O_4/PEG + polycaprolactone	–	Radiotherapy	CT imaging MRI	Sun [141]

applied in tumor therapy under the control of multiphoton spectroscopy [152]. DNA-coated [153] and BSA-coated [154] GNRs integrated with DOX were utilized in combined chemotherapy and PT therapy. The efficacy of intracellular penetration was enhanced with GNP–DNA–FA–DOX conjugates [155].

GNPs modified with DOX and aptamers for prostate-specific antigen were designed for simultaneous CT diagnosis and therapy of prostate cancer [156]. For combined therapy and positron emission tomography, GNRs were conjugated to DOX [157] or trastuzumab [158] and additionally to a cRGD targeting peptide. GNPs with attached DOX and targeting peptide A54 [159] were useful in dual chemotherapy and PT therapy.

For synchronous PT and PD therapy and for tumor cell monitoring by NIR imaging, GNRs were coupled to indocyanine green and antiepidermal growth factor receptor antibodies [160]. GNPs integrated with tumor-specific antibodies and with phthalocyanine, a PD dye, ensured successful PD therapy [161].

GNPs coated simultaneously with an arginylglycylaspartic acid (RGD peptide) and fluorochrome-labeled heparin were effective in penetrating metastatic cells, allowing them to be visualized by FL microscopy, and caused cell death owing to the apoptotic action of heparin [162]. Gold–silver nanorods conjugated simultaneously to rhodamine 6G and phage fusion protein specific for colorectal cancer cells were used for the FL imaging and PT therapy of tumors [163].

A GNSt-based multifunctional conjugate was developed for NIR imaging and combined PT, PD, and chemotherapy of tumors [164]. The GNSts were functionalized with three ligands: a targeting peptide, DOX, and indocyanine green.

For GNR labeling, Huang *et al.* [165] also employed three types of probing molecules. These included (1) a single-chain variable fragment peptide that recognizes the epidermal growth factor receptor, (2) an amino terminal fragment peptide that recognizes the urokinase plasminogen activator receptor, and (3) a cRGD peptide that recognizes the $\alpha v \beta 3$ integrin receptor. An important result from that study was that the general effectiveness of particle delivery to the cells depended weakly on the presence of probing molecules but affected strongly the distribution of the particles in the intercellular space.

For increasing the efficacy of PT therapy of tumors and improving the intracellular penetration of drugs, a composite was synthesized, consisting of cRGD-functionalized GNPs coated with BSA–rifampicin [166]. PEG-coated multifunctionalized GNPs conjugated sequentially to targeting antibodies and peptides circulated in the bloodstream for a longer time [167] and accumulated in larger amounts in cells.

In the fairly complex construct developed by Zhang *et al.* [168], GNPs were functionalized sequentially with oligonucleotides, chelated ^{64}Cu, and the fluorophore Cy5 for positron emission tomography and FL imaging of tumor cells. For concurrent boron neutron capture therapy and FL biodetection of tumors, GNPs were conjugated to fluorescein isothiocyanate (FITC), boronophenylalanine, and FA [169], yielding a nanostructure with high therapeutic potential toward cancer cells of several lines.

Dixit *et al.* [170] modified GNPs with peptides against both the epidermal growth factor and transferrin receptors and loaded them with the photosensitizer phthalocyanine. These nanoparticles improved specificity and worked synergistically to decrease the time of maximal accumulation in human glioma cells that overexpressed two cell surface receptors, as compared to cells that overexpressed only one.

Bao *et al.* [171] synthesized antisense DNA gold nanobeacons as a hybrid *in vivo* theranostics platform for the inhibition and FL analysis of cancer cells and metastasis. Similar results, with GNRs, were reported by Chen *et al.* [172].

GNPs complexed with DOX and the immunomodulator cytosine–phosphate–guanosine were found to be highly effective for combined immunotherapy, chemotherapy, and PT therapy in experiments *in vitro* and *in vivo* [173].

Beyond target-specific drug delivery, multifunctional GNPs are often employed to transport genetic material to the cell nucleus. Efficient intracellular delivery of siRNA was achieved with GNPs–siRNA–FA [174] and with cell-penetration peptides [175]. For protecting siRNA from intracellular endonucleases and improving cellular penetration, the GNS–siRNA conjugate was additionally coated with transactivator of transcription (TAT)-lipid [176]. In all the previous examples, the targeted substances (drugs and siRNA) were the major therapeutic agents, the auxiliary substances (antibodies, aptamers, cell-penetration peptides) helped to effect the targeted transport of the NCs, and the GNPs themselves and the dyes served for diagnostic purposes and for PT and PD therapy.

Vaccine design is yet another application of multifunctionalized GNPs. Ojeda *et al.* [177] synthesized gold glyconanoparticles incorporating two tumor antigens and T-cell helper peptides in their carbohydrate shell, with potential utility in the development of anticancer vaccines (Figure 6.6).

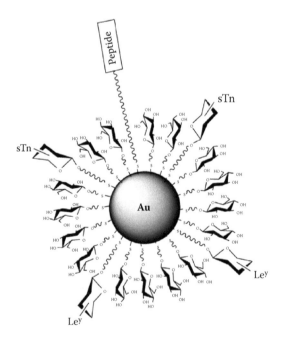

FIGURE 6.6 Scheme for a glyconanoparticle with incorporated antigens and active peptide. (Adapted from Ojeda, R., de Paz, J.L., Barrientos, A.G., Martín-Lomas, M., Penadés, S., *Carbohydr. Res.*, 342, 448–459, 2007. With permission.)

Likewise, corpuscular immunogens against other tumor types [178] and against *Streptococcus pneumoniae* [179], as well as a prototype anti-HIV drug [180], were obtained. Gold glyconanoparticles may also incorporate drugs, siRNA, fluorophores, and other ligands for diverse biomedical applications [181,182].

The data on the theranostic use of multifunctionalized GNPs are summarized in Table 6.2.

TABLE 6.2   Multifunctionalized GNPs

Type of Gold Nanoparticle	Functionalization	Therapeutic Modality	Diagnostic Modality	References
GNPs	Biotin Paclitaxel Rhodamine B	Drug delivery	FL imaging	Heo [144]
GNRs	Biotin DOX	Drug delivery PT therapy	PT imaging PA imaging	Chen [145]
GNRs	FA DOX	Drug delivery	Multi-photon FL imaging	Book Newell [152]
GNPs	DOX Aptamer	Drug delivery	CT imaging	Kim [156]
GNRs	DOX cRGD peptide ^{64}Cu-chelator	Drug delivery	Positron emission tomography CT imaging	Xiao [157]
GNRs	Indocyanine green Anti-EGFR antibodies	PT therapy PD therapy	FL imaging	Kuo [160]
GNPs	Heparin cRGD peptide HiLyte-Fluor	Drug delivery	FL imaging	Lee [162]
GNRs	Phage fusion protein Rhodamine B	PT therapy	FL imaging	Wang [163]
GNSts	DOX cRGD peptide Indocyanine green	Drug delivery PT therapy	NIR imaging	Chen [164]
GNPs	FITC Boronophenylalanine FA	Boron neutron capture therapy	FL imaging	Mandal [169]
GNPs	Epidermal growth factor peptide Transferrin peptide Phthalocyanine 4	PD therapy	FL imaging	Dixit [170]
GNPs	Antisense DNA Cy3	Gene therapy	FL imaging	Bao [171]
GNRs	Antisense DNA Fluorescein	Gene therapy	FL imaging	Chen [172]
GNPs	BSA siRNA Carboxyfluorescein	Gene therapy	FL imaging CT imaging	Wang [183]
GNPs	DOX Fucoidan	Drug delivery	PA imaging	Manivasagan [184]

6.4 Multifunctionalized Composite GNPs and Au NCls

In this section, we consider multifunctionalized composite GNPs—composite nanoparticles conjugated to several functional probes—which we believe to offer the most interest and promise. There is still a shortage of data on such nanostructures, but the possibilities they present for theranostics are impressive. In addition, we discuss briefly a new type of theranostic agent—multifunctionalized fluorescent atomic GNCls.

Salem and Searson [185] were among the first to report a successful application of multifunctionalized composites. Two-segment gold–nickel nanorods were synthesized. Plasmid DNA was tethered to the nickel segment, whereas transferrin tagged with the FL dye rhodamine was bound to the gold segment. The resultant NC permitted effective transfection of plasmid DNA to HEK293 cells. The transferrin-facilitated transfection was monitored by confocal microscopy by identifying the rhodamine that had entered into the cell. Additionally, it became possible to manipulate the transfection with magnetic fields at the cost of the magnetic properties of the nickel segment. This nanostructure is fairly promising for transgene generation, gene therapy, and DNA vaccination.

Another dual-metal nanorod system consisting of gold and nickel enabled dual-FL imaging and delivery of siRNA for cancer treatment [186]. FITC-labeled luteinizing hormone-releasing hormone peptides were attached to the surface of a nickel block to specifically bind to a breast cancer cell line, MCF-7. The gold block was modified with tetramethylrhodamine-labeled thiolated siRNA in order to knock down the vascular endothelial growth factor protein to inhibit cancer growth. These two-component nanorods actively targeted, and were internalized into, MCF-7 cells to induce apoptosis through RNA interference.

Wang *et al.* [187] and Zhang *et al.* [188] developed mesoporous titania-based yolk–shell nanoparticles as a multifunctional therapeutic platform for SERS imaging and chemo- and PT treatment. The yolk–shell nanoparticles are a special type of core–shell nanoparticle with interstitial hollow spaces that allow the core to move to the region confined by the shell section. The hollow space will be suitable for drug loading and SERS generation. In these GNRs/SiO$_2$/TiO$_2$ or GNRs/void/TiO$_2$ yolk–shell NCs, the NIR light absorbing GNR inner core serves as a SERS substrate (4-mercaptopyridine or 3,3′-diethylthiadicarbocyanine iodine) and PT agent, whereas the mesoporous titania outer shell can be loaded with DOX and a FL dye (flavin mononucleotide or 5-carboxyfluorescein diacetate).

A novel multifunctional nanoassembly, consisting of GNPs stabilized with folate- and FITC-preconjugated PAMAM dendrimer, was used for combined detection of tumor cells by flow cytometry, confocal microscopy, inductively coupled plasma mass spectrometry, and CT [189,190].

GNPs encased within a polymeric capsule and conjugated simultaneously with a Raman reporter, anti-HER2 antibodies, and DOX were proposed for targeted drug delivery and SERS imaging [191]. GNSts coated with TAT peptide and methylene blue-encapsulated silica aided in PD therapy and SERS analysis [192] and silica-coated GNRs in complex with protoporphyrin IX and a Raman reporter were employed in PD therapy, SERS analysis, and FL imaging [193]. Another nanoformulation, GNRs/SiO$_2$, had the merit of enhanced DOX- and rhodamine isothiocyanate-loading capacity owing to the presence of a mesoporous silica coat [194]. In that study, the combined functionalities of simultaneous cell imaging and drug delivery of the synthesized NCs were demonstrated. Jiang *et al.* [195] used GNRs@mesoporous SiO_2/rhodamine B isothiocyanate/DOX composite nanoparticles for FL imaging photocontrolled drug release, and PT therapy for cancer cells.

DOX-loaded composite nanomicelles formed from GNPs with a folate-modified amphiphilic block copolymer coating were prepared by Prabaharan *et al.* [196] as a carrier for targeted DOX delivery to tumor cells. Zhang *et al.* [197] reported success in using their theranostic nanoplatform (GNPs with silica shells doped with silver nanoparticles, an aptamer, and a PD dye) for target-specific nanoparticle delivery to cancer cells followed by combined PD and PT therapy.

The group of Naomi Halas of Rice University has made a large contribution to the development of multifunctionalized composite GNPs and their application in theranostics. In 2010, they constructed a sophisticated NC, in which GNSs on silica cores were wrapped in a silica epilayer with incorporated magnetic nanoparticles and were coupled with antitumor antibodies and indocyanine green [198]. The four modalities of the resultant nanocomplex—the plasmon resonant peculiarities of gold, the magnetic properties of Fe_3O_4 nanoparticles, the targeting function of the antibodies, and the PD properties of the dye—enabled its use in both diagnosis (FL methods, magnetic resonance tomography) and therapy (PT and PD) of tumors. Later, a similar NC (with antibodies conjugated through the avidin–biotin system; Figure 6.7) was tested *in vivo* on mice with breast cancer cell xenografts [199]. Its diagnostic efficacy was confirmed by NIR FL and magnetic resonance tomography. The authors determined nanoparticle biodistribution in different organs and tissues at 72 h postinjection, showed selective accumulation of nanocomplexes in tumors, and performed successful PT therapy. In their subsequent work, the obtained nanocomplexes proved of benefit for the treatment of ovarian cancer [200].

Liu *et al.* [201] designed nanorattles consisting of several silica spheres loaded with the antitumor agent docetaxel and surrounded by a PEGylated gold shell. These nanorattles

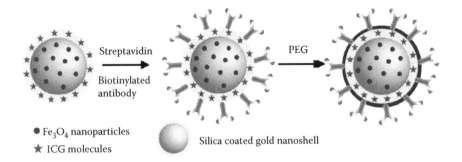

FIGURE 6.7 Scheme for the preparation of a multifunctionalized composite nanocomplex. (Adapted from Bardhan, R., Chen, W., Bartels, M., Perez-Torres, C., Botero, M.F., McAninch, R.W., Contreras, A., Schiff, R., Pautler, R.G., Halas, N.J., Joshi, A., *Nano Lett.*, 10, 4920–4928, 2010. With permission.)

had good biocompatibility and high potential for concurrent chemotherapy and PT therapy. The authors tested their "magic bullet" *in vitro* and *in vivo* for the treatment of hepatocellular carcinomas, demonstrating a pronounced synergistic effect of the NCs. Hu *et al.* [202] performed SERS imaging, targeted drug delivery, and PT therapy with nanorattles made of GNCs loaded with the Raman reporter *p*-aminothiophenol and coated with hollow silica shells that were functionalized simultaneously with a cell-penetrating peptide and DOX.

A unique nanoprobe with five functional modalities was reported by Vo-Dinh's group [203]. It consisted of PEGylated GNSts that were conjugated to *p*-mercaptobenzoic acid (SERS reporter molecule) and enclosed in silica shells linked with a gadolinium complex. This design made it possible to utilize the nanoprobe in SERS, magnetic resonance tomography, CT, two-photon luminescence, and PT therapy (Figure 6.8).

In another intriguing nanostructure, gold nanoprisms were enclosed in paclitaxel-doped folate-linked biodegradable gelatin capsules [204]. Paclitaxel was effective in penetrating the tumor cell interior, causing cell death. Of no less interest is the theranostic nanoplatform proposed by Topete *et al.* [205], in which the surface of biodegradable poly(lactide-co-glycolide) nanoparticles with encapsulated DOX was coated with a gold shell and was functionalized with indocyanine green- and folate-conjugated human serum albumin. This virus-like nanoplatform enabled FL imaging *in vivo*, as well as simultaneous targeted chemotherapy and PT therapy. *In vivo*, it exhibited a high synergistic effect in experiments with mice implanted with breast cancer cells.

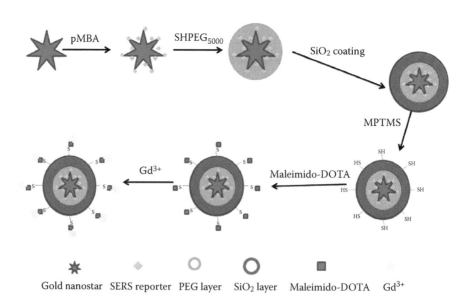

FIGURE 6.8 Schematic representation of the basic steps to fabricate a plasmonic NCs with five theranostic modalities: SERS, MRI, CT, two-photon luminescence imaging, and PT therapy. (Adapted from Liu, Y., Chang, Z., Yuan, H., Fales, A.M., Vo-Dinh, T., *Nanoscale*, 5, 12126–12131, 2013. With permission.)

Hao *et al.* [206] fabricated gold-nanoshelled docetaxel-loaded poly(lactide-co-glycolide) nanoparticles. A tumor-targeting peptide, angiopep-2, was introduced onto the nanoshell through the Au–S bond, achieving drug delivery with active targeting capability. This novel system afforded combined chemotherapy and thermal therapy for cancer, as well as showing potential x-ray imaging ability.

Navarro *et al.* [207] described multicharged GNP–block copolymer–chromophore conjugates as efficient nanocarriers for the delivery of ultrahigh loads of light-activated theranostic agents for FL imaging and PD therapy of cancer cells. Zeng *et al.* [208] reported the synthesis and characterization of a multifunctional theranostic agent for targeted MRI/computed x-ray tomography dual-mode imaging and PT therapy of hepatocellular carcinoma. These NCs were characterized as having a core–shell structure with GNPs/polydopamine as the inner core, indocyanine green, which is electrostatically absorbed onto the surface of polydopamine, as the PT therapeutic agent, and lipids modified with gadolinium–1,4,7,10-tetraacetic acid and lactobionic acid, which is self-assembled on the outer surface, as the shell.

Zhao *et al.* [209] fabricated gold nanochains by simple physical mixing to assemble citrate-stabilized GNPs into nanochains by using hyaluronic acid and hydrocaffeic acid conjugates as templates. Raman reporters and photosensitizers were conjugated onto the surface of the nanochains for multiplex detection and PD therapy. Finally, Yuan *et al.* [210] developed a combination of monoclonal antibodies, FA, and miR-122 (an RNA genetic drug)-loaded GNPs on graphene NCs for drug delivery, PT therapy, and imaging of tumor cells. Finally, Chen *et al.* [211] prepared a multifunctional NC to serve as a cancer cell tracking agent through Raman imaging and a PT therapy agent for cancer therapy. The NC was constructed by mesoporous silica self-assembly on the reduced graphene oxide nanosheets with nanogap-aligned GNPs encapsulated and arranged inside the nanochannels of the mesoporous silica layer. Functionalization was with rhodamine 6G as a Raman reporter and with anti-EGFR antibodies. According to the authors, the resultant NC is an excellent new theranostic nanosystem with cell targeting, cell tracking, and PT therapy capabilities.

Noble metal NCls are a new type of fluorophores [212–215] that have attracted significant attention owing to their advantageous photophysical properties, as compared to small-molecule dyes, fluorescent proteins, quantum dots, and upconverting and dye-doped nanoparticles [216–219]. Today, highly fluorescent stable NCls can be fabricated with atomically precise sizes and tunable optical properties [220–222].

For biomedical applications, GNCls have been synthesized using biocompatible capping ligands, including DNA [223], glutathione (GSH) [224], BSA [225], and other proteins (see, *e.g.*, [226] and references therein). Moreover, $HAuCl_4$ can be used as a precursor without any reducing agent to obtain fluorescent GNCls under native conditions through *in vivo* biosynthesis [227,228]. Protein-based protocols hold particular promise, as proteins act as reducing and stabilizing agents that produce water-soluble and biocompatible GNCls under mild reaction conditions. Protein-capped GNCls retain their FL in a wide range of pH and can easily be conjugated with target and drug molecules.

Thus, fluorescent GNCls can be considered a specific type of NC in which the Au atoms are intrinsically combined with polymer capping agents (DNA, GSH, BSA, PAMAM, *etc.*) to produce intense emitted light, whereas the Au atomic core is not FL itself or has a fairly low quantum yield (less than 0.1%). On the other hand, owing to their multivalent functionalization, such NCs belong to a particular kind of multifunctional-ized Au-based composite, and for this reason, they are discussed in this section.

Most published applications of GNCls have been focused on sensing (heavy metal ions, inorganic anions, small biomolecules, proteins, *etc.*) [229], bioimaging *in vitro* and *in vivo* [229,230], and catalysis [231,232]. However, in contrast to the numerous stud-ies on multifunctional composites based on plasmonic nanoparticles [12,20] there have been few reports describing analogous GNCl-based multifunctional theranostic NCs. Furthermore, those studies were aimed mainly at cancer theranostics. One of the first investigations dealing with multifunctional GNCls was conducted by Retnakumari *et al.* [233], who prepared BSA-protected GNCls conjugated with FA (Au–BSA–FA) for receptor-targeted detection of oral squamous cell carcinoma (KB) and breast adeno-carcinoma MCF-7 cells. Similarly, Wang *et al.* [234] synthesized BSA-protected GNCls, which were conjugated with herceptin to achieve specific targeting and nuclear localiza-tion in the overexpressed ErbB2 receptors. Perhaps the first reports on the theranostic application of GNCs were published by Chen *et al.* [235,236], in which Au–BSA NCls were additionally loaded with FA, methionine, DOX, or the NIR FL dye MPA. Ding and Tian [237] applied Au–BSA–FITC–FA NCls to specific bioimaging and biosens-ing of cancer cells, in which GNCls produced a reference FL signal, FITC allowed the monitoring of the pH, and FA acted as targeting molecules. Alongside GNPs, GNCls find expanding applications in theranostics [229], *e.g.*, as part of multifunctional GNCls based on transferrin-modified polymeric micelles, which were developed for targeted docetaxel delivery and for tumor bioimaging [238]. Ding *et al.* [239] presented the devel-opment of novel protein–gold hybrid nanocubes, which were assembled with GNCls, BSA, and tryptophan as building blocks. These NCs were loaded with rhodamine 6G and DOX. The protein–metal hybrid nanocubes can serve as a new type of dual-purpose tool: a blue-emitting cell marker in bioimaging investigation and a nanocarrier in drug delivery studies.

An intriguing structure was proposed by Hembury *et al.* [240]. GNPs of ~7-nm diameter were placed inside hollow silica shells hosting GNCls (gold quantum dots) within their mes-opores. The NCls had a diameter of <2 nm and expressed magnetic and FL properties. The gold–silica quantum rattles so produced had a total size of ~150 nm and were used for deliv-ery, into tumor cells, of DOX conjugated to their surface; PT therapy; and FL, PA, and MRI.

Zhang *et al.* [241] recently developed multifunctional theranostic probes comprising GSH-capped GNCls; after these were further coupled to FA and PEG, the photosensi-tizer chlorin e6 (Ce6) was trapped within PEG networks (Figure 6.9).

Antibiotic-resistant pathogenic bacteria (*e.g.*, methycillin-resistant *S. aureus*) [242] is a serious problem for many clinics [243,244]. In recent years, several plasmonic NCs have been developed for simultaneous detection and PD and PT antimicrobial inactivation (see [69] and references therein). Figure 6.10 illustrates the fabrication of Au–BSA multifunc-tional NCl functionalized with targeting molecules (human antistaphylococcal immuno-globulin, antiSAIgG) and the PD dye Photosens (PS) for selective detection and effective PD inactivation of both methycillin-resistant and methicillin-sensitive *S. aureus* [245].

The data on the theranostic use of multifunctionalized composite GNPs and GNCls are summarized in Table 6.3.

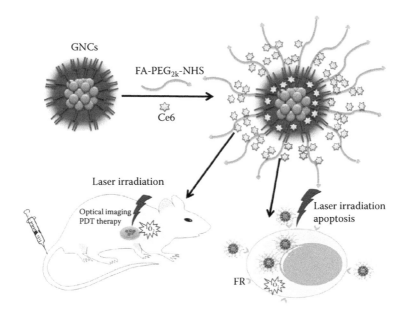

FIGURE 6.9 Fabrication of GNCl-based nanoprobes for *in vitro* and *in vivo* application. As-prepared GSH-capped red-emitting GNCls were further functionalized with FA, PEG_{2K}-NHS, and PD agent chlorine e6. These multifunctionalized NCs have been used for FL imaging and PD therapy *in vivo*. (Adapted from Dykman, L.A., Khlebtsov, N.G., *Biomaterials*, 108, 13–34, 2016. With permission.)

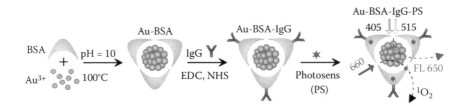

FIGURE 6.10 Scheme for the preparation of Au–BSA–IgG–PS complexes combining FL and PD properties under (405, 515 nm) and 660 nm excitation, respectively. Here, EDC, NHS and IgG designate 1-ethyl-3-[3-dimethylaminopropyl]carbodiimide hydrochloride, *N*-hydroxysuccinimide, and human antistaphylococcal immunoglobulin, respectively. Note that 405 and 515 nm correspond to the excitation spectrum maxima, whereas the FL excitation band is between 350 and 600 nm. (Adapted from Khlebtsov, B., Tuchina, E., Tuchin, V., Khlebtsov, N., *RSC Adv.*, 5, 61639–61649, 2015. With permission.)

TABLE 6.3 Multifunctionalized Composite GNPs

Composite	Functionalization	Therapeutic Modality	Diagnostic Modality	References
Au-Ni nanorods	FITC-labeled luteinizing hormone-releasing hormone peptide Tetramethylrhodamine-labeled siRNA	Drug delivery	Dual FL imaging	Choi [186]
GNRs/SiO$_2$/TiO$_2$	4-Mercaptopyridine Flavin mononucleotide DOX	Drug delivery PT therapy	FL SERS	Wang [187]
GNRs/void/TiO$_2$	3,3′-diethylthiadicarbocyanine iodine 5-Carboxyfluorescein diacetate DOX	Drug delivery PT therapy	SERS imaging FL imaging	Zhang [188]
GNRs/PAMAM	α-tocopheryl succinate FITC FA	Drug delivery	FL imaging CT imaging	Zhu [190]
GNPs/PEG/methyl methacrylate/4-vinylpyridine	Raman reporter Anti-HER2 antibodies DOX	Drug delivery	SERS imaging	Song [191]
GNSts/SiO$_2$	Methylene blue TAT peptide	PD therapy	SERS spectroscopy	Yuan [192]
GNRs/SiO$_2$	Protoporphyrin IX Raman reporter	PD therapy	FL imaging SERS spectroscopy	Zhang [193]
GNRs/SiO$_2$	DOX Rhodamine isothiocyanate	Drug delivery	FL imaging	Wang [194]
GNRs/SiO$_2$	DOX Rhodamine isothiocyanate	Drug delivery PT therapy	FL imaging	Jiang [195]
GNSs/SiO$_2$/Fe$_3$O$_4$	Indocyanine green Anti-HER2 antibodies	PD therapy PT therapy	FL imaging MRI	Bardhan [198]

(Continued)

TABLE 6.3 (CONTINUED) Multifunctionalized Composite GNPs

Composite	Functionalization	Therapeutic Modality	Diagnostic Modality	References
GNPs/Fe nanoparticles/PEO-*b*-PCL	Paclitaxel	Drug delivery PT therapy	MRI	Zhang [246]
GNCs/SiO$_2$	DOX TAT peptide Raman reporter	Drug delivery PT therapy	SERS imaging	Hu [202]
GNSts/SiO$_2$	Raman reporter Maleimide-DOTA+ Gd^{3+}	PT therapy	SERS spectroscopy Magnetic resonance, CT and two-photon luminescence imaging	Liu [203]
Poly(lactic-co-glycolic acid) nanoparticles/DOX/Au shell	Human serum albumin Indocyanine green FA	Drug delivery PD therapy PT therapy	FL imaging	Topete [205]
Poly(lactic-co-glycolic acid) nanoparticles/Au shell	Docetaxel Angiopep-2	Drug delivery PT therapy	X-ray imaging	Hao [206]
GNPs/poly(NAM-co-NAS)	Lucifer Yellow Dibromobenzene	PD therapy	FL imaging	Navarro [207]
GNPs/polydopamine/lipid	Indocyanine green Gadolinium–1,4,7,10-tetraacetic acid Lactobionic acid	PD therapy PT therapy	MRI CT imaging	Zeng [208]
Hyaluronic acid/hydrocaffeic acid/GNPs	Raman reporter Pheophorbide	PD therapy	SERS imaging	Zhao [209]
GNPs/graphene oxide	RNA genetic drug (miR-122) Monoclonal antibodies FA	Drug delivery PT therapy	Transmission electron microscopy (TEM) imaging SEM imaging	Yuan [210]
D-α-tocopheryl polyethylene glycol 1000 succinate/GNCls	Docetaxel Transferrin	Drug delivery	FL imaging	Muthu [238]

(Continued)

TABLE 6.3 (CONTINUED) Multifunctionalized Composite GNPs

Composite	Functionalization	Therapeutic Modality	Diagnostic Modality	References
BSA/tryptophan/GNCls	DOX Rhodamine 6G	Drug delivery	FL imaging	Ding [239]
GNPs/SiO$_2$/GNCls	DOX	Drug delivery PT therapy	FL imaging MRI PA imaging	Hembury [240]
GNRs/SiO$_2$/GNCls	Ce6 FA Cell-penetrating peptide	PD therapy PT therapy	SERS spectroscopy Dark-field and FL imaging	Li [247]
SiO$_2$/Fe$_2$O$_3$/GNPs	Boronic acid FA DOX	Drug delivery	MRI CT imaging	Tseng [248]
GNPs/poly-β-benzyl-L-aspartate/PEG	FA Verteporfin	PD therapy	TEM imaging SEM imaging FL imaging	Zhao [249]

6.5 Concluding Remarks

Owing to the development and improvement of chemical synthesis technologies for GNPs over the past decade, a huge variety of particles with required sizes, shapes, structures, and optical properties are now available to researchers. Furthermore, the current agenda is to perform primary modeling of a nanoparticle with the desired properties and to subsequently develop a procedure for the synthesis of the modeled nanostructure. From the standpoint of medical applications, it is crucial to develop efficient technologies for GNP functionalization with different classes of molecules that provide stability of nanoparticles *in vivo*, their biospecific interaction with biological targets, and consequently, efficient delivery of drugs or diagnostic markers.

It is now generally recognized that GNP conjugates can serve as excellent labels for use in bioimaging, which can be implemented by various technologies, including optical imaging, PA imaging, SERS, coherent anti-Stokes Raman spectroscopy (CARS), and CT imaging, among others. Along with the reported cases of clinical diagnosis of cancer, Alzheimer's disease, HIV, hepatitis, tuberculosis, diabetes, and other diseases, new diagnostic applications of GNPs are to be expected.

The nanoparticle-aided targeted delivery of DNA, antigens, and drugs seems one of the most promising areas in biomedicine. Functionalization of GNPs with molecular vectors to the receptors of cancer cells significantly increases GNP delivery to target cells. Accordingly, when GNPs are additionally loaded with an anticancer agent, the risk of side effects, typical of most effective anticancer drugs, is reduced. In addition to chemotherapy, such conjugates can be used for the thermal treatment of tumors by application of optical radiation, radio waves, and an alternating magnetic field.

Analysis of the published data shows that a universal carrier, which might be used for all types of materials and biological targets, is unlikely to be developed. A more probable and appropriate approach would be to design a carrier that would be optimized both with respect to the load with a specific substance and with respect to the efficiency of delivery to a specific target. In particular, the stability of the carrier conjugate in the bloodstream and the weak interaction with nontarget and immune cells should be combined with effective penetration into the target cells. It is quite possible that such properties can be obtained by using dynamically controlled, rather than static, nanosystems, which can be "switched over" to perform specific functions in response to an optical, a magnetic, or an acoustic signal. In connection with this, great hopes are being pinned on rapid progress in technologies of synthesizing multifunctional NCs, which combine controlled physical properties (magnetic, optical, PD, radioactive, *etc.*) with advanced technologies of molecular surface targeting. It is such structures that can ensure progress in theranostics, which enables targeted delivery of nanoparticles both for visualization and for therapy. Multifunctional nanoparticles enable several treatment agents to be delivered simultaneously, which leads to effective combined therapy of cancer.

Plasmonic PT laser therapy of cancer with GNPs, first described in 2003, has by now moved to the stage of clinical testing. The available experimental data indicate that further progress in nanooncology is expected to be made by combining different technologies, including photodynamics, chemotherapy, gene therapy, and other approaches of this type. The delivery of genetic material that can suppress the aggressive expression of cancer cells and their metastasis, in combination with other therapeutic nanobio-technologies and surgical approaches, may be the most efficient direction to take. In addition, the potential of using multifunctional composites as a platform for the design of nanovaccines is expanding.

This chapter has highlighted various uses of multifunctional GNPs in theranostics. It is the presence of a multitude of functions, determined either by different nanoparticle compositions (NCs), by different functional groups on their surface (multifunctional-ized GNPs), or by a combination of these properties that explains why these NCs are in such active use in theranostics. Multifunctionality ensures both their diagnostic and their therapeutic (chemical and physical action) applications. Of note, multifunctional-ity is often manifested as a synergistic effect of GNPs both *in vitro* and *in vivo*. We believe that multifunctional GNPs hold great diagnostic and treatment potential in diverse bio-medical studies, most importantly in practical personalized medicine.

An important direction in theranostics is the development of methods affording con-trolled release of the payloads delivered by multifunctional NCs to biotargets. Control of this type is usually achieved in a passive or active way. In the former case, the payload is released upon a change in physical–chemical conditions in the local area of the biotar-get. Examples of passive stimuli are pH changes and redox potential changes. Another possibility is the payload release caused by the action of a specific enzyme. For the active release of delivered substances, an external stimulus (light, ultrasound, oscillating mag-netic field, *etc.*) is used to change the NC structure so as to ensure that the therapeutic cargo is delivered to the external local environment of the NC.

The advances in genomics and proteomics have led to an increase in information about the molecular biomarkers of various types of cancer. This information will help to create new multifunctional GNPs with the ability to identify target tumor cells and to act on them with greater accuracy and specificity. The new, more complex functions of multimodal GNPs will make possible an early diagnosis of diseases and monitoring of treatment of patients in real time.

We conclude this section with a special attention to small-sized (about 1–2 nm) fluores-cent GNCls and multimodal composites based on these robust and efficient luminescent nanoparticles. By contrast to recognized great potential of GNPs, related multimodal NCs and numerous reports published during last 10–15 years, intense studies of unique properties of GNCls and cluster-based composites have just recently started to emerge [218,219,250]. One may expect a rapid progress in synthesis protocols and promising theranostic applications of NCl composites in the nearest future.

To summarize, it is hoped that this chapter will motivate chemists to create new effi-cient and robust toolkits for controlled synthesis of multifunctional and biocompatible NCs. These new structures will create new challenging tasks for plasmonic commu-nity to simulate their electromagnetic responses. In particular, a broad class of weak processes such as SERS, CARS, metal-enhanced FL, antigen–antibody interactions, and detecting of low concentrations of small molecules benefit directly from the enhanced local fields that can be generated by new nanostructures. Accordingly, the chemical and biological sensing technologies will benefit from the progress in the development of multifunctional nanostructures. However, our main expectation is that cited and discussed publications will stimulate the nanobiotechnology community to create new imaging, sensing, and theranostic technologies. These applications should be based on desired properties that might exist rather than choosing biomedical applications for nanostructures that currently exist.

References

1. Sanchez C., Belleville P., Popall M., Nicole L. Applications of advanced hybrid organic-inorganic nanomaterials: From laboratory to market // *Chem. Soc. Rev.* 2011. V. 40. P. 696–753.
2. Lammers T., Aime S., Hennink W.E., Storm G., Kiessling F. Theranostic nanomedicine // *Acc. Chem. Res.* 2011. V. 44. P. 1029–1038.
3. *Handbook of Nanobiomedical Research: Fundamentals, Applications and Recent Developments* / Ed. Torchilin V. Singapore: World Scientific. 2014.
4. Funkhouser J. Reinventing pharma: The theranostic revolution // *Curr. Drug. Discov.* 2002. V. 2. P. 17–19.
5. Picard F.J., Bergeron M.G. Rapid molecular theranostics in infectious diseases // *Drug Discov. Today.* 2002. V. 7. P. 1092–1101.
6. Lukianova-Hleb E., Hanna E.Y., Hafner J.H., Lapotko D.O. Tunable plasmonic nanobubbles for cell theranostics // *Nanotechnology.* 2010. V. 21. 085102.
7. Warner S. Diagnostics + therapy = theranostics // *Scientist.* 2004. V. 18. P. 38–39.
8. Kelkar S.S., Reineke T.M. Theranostics: Combining imaging and therapy // *Bioconjug. Chem.* 2011. V. 22. P. 1879–1903.
9. Sailor M.J., Park J.-H. Hybrid nanoparticles for detection and treatment of cancer // *Adv. Mater.* 2012. V. 24. P. 3779–3802.
10. Chen W., Zhang S., Yu Y., Zhang H., He Q. Structural-engineering rationales of gold nanoparticles for cancer theranostics // *Adv. Mater.* 2016. V. 28. P. 8567–8585.
11. Loo C., Lowery A., Halas N., West J., Drezek R. Immunotargeted nanoshells for integrated cancer imaging and therapy // *Nano Lett.* 2005. V. 5. P. 709–711.
12. Bardhan R., Lal S., Joshi A., Halas N.J. Theranostic nanoshells: From probe design to imaging and treatment of cancer // *Acc. Chem. Res.* 2011. V. 44. P. 936–946.
13. Urries I., Muñoz C., Gomez L., Marquina C., Sebastian V., Arruebo M., Santamaria J. Magneto-plasmonic nanoparticles as theranostic platforms for magnetic resonance imaging, drug delivery and NIR hyperthermia applications // *Nanoscale.* 2014. V. 6. P. 9230–9240.
14. Jiao P.F., Zhou H.Y., Chen L.X. Yan B. Cancer-targeting multifunctionalized gold nanoparticles in imaging and therapy // *Curr. Med. Chem.* 2011. V. 18. P. 2086–2102.
15. Liang R., Wei M., Evans D.G., Duan X. Inorganic nanomaterials for bioimaging, targeted drug delivery and therapeutics // *Chem. Commun. (Camb.).* 2014. V. 50. P. 14071–14081.
16. Kim J., Piao Y., Hyeon T. Multifunctional nanostructured materials for multimodal imaging, and simultaneous imaging and therapy // *Chem. Soc. Rev.* 2009. V. 38. P. 372–390.
17. Sotiriou G.A. Biomedical applications of multifunctional plasmonic nanoparticles // *Wiley Interdiscip. Rev. Nanomed. Nanobiotechnol.* 2013. V. 5. P. 19–30.

18. Bao G., Mitragotri S., Tong S. Multifunctional nanoparticles for drug delivery and molecular imaging // *Annu. Rev. Biomed. Eng.* 2013. V. 15. P. 253–282.

19. Jin Y. Multifunctional compact hybrid au nanoshells: A new generation of nano-plasmonic probes for biosensing, imaging, and controlled release // *Acc. Chem. Res.* 2014. V. 47. P. 138–148.

20. Cabral R.M., Baptista P.V. Anti-cancer precision theranostics: A focus on multi-functional gold nanoparticles // *Expert Rev. Mol. Diagn.* 2014. V. 14. P. 1041–1052.

21. Sahay R., Reddy V.J., Ramakrishna S. Synthesis and applications of multifunc-tional composite nanomaterials // *Int. J. Mech. Mater. Eng.* 2014. V. 9. 25.

22. Webb J.A., Bardhan R. Emerging advances in nanomedicine with engineered gold nanostructures // *Nanoscale.* 2014. V. 6. P. 2502–2530.

23. Lim E.-K., Kim T., Paik S., Haam S., Huh Y.-M., Lee K. Nanomaterials for theranostics: Recent advances and future challenges // *Chem. Rev.* 2015. V. 115. P. 327–394.

24. Chen Q., Ke H., Dai Z., Liu Z. Nanoscale theranostics for physical stimulus-responsive cancer therapies // *Biomaterials.* 2015. V. 73. P. 214–230.

25. El-Toni A.M., Habila M.A., Labis J.P., ALOthman Z.A., Alhoshan M., Elzatahry A.A., Zhang F. Design, synthesis and applications of core-shell, hollow core, and nanorattle multifunctional nanostructures // *Nanoscale.* 2016. V. 8. P. 2510–2531.

26. Shahbazi R., Ozpolat B., Ulubayram K. Oligonucleotide-based theranostic nanoparticles in cancer therapy // *Nanomedicine (Lond.).* 2016. V. 11. P. 1287–1308.

27. Sounderya N., Zhang Y. Use of core/shell structured nanoparticles for biomedical applications // *Recent Patents Biomed. Eng.* 2008. V. 1. P. 34–42.

28. Cortie M.B., McDonagh A.M. Synthesis and optical properties of hybrid and alloy plasmonic nanoparticles // *Chem. Rev.* 2011. V. 111. P. 3713–3735.

29. Pereira S.O., Barros-Timmons A., Trindade T. Biofunctionalisation of colloi-dal gold nanoparticles via polyelectrolytes assemblies // *Colloid Polym. Sci.* 2014. V. 292. P. 33–50.

30. Bielinska A., Eichman J.D., Lee I., Baker J.R., Jr., Balogh, L. Imaging {Au0-PAMAM} gold-dendrimer nanocomposites in cells // *J. Nanopart. Res.* 2002. V. 4. P. 395–403.

31. Kojima C., Umeda Y., Ogawa M., Harada A., Magata Y., Kono K. X-ray computed tomography contrast agents prepared by seeded growth of gold nanoparticles in PEGylated dendrimer // *Nanotechnology.* 2010. V. 21. 245104.

32. Balogh L.P., Nigavekar S.S., Cook A.C., Minc L., Khan M.K. Development of dendrimer-gold radioactive nanocomposites to treat cancer microvasculature // *PharmaChem.* 2003. V. 2. P. 94–99.

33. Li X., Takashima M., Yuba E., Harada A., Kono K. PEGylated PAMAM dendrimer-doxorubicin conjugate-hybridized gold nanorod for combined photothermal-chemotherapy // *Biomaterials.* 2014. V. 35. P. 6576–6584.

34. Young J.K., Figueroa E.R., Drezek R.A. Tunable nanostructures as photothermal theranostic agents // *Ann. Biomed. Eng.* 2012. V. 40. P. 438–459.

35. Doane T.L., Burda C. The unique role of nanoparticles in nanomedicine: Imaging, drug delivery and therapy // *Chem. Soc. Rev.* 2012. V. 41. P. 2885–2911.

36. Esumi K., Takei N., Yoshimura T. Antioxidant-potentiality of gold–chitosan nanocomposites // *Colloids Surf. B.* 2003. V. 32. P. 117–123.

37. Ding L., Hao C., Xue Y., Ju H. A bio-inspired support of gold nanoparticles-chitosan nanocomposites gel for immobilization and electrochemical study of K562 leukemia cells // *Biomacromolecules.* 2007. V. 8. P. 1341–1346.

38. Zhao X.J., Mai Z.B., Kang X.H., Dai Z., Zou X.Y. Clay-chitosan-gold nanoparticle nanohybrid: Preparation and application for assembly and direct electrochemistry of myoglobin // *Electrochim. Acta.* 2008. V. 53. P. 4732–4739.

39. Hu Y., Chen Q., Ding Y., Li R., Jiang X., Liu B. Entering and lighting up nuclei using hollow chitosan–gold hybrid nanospheres // *Adv. Mater.* 2009. V. 21. P. 3639–3643.

40. Chen R., Zheng X., Qian H., Wang X., Wang J., Jiang X. Combined near-IR photothermal therapy and chemotherapy using gold-nanorod/chitosan hybrid nanospheres to enhance the antitumor effect // *Biomater. Sci.* 2013. V. 3. P. 285–293.

41. Chen R., Wang X., Yao X., Zheng X., Wang J., Jiang X. Near-IR-triggered photothermal / photodynamic dual-modality therapy system via chitosan hybrid nanospheres // *Biomaterials.* 2013. V. 34. P. 8314–8322.

42. Cho H.-J., Oh J., Choo M.-K., Ha J.-I., Park Y., Maeng H.-J. Chondroitin sulfate-capped gold nanoparticles for the oral delivery of insulin // *Int. J. Biol. Macromol.* 2014. V. 63. P. 15–20.

43. Pan D.P.J., Pramanik M., Senpan A., Ghosh S., Wickline S.A., Wang L.V., Lanza G.M. Near infrared photoacoustic detection of sentinel lymph nodes with gold nanobeacons // *Biomaterials.* 2010. V. 31. P. 4088–4093.

44. Tam J.M., Tam J.O., Murthy A., Ingram D.R., Ma L.L., Travis K., Johnston K.P., Sokolov K.V. Controlled assembly of biodegradable plasmonic nanoclusters for near-infrared imaging and therapeutic applications // *ACS Nano.* 2010. V. 4. P. 2178–2184.

45. Song J., Pu L., Zhou J., Duan B., Duan H. Biodegradable theranostic plasmonic vesicles of amphiphilic gold nanorods // *ACS Nano.* 2013. V. 7. P. 9947–9960.

46. Ke H., Wang J., Dai Z., Jin Y., Qu E., Xing Z., Guo C., Yue X., Liu J. Gold-nanoshelled microcapsules: A theranostic agent for ultrasound contrast imaging and photothermal therapy // *Angew. Chem. Int. Ed.* 2011. V. 50. P. 3017–3021.

47. Huang P., Lin J., Li W., Rong P., Wang Z., Wang S., Wang X., Sun X., Aronova M., Niu G., Leapman R.D., Nie Z., Chen X. Biodegradable gold nanovesicles with an ultrastrong plasmonic coupling effect for photoacoustic imaging and photothermal therapy // *Angew. Chem. Int. Ed.* 2013. V. 52. P. 13958–13964.

48. Lin J., Wang S., Huang P., Wang Z., Chen S., Niu G., Li W., He J., Cui D., Lu G., Chen X., Nie Z. Photosensitizer-loaded gold vesicles with strong plasmonic coupling effect for imaging-guided photothermal/photodynamic therapy // *ACS Nano.* 2013. V. 7. P. 5320–5329.

49. Deng H., Dai F., Ma G., Zhang X. Theranostic gold nanomicelles made from biocompatible comb-like polymers for thermochemotherapy and multifunctional imaging with rapid clearance // *Adv. Mater.* 2015. V. 27. P. 3645–3653.

50. Gao L., Fei J., Zhao J., Li H., Cui Y., Li J. Hypocrellin-loaded gold nanocages with high two-photon efficiency for photothermal/photodynamic cancer therapy *in vitro* // *ACS Nano.* 2012. V. 6. P. 8030–8040.

51. Vankayala R., Lin C.-C., Kalluru P., Chiang C.-S., Hwang K.C. Gold nanoshells-mediated bimodal photodynamic and photothermal cancer treatment using ultra-low doses of near infra-red light // *Biomaterials*. 2014. V. 35. P. 5527–5538.

52. Liao J., Li W., Peng J., Yang Q., Li H., Wei Y., Zhang X., Qian Z. Combined cancer photothermal-chemotherapy based on doxorubicin/gold nanorod-loaded poly-mersomes // *Theranostics*. 2015. V. 5. P. 345–356.

53. Wu G., Mikhailovsky A., Khant H.A., Zasadzinski, J.A. Synthesis, characterization, and optical response of gold nanoshells used to trigger release from lipo-somes // *Methods Enzymol*. 2009. V. 464. P. 279–307.

54. Jin Y.D., Gao X.H. Spectrally tunable leakage-free gold nanocontainers // *J. Am. Chem. Soc.* 2009. V. 131. P. 17774–17776.

55. Liu Z., Ye B., Jin M., Chen H., Zhong H., Wang X., Guo Z. Dye-free near-infrared surface-enhanced Raman scattering nanoprobes for bioimaging and high-performance photothermal cancer therapy // *Nanoscale*. 2015. V. 7. P. 6754–6761.

56. Park H., Yang J., Seo S., Kim K., Suh J., Kim D., Haam S., Yoo K.-H. Multifunctional nanoparticles for photothermally controlled drug delivery and magnetic reso-nance imaging enhancement // *Small*. 2008. V. 4. P. 192–196.

57. Lee S.-M., Park H., Yoo K.-H. Synergistic cancer therapeutic effects of locally delivered drug and heat using multifunctional nanoparticles // *Adv. Mater.* 2010. V. 22. P. 4049–4053.

58. Wang Y.-H., Chen S.-P., Liao A.-H., Yang Y.-C., Lee C.-R., Wu C.-H., Wu P.-C., Liu T.-M., Wang C.-R.C., Li P.-C. Synergistic delivery of gold nanorods using multifunctional microbubbles for enhanced plasmonic photothermal therapy // *Sci. Rep.* 2014. V. 4. 5685.

59. Peralta D.V., He J., Wheeler D.A., Zhang J.Z., Tarr M.A. Encapsulating gold nano-materials into size-controlled human serum albumin nanoparticles for cancer therapy platforms // *J. Microencapsul*. 2014. V. 31. P. 824–831.

60. Simon T., Potara M., Gabudean A.-M., Licarete E., Banciu M., Astilean S. Designing theranostic agents based on Pluronic stabilized gold nanoaggregates loaded with Methylene Blue for multimodal cell imaging and enhanced photody-namic therapy // *ACS Appl. Mater. Interfaces*. 2015. 7. P. 16191–16201.

61. Costa Lima S.A., Reis S. Temperature-responsive polymeric nanospheres contain-ing methotrexate and gold nanoparticles: A multi-drug system for theranostic in rheumatoid arthritis // *Colloids Surf. B*. 2015. V. 133. P. 378–387.

62. Huang C.-C., Liu T.-M. Controlled Au-polymer nanostructures for multiphoton imaging, prodrug delivery, and chemo-photothermal therapy platforms // *ACS Appl. Mater. Interfaces*. 2015. V. 7. P. 25259–25269.

63. Park G.-S., Kwon H., Kwak D.W., Park S.Y., Kim M., Lee J.-H., Han H., Heo S., Li X.S., Lee J.H., Kim Y.H., Lee J.-G., Yang W., Cho H.Y., Kim S.K., Kim K. Full surface embedding of gold clusters on silicon nanowires for efficient capture and photothermal therapy of circulating tumor cells // *Nano Lett*. 2012. V. 12. P. 1638–1642.

64. Botella P., Ortega Í., Quesada M., Madrigal R.F., Muniesa C., Fimia A., Fernández E., Corma A. Multifunctional hybrid materials for combined photo and chemo-therapy of cancer // *Dalton Trans*. 2012. V. 41. P. 9286–9296.

65. Zhang Z., Wang L., Wang J., Jiang X., Li X., Hu Z., Ji Y., Wu X., Chen C. Mesoporous silica-coated gold nanorods as a light-mediated multifunctional theranostic platform for cancer treatment // *Adv. Mater.* 2012. V. 24. P. 1418–1423.

66. Shen S., Tang H., Zhang X., Ren J., Pang Z., Wang D., Gao H., Qian Y., Jiang X., Yang W. Targeting mesoporous silica-encapsulated gold nanorods for chemo-photothermal therapy with near-infrared radiation // *Biomaterials.* 2013. V. 34. P. 3150–3158.

67. Monem A.S., Elbialy N., Mohamed N. Mesoporous silica coated gold nanorods loaded doxorubicin for combined chemo–photothermal therapy // *Int. J. Pharm.* 2014. V. 470. P. 1–7.

68. Ouhenia-Ouadahi K., Yasukuni R., Yu P., Laurent G., Pavageau C., Grand J., Guérin J., Léaustic A., Félidj N., Aubard J., Nakatani K., Métivier R. Photochromic–fluorescent–plasmonic nanomaterials: Towards integrated three component photoactive hybrid nanosystems // *Chem. Commun.* 2014. V. 50. P. 7299–7302.

69. Khlebtsov B.N., Tuchina E.S., Khanadeev V.A., Panfilova E.V., Petrov P.O., Tuchin V.V., Khlebtsov N.G. Enhanced photoinactivation of *Staphylococcus aureus* with nanocomposites containing plasmonic particles and hematoporphyrin // *J. Biophotonics.* 2013. V. 6. P. 338–351.

70. Khlebtsov B., Panfilova E., Khanadeev V., Bibikova O., Terentyuk G., Ivanov A., Rumyantseva V., Shilov I., Ryabova A., Loshchenov V., Khlebtsov N. Nanocomposites containing silica-coated gold-silver nanocages and Yb-2,4-dimethoxyhematoporphyrin: Multifunctional capability of IR-luminescence detection, photosensitization, and photothermolysis // *ACS Nano.* 2011. V. 5. P. 7077–7089.

71. Khlebtsov B.N., Panfilova E.V., Khanadeev V.A., Markin A.V., Terentyuk G.S., Rumyantseva V.D., Ivanov A.V., Shilov I.P., Khlebtsov N.G. Composite multifunctional nanoparticles based on silica-coated gold-silver nanocages functionalized by Yb-hematoporphyrin // *Nanotechnol. Russia.* 2011. V. 6. P. 496–503.

72. Terentyuk G., Panfilova E., Khanadeev V., Chumakov D., Genina E., Bashkatov A., Tuchin V., Bucharskaya A., Maslyakova G., Khlebtsov N., Khlebtsov B. Gold nanorods with hematoporphyrin-loaded silica shell for dual-modality photodynamic and photothermal treatment of tumors *in vivo* // *Nano Res.* 2014. V. 7. P. 325–337.

73. Huang P., Bao L., Zhang C., Lin J., Luo T., Yang D., He M., Li Z., Gao G., Gao B., Fu S., Cui D. Folic acid-conjugated silica-modified gold nanorods for X-ray/CT imaging-guided dual-mode radiation and photo-thermal therapy // *Biomaterials.* 2011. V. 32. P. 9796–9809.

74. Ayala-Orozco C., Urban C., Knight M.W., Urban A.S., Neumann O., Bishnoi S.W., Mukherjee S., Goodman A.M., Charron H., Mitchell T., Shea M., Roy R., Nanda S., Schiff R., Halas N.J., Joshi A. Au nanomatryoshkas as efficient near-infrared photothermal transducers for cancer treatment: Benchmarking against nanoshells // *ACS Nano.* 2014. V. 8. P. 6372–6381.

75. Wu C.-H., Cook J., Emelianov S., Sokolov K. Multimodal magneto-plasmonic nanoclusters for biomedical applications // *Adv. Funct. Mater.* 2014. V. 24. P. 6862–6871.

76. Stoeva S.I., Huo F., Lee J.-S., Mirkin C.A. Three-layer composite magnetic nanoparticle probes for DNA // *J. Am. Chem. Soc.* 2005. V. 127. P. 15362–15363.

77. Kim J., Park S., Lee J.E., Jin S.M., Lee J.H., Lee I.S., Yang I., Kim J.S., Kim S.K., Cho M.H., Hyeon T. Designed fabrication of multifunctional magnetic gold nanoshells and their application to magnetic resonance imaging and photothermal therapy // *Angew. Chem. Int. Ed.* 2006. V. 45. P. 7754–7758.

78. Dykman L.A., Khlebtsov N.G. Multifunctional gold-based nanocomposites for theranostics // *Biomaterials.* 2016. V. 108. P. 13–34.

79. Larson T.A., Bankson J., Aaron J., Sokolov K. Hybrid plasmonic magnetic nanoparticles as molecular specific agents for MRI/optical imaging and photothermal therapy of cancer cells // *Nanotechnology.* 2007. V. 18. 325101.

80. Zhou T., Wu B., Xing D. Bio-modified Fe_3O_4 core/Au shell nanoparticles for targeting and multimodal imaging of cancer cells // *J. Mater. Chem.* 2012. V. 22. P. 470–477.

81. Kim D., Kim J.W., Jeong Y.Y., Jon S. Antibiofouling polymer coated gold@iron oxide nanoparticle (GION) as a dual contrast agent for CT and MRI // *Bull. Korean Chem. Soc.* 2009. V. 30. P. 1855–1857.

82. Kim D., Yu M.K., Lee T.S., Park J.J., Jeong Y.Y., Jon S. Amphiphilic polymer-coated hybrid nanoparticles as CT/MRI dual contrast agents // *Nanotechnology.* 2011. V. 22. 155101.

83. Jin Y.D., Jia C.X., Huang S.-W., O'Donnell M., Gao X.H. Multifunctional nanoparticles as coupled contrast agents // *Nat. Commun.* 2010. V. 1. 41.

84. Li Z., Yin S., Cheng L., Yang K., Li Y., Liu Z. Magnetic targeting enhanced theranostic strategy based on multimodal imaging for selective ablation of cancer // *Adv. Funct. Mater.* 2014. V. 24. P. 2312–2321.

85. Wang C., Chen J., Talavage T., Irudayaraj J. Gold nanorod/Fe_3O_4 nanoparticle "nano-pearl-necklace" for simultaneous targeting, dual-mode imaging and photothermal ablation of cancer cells // *Angew. Chem. Int. Ed.* 2009. V. 48. P. 2759–2763.

86. Mohammad F., Balaji G., Weber A., Uppu R.M., Kumar C.S.S.R. Influence of gold nanoshell on hyperthermia of superparamagnetic iron oxide nanoparticles // *J. Phys. Chem. C.* 2010. V. 114. P. 19194–19201.

87. Kirui D.K., Rey D.A., Batt C.A. Gold hybrid nanoparticles for targeted phototherapy and cancer imaging // *Nanotechnology.* 2010. V. 21. 105105.

88. Feng W., Zhou X., Nie W., Chen L., Qiu K., Zhang Y., He C. Au/polypyrrole@Fe_3O_4 nanocomposites for MR/CT dual-modal imaging guided-photothermal therapy: An *in vitro* study // *ACS Appl. Mater. Interfaces.* 2015. V. 7. P. 4354–4367.

89. Li J., Hu Y., Yang J., Wei P., Sun W., Shen M., Zhang G., Shi X. Hyaluronic acid-modified Fe_3O_4@Au core/shell nanostars for multimodal imaging and photothermal therapy of tumors // *Biomaterials.* 2015. V. 38. P. 10–21.

90. Ji X., Shao R., Elliott A.M., Stafford R.J., Esparza-Coss E., Bankson J.A., Liang G., Luo Z.-P., Park K., Markert J.T., Li C. Bifunctional gold nanoshells with a superparamagnetic iron oxide–silica core suitable for both MR imaging and photothermal therapy // *J. Phys. Chem. C.* 2007. V. 111. P. 6245–6251.

91. Hu X.G., Wei C.-W., Xia J.J., Pelivanov I., O'Donnell M., Gao X.H. Trapping and photoacoustic detection of CTCs at the single cell per milliliter level with magneto-optical coupled nanoparticles // *Small.* 2013. V. 9. P. 2046–2052.

92. Chao X., Shi F., Zhao Y.Y., Li K., Peng M.L., Chen C., Cui Y.L. Cytotoxicity of Fe_3O_4/Au composite nanoparticles loaded with doxorubicin combined with magnetic field // *Pharmazie.* 2010. V. 65. P. 500–504.

93. Kayal S., Ramanujan R.V. Anti-cancer drug loaded irongold coreshell nanoparticles (Fe@Au) for magnetic drug targeting // *J. Nanosci. Nanotechnol.* 2010. V. 10. P. 5527–5539.

94. Ma M., Chen H., Chen Y., Wang X., Chen F., Cui X., Shi J. Au capped magnetic core/mesoporous silica shell nanoparticles for combined photothermo-/chemotherapy and multimodal imaging // *Biomaterials.* 2012. V. 33. P. 989–998.

95. Lee J., Yang J., Ko H., Oh S.J., Kang J., Son J.-H., Lee K., Lee S.-W., Yoon H.-G., Suh J.-S., Huh Y.-M., Haam S. Multifunctional magnetic gold nanocomposites: Human epithelial cancer detection via magnetic resonance imaging and localized synchronous therapy // *Adv. Funct. Mater.* 2008. V. 18. P. 258–264.

96. Melancon M.P., Elliott A., Ji X., Shetty A., Yang Z., Tian M., Taylor B., Stafford R.J., Li C. Theranostics with multifunctional magnetic gold nanoshells: Photothermal therapy and t2* magnetic resonance imaging // *Invest. Radiol.* 2011. V. 46. P. 132–140.

97. Cheng L., Yang K., Li Y., Zeng X., Shao M., Lee S.-T., Liu Z. Multifunctional nanoparticles for upconversion luminescence/MR multimodal imaging and magnetically targeted photothermal therapy // *Biomaterials.* 2012. V. 33. P. 2215–2222.

98. Sotiriou G.A., Starsich F., Dasargyri A., Wurnig M.C., Krumeich F., Boss A., Leroux J.-C., Pratsinis S.E. Photothermal killing of cancer cells by the controlled plasmonic coupling of silica-coated Au/Fe_2O_3 nanoaggregates // *Adv. Funct. Mater.* 2014. V. 24. P. 2818–2827.

99. Kim D.H., Rozhkova E.A., Rajh T., Bader S.D., Novosad, V. Synthesis of hybrid gold/iron oxide nanoparticles in block copolymer micelles for imaging, drug delivery, and magnetic hyperthermia // *IEEE Trans. Magn.* 2009. V. 45. P. 4821–4824.

100. Yang H.W., Liu H.L., Li M.L., Hsi I.W., Fan C.T., Huang C.Y., Lu Y.J., Hua M.Y., Chou H.Y., Liaw J.W., Ma C.C., Wei K.C. Magnetic gold-nanorod/PNIPAAmMA nanoparticles for dual magnetic resonance and photoacoustic imaging and targeted photothermal therapy // *Biomaterials.* 2013. V. 34. P. 5651–5660.

101. Carril M., Fernández I., Rodríguez J., García I., Penadés S. Gold-coated iron oxide glyconanoparticles for MRI, CT, and US multimodal imaging // *Part. Part. Syst. Charact.* 2014. V. 31. P. 81–87.

102. Huang W.C., Tsai P.-J., Chen Y.-C. Multifunctional Fe_3O_4@Au nanoeggs as photothermal agents for selective killing of nosocomial and antibiotic-resistant bacteria // *Small.* 2009. V. 4. P. 51–56.

103. Fan Z., Shelton M., Singh A.K., Senapati D., Khan S.A., Ray P.C. Multifunctional plasmonic shell-magnetic core nanoparticles for targeted diagnostics, isolation, and photothermal destruction of tumor cells // *ACS Nano.* 2012. V. 6. P. 1065–1073.

104. Chen W., Xu N., Xu L., Wang L., Li Z., Ma W., Zhu Y., Xu C., Kotov N.A. Nanoparticle assemblies for cancer therapy and diagnostics (theranostics) // *Macromol. Rapid Commun.* 2010. V. 31. P. 228–236.

105. Cheng F.-Y., Su C.-H., Wu P.-C., Yeh C.-S. Multifunctional polymeric nanoparticles for combined chemotherapeutic and near-infrared photothermal cancer therapy *in vitro* and *in vivo* // *Chem. Commun.* 2010. V. 46. P. 3167–3169.

106. Ke H., Yue X., Wang J., Xing S., Zhang Q., Dai Z., Tian J., Wang S., Jin Y. Gold nanoshelled liquid perfluorocarbon nanocapsules for combined dual modal ultrasound/CT imaging and photothermal therapy of cancer // *Small.* 2014. V. 10. P. 1220–1227.

107. Ke H., Wang J., Tong S., Jin Y., Wang S., Qu E., Bao G., Dai Z. Gold Nanoshelled Liquid perfluorocarbon magnetic nanocapsules: A nanotheranostic platform for bimodal ultrasound/magnetic resonance imaging guided photothermal tumor ablation // *Theranostics.* 2014. V. 4. P. 12–23.

108. Baek S.M., Singh R.K., Kim T.-H., Seo J.-W., Shin U.S., Chrzanowski W., Kim H.-W. Triple hit with drug carriers: pH- and temperature-responsive theranostics for multimodal chemo- and photothermal-therapy and diagnostic applications // *ACS Appl. Mater. Interfaces.* 2016. V. 8. P. 8967–8979.

109. Kim J.-W., Galanzha E.I., Shashkov E.V., Moon H.-M., Zharov V.P. Golden carbon nanotubes as multimodal photoacoustic and photothermal high-contrast molecular agents // *Nat. Nanotechnol.* 2009. V. 4. P. 688–694.

110. Modugno G., Ménard-Moyon C., Prato M., Bianco A. Carbon nanomaterials combined with metal nanoparticles for theranostic applications // *Br. J. Pharm.* 2015. V. 172. P. 975–991.

111. Minati L., Antonini V., Dalla Serra M., Speranza G. Multifunctional branched gold–carbon nanotube hybrid for cell imaging and drug delivery // *Langmuir.* 2012. V. 28. P. 15900–15906.

112. Ma X., Qu Q., Zhao Y., Luo Z., Zhao Y., Ng K.W., Zhao Y. Graphene oxide wrapped gold nanoparticles for intracellular Raman imaging and drug delivery // *J. Mater. Chem.* B. 2013. V. 1. P. 6495–6500.

113. Wang X., Wang C., Cheng L., Lee S.T., Liu Z. Noble metal coated single-walled carbon nanotubes for applications in surface enhanced Raman scattering imaging and photothermal therapy // *J. Am. Chem. Soc.* 2012. V. 134. P. 7414–7422.

114. Tchounwou C., Sinha S.S., Viraka Nellore B.P., Pramanik A., Kanchanapally R., Jones S., Chavva S.R., Ray P.C. Hybrid theranostic platform for second near-IR window light triggered selective two-photon imaging and photothermal killing of targeted melanoma cells // *ACS Appl. Mater. Interfaces.* 2015. V. 7. P. 20649–20656.

115. Shi X., Gong H., Li Y., Wang C., Cheng L., Liu Z. Graphene-based magnetic plasmonic nanocomposite for dual bioimaging and photothermal therapy // *Biomaterials.* 2013. V. 34. P. 4786–4793.

116. Kim Y.-K., Na H.-K., Kim S., Jang H., Chang S.-J., Min D.-H. One-pot synthesis of multifunctional Au@graphene oxide nanocolloid core@shell nanoparticles for Raman bioimaging, photothermal, and photodynamic therapy // *Small.* 2015. V. 11. P. 2527–2535.

117. Maji S.K., Mandal A.K., Nguyen K.T., Borah P., Zhao Y. Cancer cell detection and therapeutics using peroxidase-active nanohybrid of gold nanoparticle-loaded mesoporous silica-coated graphene // *ACS Appl. Mater. Interfaces.* 2015. V. 7. P. 9807–9816.

118. Chen H., Liu F., Lei Z., Ma L., Wang Z. Fe_2O_3@Au core@shell nanoparticle–graphene nanocomposites as theranostic agents for bioimaging and chemo-photothermal synergistic therapy // *RSC Adv.* 2015. V. 5. P. 84980–84987.

119. Wang Y., Yan B., Chen L. SERS tags: Novel optical nanoprobes for bioanalysis // *Chem. Rev.* 2013. V. 113. P. 1391–1428.

120. Granger J.H., Schlotter N.E., Crawford A.C., Porter M.D. Prospects for point-of-care pathogen diagnostics using surface-enhanced Raman scattering (SERS) // *Chem. Soc. Rev.* 2016. V. 45. P. 3865–3882.

121. Lim D.K., Jeon K.S., Hwang J.H., Kim H., Kwon S., Suh Y.D., Nam J.M. Highly uniform and reproducible surface-enhanced Raman scattering from DNA-tailorable nanoparticles with 1-nm interior gap // *Nat. Nanotechnol.* 2011. V. 6. P. 452–460.

122. Gandra N., Hendargo H.C., Norton S.J., Fales A.M., Palmer G.M., Vo-Dinh T. Tunable and amplified Raman gold nanoprobes for effective tracking (TARGET): *in vivo* sensing and imaging // *Nanoscale.* 2016. V. 8. P. 8486–8494.

123. Kang J.W., So P.T.C., Dasari R.R., Lim D.-K. High resolution live cell Raman imaging using subcellular organelle-targeting SERS-sensitive gold nanoparticles with highly narrow intra-nanogap // *Nano Lett.* 2015. V. 15. P. 1766–1772.

124. Feng Y., Wang Y., Wang H., Chen T., Tay Y.Y., Yao L., Yan Q., Li S., Chen H. Engineering "hot" nanoparticles for surface-enhanced Raman scattering by embedding reporter molecules in metal layers // *Small.* 2012. V. 8. P. 246–251.

125. Gandra N., Singamaneni S. Bilayered Raman-intense gold nanostructures with hidden tags (BRIGHTS) for high-resolution bioimaging // *Adv. Mater.* 2013. V. 25. P. 1022–1027.

126. Oh J.-W., Lim D.-K., Kim G.-H., Suh Y.D., Nam J.-M. Thiolated DNA-based chemistry and control in the structure and optical properties of plasmonic nanoparticles with ultrasmall interior nanogap // *J. Am. Chem. Soc.* 2014. V. 136. P. 14052–14059.

127. Song J., Duan B., Wang C., Zhou J., Pu L., Fang Z., Wang P., Lim T.T., Duan H. SERS-encoded nanogapped plasmonic nanoparticles: Growth of metallic nanoshell by templating redox-active polymer brushes // *J. Am. Chem. Soc.* 2014. V. 136. P. 6838–6841.

128. Zhao B., Shen J., Chen S., Wang D., Li F., Mathur S., Song S., Fan C. Gold nanostructures encoded by non-fluorescent small molecules in polya-mediated nanogaps as universal SERS nanotags for recognizing various bioactive molecules // *Chem. Sci.* 2014. V. 5. P. 4460–4466.

129. Lin L., Zapata M., Xiong M., Liu Z., Wang S., Xu H., Borisov A.G., Gu H., Nordlander P., Aizpurua J., Ye J. Nanooptics of plasmonic nanomatryoshkas: Shrinking the size of a core–shell junction to subnanometer // *Nano Lett.* 2015. V. 15. P. 6419–6428.

130. Lin L., Gu H., Ye J. Plasmonic multi-shell nanomatryoshka particles as highly tunable SERS tags with built-in reporters // *Chem. Commun.* 2015. V. 51. P. 17740–17743.

131. Lee J.-H., Oh J.-W., Nam S.H., Cha Y.S., Kim G.-H., Rhim W.-K., Kim N.H., Kim J., Sang Han W., Suh Y.D., Nam J.-M. Synthesis, optical properties, and multiplexed Raman bio-imaging of surface roughness-controlled nanobridged nanogap particles // *Small*. 2016. V. 12. P. 4726–4734.

132. Ngo H.T., Gandra N., Fales A.M., Taylor S.M., Vo-Dinh T. Sensitive DNA detection and SNP discrimination using ultrabright SERS nanorattles and magnetic beads for malaria diagnostics // *Biosens. Bioelectron*. 2016. V. 81. P. 8–14.

133. Gandra N., Portz C., Singamaneni S. Multifunctional plasmonic nanorattles for spectrum-guided locoregional therapy // *Adv. Mater*. 2014. V. 26. P. 424–429.

134. Liu K.-K., Tadepalli S., Tian L., Singamaneni S. Size-dependent surface enhanced Raman scattering activity of plasmonic nanorattles // *Chem. Mater*. 2015. V. 27. P. 5261–5270.

135. Khlebtsov N.G., Khlebtsov B.N. Optimal design of gold nanomatryoshkas with embedded Raman reporters // *J. Quant. Spectrosc. Radiat. Transfer*. 2017. V. 190. P. 89–102.

136. Suarasan S., Focsan M., Potara M., Soritau O., Florea A., Maniu D., Astilean S. A doxorubicin-incorporated nanotherapeutic delivery system based on gelatin-coated gold nanoparticles: Formulation, drug release and multimodal imaging of cellular internalization // *ACS Appl. Mater. Interfaces*. 2016. V. 8. P. 22900–22913.

137. Cao F., Yao Q., Yang T., Zhang Z., Han Y., Feng J., Wang X.-H. Marriage of antibody–drug conjugate with gold nanorods to achieve multi-modal ablation of breast cancer cells and enhanced photoacoustic performance // *RSC Adv*. 2016. V. 6. P. 46594–46606.

138. Khafaji M., Vossoughi M., Hormozi-Nezhad M.R., Dinarvand R., Börrnert F., Irajizad A. A new bifunctional hybrid nanostructure as an active platform for photothermal therapy and MR imaging // *Sci. Rep*. V. 6. 27847.

139. Song J., Wang F., Yang X., Ning B., Harp M.G., Culp S.H., Hu S., Huang P., Nie L., Chen J., Chen X. Gold nanoparticle coated carbon nanotube ring with enhanced Raman scattering and photothermal conversion property for theranostic applications // *J. Am. Chem. Soc*. 2016. V. 138. P. 7005–7015.

140. Jia Q., Ge J., Liu W., Liu S., Niu G., Guo L., Zhang H., Wang P. Gold nanorod@silica-carbon dots as multifunctional phototheranostics for fluorescent and photoacoustic imaging-guided synergistic photodynamic/photothermal therapy // *Nanoscale*. 2016. V.8. P. 13067–13077.

141. Sun L., Joh D.Y., Al-Zaki A., Stangl M., Murty S., Davis J.J., Baumann B.C., Alonso-Basanta M., Kao G.D., Tsourkas A., Dorsey J.F. Theranostic application of mixed gold and superparamagnetic iron oxide nanoparticle micelles in glioblastoma multiforme // *J. Biomed. Nanotechnol*. 2016. V. 12. P. 347–356.

142. Mukherjee P., Bhattacharya R., Mukhopadhyay D. Gold nanoparticles bearing functional anti-cancer drug and anti-angiogenic agent: A "2 in 1" system with potential application in cancer therapeutics // *J. Biomed. Nanotechnol*. 2005. V. 1. P. 224–228.

143. Paciotti G.F., Kingston D.G.I., Tamarkin L. Colloidal gold nanoparticles: A novel nanoparticle platform for developing multifunctional tumor-targeted drug delivery vectors // *Drug Dev. Res*. 2006. V. 67. P. 47–54.

144. Heo D.N., Yang D.H., Moon H.-J., Lee J.B., Bae M.S., Lee S.C., Lee W.J., Sun I.-C., Kwon I.K. Gold nanoparticles surface-functionalized with paclitaxel drug and biotin receptor as theranostic agents for cancer therapy // *Biomaterials*. 2012. V. 33. P. 856–866.

145. Chen W.-H., Yang C.-X., Qiu W.-X., Luo G.-F., Jia H.-Z., Lei Q., Wang X.-Y., Liu G., Zhuo R.-X., Zhang X.-Z. Multifunctional theranostic nanoplatform for cancer combined therapy based on gold nanorods // *Adv. Healthc. Mater.* 2015. V. 4. P. 2247–2259.

146. Bhattacharyya S., Khan J.A., Curran G.L., Robertson J.D., Bhattacharya R., Mukherjee P. Efficient delivery of gold nanoparticles by dual receptor targeting // *Adv. Mater.* 2011. V. 23. P. 5034–5038.

147. Patra C.R., Bhattacharya R., Mukherjee P. Fabrication and functional characterization of gold nanoconjugates for potential application in ovarian cancer // *J. Mater. Chem.* 2010. V. 20. P. 547–554.

148. Hosta-Rigau L., Olmedo I., Arbiol J., Cruz L.J., Kogan M.J., Albericio F. Multifunctionalized gold nanoparticles with peptides targeted to gastrin-releasing peptide receptor of a tumor cell line // *Bioconjug. Chem.* 2010. V. 21. P. 1070–1078.

149. Kumar A., Ma H., Zhang X., Huang K., Jin S., Liu J., Wei T., Cao W., Zou G., Liang X.-J. Gold nanoparticles functionalized with therapeutic and targeted peptides for cancer treatment. *Biomaterials*. 2012. V. 33. P. 1180–1189.

150. Kang B., Mackey M.A., El-Sayed M.A. Nuclear targeting of gold nanoparticles in cancer cells induces DNA damage, causing cytokinesis arrest and apoptosis // *J. Am. Chem. Soc.* 2010. V. 132. P. 1517–1519.

151. You J., Zhang R., Xiong C., Zhong M., Melancon M., Gupta S., Nick A.M., Sood A.K., Li C. Effective photothermal chemotherapy using doxorubicinloaded gold nanospheres that target EphB4 receptors in tumors // *Cancer Res.* 2012. V. 72. P. 4777–4786.

152. Book Newell B., Wang Y., Irudayaraj J. Multifunctional gold nanorod theragnostics probed by multi-photon imaging // *Eur. J. Med. Chem.* 2012. V. 48. P. 330–337.

153. Wang D., Xu Z., Yu H., Chen X., Feng B., Cui Z., Lin B., Yin Q., Zhang Z., Chen C., Wang J., Zhang W., Li Y. Treatment of metastatic breast cancer by combination of chemotherapy and photothermal ablation using doxorubicin-loaded DNA wrapped gold nanorods // *Biomaterials*. 2014. V. 35. P. 8374–8384.

154. Chen H., Chi X., Li B., Zhang M., Ma Y., Achilefud S., Gu Y. Drug loaded multi-layered gold nanorods for combined photothermal and chemotherapy // *Biomater. Sci.* 2014. V. 2. P. 996–1006.

155. Alexander C.M., Hamner K.L., Maye M.M., Dabrowiak J.C. Multifunctional DNA-gold nanoparticles for targeted doxorubicin delivery // *Bioconjug. Chem.* 2014. V. 25. P. 1261–1271.

156. Kim D., Jeong Y.Y., Jon S. A drug-loaded aptamer-gold nanoparticle bioconjugate for combined CT imaging and therapy of prostate cancer // *ACS Nano*. 2010. V. 4. P. 3689–3696.

157. Xiao Y., Hong H., Matson V.Z., Javadi A., Xu W., Yang Y., Zhang Y., Engle J.W., Nickles R.J., Cai W., Steeber D.A., Gong S. Gold nanorods conjugated with doxorubicin and cRGD for combined anti-cancer drug delivery and PET imaging // *Theranostics*. 2012. V. 2. P. 757–768.

158. Avvakumova S., Colombo M., Tortora P., Prosperi D. Biotechnological approaches toward nanoparticle biofunctionalization // *Trends Biotechnol.* 2014. V. 32. P. 11–20.

159. Liang Z., Li X., Xie Y., Liu S. 'Smart' gold nanoshells for combined cancer chemotherapy and hyperthermia // *Biomed. Mater.* 2014. V. 9. 025012.

160. Kuo W.S., Chang C.N., Chang Y.T., Yang M.H., Chien Y.H., Chen S.J., Yeh C.S. Gold nanorods in photodynamic therapy, as hyperthermia agents, and in near-infrared optical imaging // *Angew. Chem. Int. Ed.* 2010. V. 49. P. 2711–2715.

161. Stuchinskaya T., Moreno M., Cook M.J., Edwards D.R., Russell D.A. Targeted photodynamic therapy of breast cancer cells using antibody-phthalocyanine-gold nanoparticle conjugates // *Photochem. Photobiol. Sci.* 2011. V. 10. P. 822–831.

162. Lee K., Lee H., Bae K.H., Park T.G. Heparin immobilized gold nanoparticles for targeted detection and apoptotic death of metastatic cancer cells // *Biomaterials.* 2010. V. 31. P. 6530–6536.

163. Wang F., Liu P., Sun L., Li C., Petrenko V.A., Liu A. Bio-mimetic nanostructure self-assembled from Au@Ag heterogeneous nanorods and phage fusion proteins for targeted tumor optical detection and photothermal therapy // *Sci. Rep.* 2014. V. 4. 6808.

164. Chen H., Zhang X., Dai S., Ma Y., Cui S., Achilefu S., Gu Y. Multifunctional gold nanostar conjugates for tumor imaging and combined photothermal and chemotherapy // *Theranostics.* 2013. V. 3. P. 633–649.

165. Huang X., Peng X., Wang Y., Wang Y., Shin D.M., El-Sayed M.A., Nie S. A reexamination of active and passive tumor targeting by using rod-shaped gold nanocrystals and covalently conjugated peptide ligands // *ACS Nano.* 2010. V. 4. P. 5887–5896.

166. Ali M.R.K., Panikkanvalappil S.R., El-Sayed M.A. Enhancing the efficiency of gold nanoparticles treatment of cancer by increasing their rate of endocytosis and cell accumulation using rifampicin // *J. Am. Chem. Soc.* 2014. V. 136. P. 4464–4467.

167. Kumar S., Harrison N., Richards-Kortum R., Sokolov K. Plasmonic nanosensors for imaging intracellular biomarkers in live cells // *Nano Lett.* 2007. V. 7. P. 1338–1343.

168. Zhang Z., Liu Y., Jarreau C., Welch M.J., Taylor J.-S.A. Nucleic acid-directed self-assembly of multifunctional gold nanoparticle imaging agents // *Biomater. Sci.* 2013. V. 1. P. 1055–1064.

169. Mandal S., Bakeine G.J., Krol S., Ferrari C., Clerici A.M., Zonta C., Cansolino L., Ballarini F., Bortolussi S., Stella S., Protti N., Bruschi P., Altieri S. Design, development and characterization of multi-functionalized gold nanoparticles for bio-detection and targeted boron delivery in BNCT applications // *Appl. Radiat. Isot.* 2011. V. 69. P. 1692–1697.

170. Dixit S., Miller K., Zhu Y., McKinnon E., Novak T., Kenney M.E., Broome A.M. Dual receptor-targeted theranostic nanoparticles for localized delivery and activation of photodynamic therapy drug in glioblastomas // *Mol. Pharm.* 2015. V. 12. P. 3250–3260.

171. Bao C., Conde J., Curtin J., Artzi N., Tian F., Cui D. Bioresponsive antisense DNA gold nanobeacons as a hybrid *in vivo* theranostics platform for the inhibition of cancer cells and metastasis // *Sci. Rep.* 2015. V. 5. 12297.

172. Chen S., Li Q., Xu Y., Li H., Ding X. Gold nanorods bioconjugates for intracellular delivery and cancer cell apoptosis // *J. Lab. Autom.* 2015. V. 20. P. 418–422.

173. Tao Y., Ju E., Liu Z., Dong K., Ren J., Qu X. Engineered, self-assembled near-infrared photothermal agents for combined tumor immunotherapy and chemo-photothermal therapy // *Biomaterials.* 2014. V. 35. P. 6646–6656.

174. Lu W., Zhang G., Zhang R., Flores L.G., Huang Q., Gelovani J.G., Li C. Tumor site-specific silencing of NF-kappaB p65 by targeted hollow gold nanosphere-mediated photothermal transfection // *Cancer Res.* 2010. V. 70. P. 3177–3188.

175. Conde J., Ambrosone A., Sanz V., Hernandez Y., Marchesano V., Tian F., Child H., Berry C.C., Ibarra M.R., Baptista P.V., Tortiglione C., de la Fuente J.M. Design of multifunctional gold nanoparticles for *in vitro* and *in vivo* gene silencing // *ACS Nano.* 2012. V. 6. P. 8316–8324.

176. Braun G.B., Pallaoro A., Wu G., Missirlis D., Zasadzinski J.A., Tirrell M., Reich N.O. Laser-activated gene silencing via gold nanoshell-siRNA conjugates // *ACS Nano.* 2009. V. 3. P. 2007–2015.

177. Ojeda R., de Paz J.L., Barrientos A.G., Martín-Lomas M., Penadés S. Preparation of multifunctional glyconanoparticles as a platform for potential carbohydrate-based anticancer vaccines // *Carbohydr. Res.* 2007. V. 342. P. 448–459.

178. Brinãs R.P., Sundgren A., Sahoo P., Morey S., Rittenhouse-Olson K., Wilding G.E., Deng W., Barchi J.J., Jr. Design and synthesis of multifunctional gold nanoparticles bearing tumor-associated glycopeptide antigens as potential cancer vaccines // *Bioconjug. Chem.* 2012. V. 23. P. 1513–1523.

179. Safari D., Marradi M., Chiodo F., Dekker H.A.T., Shan Y., Adamo R., Oscarson S., Rijkers G.T., Lahmann M., Kamerling J.P., Penadés S., Snippe H. Gold nanoparticles as carriers for a synthetic *Streptococcus pneumoniae* type 14 conjugate vaccine // *Nanomedicine (Lond.).* 2012. V. 7. P. 651–662.

180. Chiodo F., Marradi M., Calvo J., Yuste E., Penadés S. Glycosystems in nanotechnology: Gold glyconanoparticles as carrier for anti-HIV prodrugs // *Beilstein J. Org. Chem.* 2014. V. 10. P. 1339–1346.

181. Reichardt N.C., Martín-Lomas M., Penadés S. Glyconanotechnology. *Chem. Soc. Rev.* 2013. V. 42. P. 4358–4376.

182. Li X., Chen G. Glycopolymer-based nanoparticles: Synthesis and application // *Polym. Chem.* 2015. V. 6. P. 1417–1430.

183. Wang Z., Wu H., Shi H., Wang M., Huang C., Jia N. A novel multifunctional bio-mimetic Au@BSA nanocarrier as a potential siRNA theranostic nanoplatform // *J. Mater. Chem. B.* 2016. V. 4. P. 2519–2526.

184. Manivasagan P., Bharathiraja S., Bui N.Q., Jang B., Oh Y.-O., Lim I.G., Oh J. Doxorubicin-loaded fucoidan capped gold nanoparticles for drug delivery and photoacoustic imaging // *Int. J. Biol. Macromol.* 2016. V. 91. P. 578–588.

185. Salem A.K., Searson P.C., Leong K.W. Multifunctional nanorods for gene delivery // *Nat. Mater.* 2003. V. 2. P. 668–671.

186. Choi J.H., Oh B.K. Development of two-component nanorod complex for dual-fluorescence imaging and siRNA delivery // *J. Microbiol. Biotechnol.* 2014. V. 24. P. 1291–1299.

187. Wang Y., Chen L., Liu P. Biocompatible triplex Ag@SiO$_2$@mTiO$_2$ core-shell nanoparticles for simultaneous fluorescence-SERS bimodal imaging and drug delivery // *Chemistry*. 2012. V. 18. P. 5935–5943.

188. Zhang W., Wang Y., Sun X., Wang W., Chen L. Mesoporous titania based yolk-shell nanoparticles as multifunctional theranostic platforms for SERS imaging and chemo-photothermal treatment // *Nanoscale*. 2014. V. 6. P. 14514–14522.

189. Shi X., Wang S.H., Van Antwerp M.E., Chen X., Baker J.R., Jr. Targeting and detecting cancer cells using spontaneously formed multifunctional dendrimer-stabilized gold nanoparticles // *Analyst*. 2009. V. 134. P. 1373–1379.

190. Zhu J., Zheng L., Wen S., Tang Y., Shen M., Zhang G., Shi X. Targeted cancer theranostics using alpha-tocopheryl succinateconjugated multifunctional dendrimer-entrapped gold nanoparticles // *Biomaterials*. 2014. V. 35. P. 7635–7646.

191. Song J., Zhou J., Duan H. Self-assembled plasmonic vesicles of SERS-encoded amphiphilic gold nanoparticles for cancer cell targeting and traceable intracellular drug delivery // *J. Am. Chem. Soc.* 2012. V. 134. P. 13458–13469.

192. Yuan H., Fales A.M., Vo-Dinh T. TAT peptide-functionalized gold nanostars: Enhanced intracellular delivery and efficient NIR photothermal therapy using ultralow irradiance // *J. Am. Chem. Soc.* 2012. V. 134. P. 11358–11361.

193. Zhang Y., Qian J., Wang D., Wang Y., He S. Multifunctional gold nanorods with ultrahigh stability and tunability for *in vivo* fluorescence imaging, SERS detection, and photodynamic therapy // *Angew. Chem. Int. Ed.* 2013. V. 52. P. 1148–1151.

194. Wang T.T., Chai F., Wang C.G., Li L., Liu H.Y., Zhang L.Y., Su Z.M., Liao Y. Fluorescent hollow/rattle-type mesoporous Au@SiO$_2$ nanocapsules for drug delivery and fluorescence imaging of cancer cells // *J. Colloid Interface. Sci.* 2011. V. 358. P. 109–115.

195. Jiang Z., Dong B., Chen B., Wang J., Xu L., Zhang S., Song H. Multifunctional Au@mSiO$_2$/rhodamine B isothiocyanate nanocomposites: Cell imaging, photo-controlled drug release, and photothermal therapy for cancer cells // *Small*. 2013. V. 9. P. 604–612.

196. Prabaharan M., Grailer J.J., Pilla S., Steeber D.A., Gong S. Gold nanoparticles with a monolayer of doxorubicin-conjugated amphiphilic block copolymer for tumor-targeted drug delivery // *Biomaterials*. 2009. V. 30. P. 6065–6075.

197. Zhang Z., Liu C., Bai J., Wu C., Xiao Y., Li Y., Zheng J., Yang R., Tan W. Silver nanoparticle gated, mesoporous silica coated gold nanorods (AuNR@MS@AgNPs): Low premature release and multifunctional cancer theranostic platform // *ACS Appl. Mater. Interfaces.* 2015. V. 7. P. 6211–6219.

198. Bardhan R., Chen W., Perez-Torres C., Bartels M., Huschka R.M., Zhao L.L., Morosan E., Pautler R.G., Joshi A., Halas N.J. Nanoshells with targeted simultaneous enhancement of magnetic and optical imaging and photothermal therapeutic response // *Adv. Funct. Mater.* 2009. V. 19. P. 3901–3909.

199. Bardhan R., Chen W., Bartels M., Perez-Torres C., Botero M.F., McAninch R.W., Contreras A., Schiff R., Pautler R.G., Halas N.J., Joshi A. Tracking of multimodal therapeutic nanocomplexes targeting breast cancer *in vivo* // *Nano Lett.* 2010. V. 10. P. 4920–4928.

200. Chen W., Bardhan R., Bartels M., Perez-Torres C., Pautler R.G., Halas N.J., Joshi A. A molecularly targeted theranostic probe for ovarian cancer // *Mol. Cancer Ther.* 2010. V. 9. P. 1028–1038.

201. Liu H., Chen D., Li L., Liu T., Tan L., Wu X., Tang F. Multifunctional gold nanoshells on silica nanorattles: A platform for the combination of photothermal therapy and chemotherapy with low systemic toxicity // *Angew. Chem. Int. Ed.* 2011. V. 50. P. 891–895.

202. Hu F., Zhang Y., Chen G., Li C., Wang Q. Double-walled Au nanocage/SiO$_2$ nanorattles: Integrating SERS imaging, drug delivery and photothermal therapy // *Small.* 2015. V. 11. P. 985–993.

203. Liu Y., Chang Z., Yuan H., Fales A.M., Vo-Dinh T. Quintuple-modality (SERS-MRI-CT-TPL-PTT) plasmonic nanoprobe for theranostics // *Nanoscale.* 2013. V. 5. P. 12126–12131.

204. Movia D., Gerard V., Maguire C.M., Jain N., Bell A.P., Nicolosi V., O'Neill T., Scholz D., Gun'ko Y., Volkov Y., Prina-Mello A. A safe-by-design approach to the development of gold nanoboxes as carriers for internalization into cancer cells // *Biomaterials.* 2014. V. 35. P. 2543–2557.

205. Topete A., Alatorre-Meda M., Iglesias P., Villar-Alvarez E.M., Barbosa S., Costoya J.A., Taboada P., Mosquera V. Fluorescent drug-loaded, polymeric-based, branched gold nanoshells for localized multimodal therapy and imaging of tumoral cells // *ACS Nano.* 2014. V. 8. P. 2725–2738.

206. Hao Y., Zhang B., Zheng C., Ji R., Ren X., Guo F., Sun S., Shi J., Zhang H., Zhang Z., Wang L., Zhang Y. The tumor-targeting core-shell structured DTX-loaded PLGA@Au nanoparticles for chemo-photothermal therapy and X-ray imaging // *J. Control. Release.* 2015. V. 220. P. 545–555.

207. Navarro J.R., Lerouge F., Cepraga C., Micouin G., Favier A., Chateau D., Charreyre M.T., Lanoë P.H., Monnereau C., Chaput F., Marotte S., Leverrier Y., Marvel J., Kamada K., Andraud C., Baldeck P.L., Parola S. Nanocarriers with ultrahigh chromophore loading for fluorescence bio-imaging and photodynamic therapy // *Biomaterials.* 2013. V. 34. P. 8344–8351.

208. Zeng Y., Zhang D., Wu M., Liu Y., Zhang X., Li L., Li Z., Han X., Wei X., Liu X. Lipid-AuNPs@PDA nanohybrid for MRI/CT imaging and photothermal therapy of hepatocellular carcinoma // *ACS Appl. Mater. Interfaces.* 2014. V. 6. P. 14266–14277.

209. Zhao L., Kim T.H., Kim H.W., Ahn J.C., Kim S.Y. Surface-enhanced Raman scattering (SERS)-active gold nanochains for multiplex detection and photodynamic therapy of cancer // *Acta Biomater.* 2015. V. 20. P. 155–164.

210. Yuan Y., Zhang Y., Liu B., Wu H., Kang Y., Li M., Zeng X., He N., Zhang G. The effects of multifunctional MiR-122-loaded graphene-gold composites on drug-resistant liver cancer // *J. Nanobiotechnology.* 2015. V. 13. 12.

211. Chen Y.W., Liu T.Y., Chen P.J., Chang P.H., Chen S.Y. A high-sensitivity and low-power theranostic nanosystem for cell SERS imaging and selectively photothermal therapy using anti-EGFR-conjugated reduced graphene oxide/mesoporous silica/AuNPs nanosheets // *Small.* 2016. V. 12. P. 1458–1468.

212. Zheng J., Dickson R.M. Individual water-soluble dendrimer-encapsulated silver nanodot fluorescence // *J. Am. Chem. Soc.* 2002. V. 124. P. 13982–13983.

213. Zheng J., Petty J.T., Dickson R.M. High quantum yield blue emission from water-soluble Au8 nanodots // *J. Am. Chem. Soc.* 2003. V. 125. P. 7780–7781.

214. Zheng J., Zhang C., Dickson R.M. Highly fluorescent, water-soluble, size-tunable gold quantum dots // *Phys. Rev. Lett.* 2004. V. 93. 077402.

215. Cantelli A., Battistelli G., Guidetti G., Manzi J., di Giosia M., Montalti M. Luminescent gold nanoclusters as biocompatible probes for optical imaging and theranostics // *Dyes Pigments*. 2016. V. 135. P. 64–79.

216. Shang L., Dong S., Nienhaus G.U. Ultra-small fluorescent metal nanoclusters: Synthesis and biological applications // *Nano Today*. 2011. V. 6. P 401–418.

217. Choi S., Dickson R.M., Yu J. Developing luminescent silver nanodots for biological applications // *Chem. Soc. Rev.* 2012. V. 41. P. 1867–1891.

218. Lu Y., Chen W. Sub-nanometre sized metal clusters: From synthetic challenges to the unique property discoveries // *Chem. Soc. Rev.* 2012. V. 41. P. 3594–3623.

219. Zhang L., Wang E. Metal nanoclusters: New fluorescent probes for sensors and bioimaging // *Nano Today*. 2014. V. 9. P. 132–157.

220. Zhang P. X-ray spectroscopy of gold–thiolate nanoclusters // *J. Phys. Chem. C.* 2014. V. 118. P. 25291–25299.

221. Jin R. Atomically precise metal nanoclusters: Stable sizes and optical properties // *Nanoscale*. 2015. V. 7. P. 1549–1565.

222. Yu P., Wen X., Toh Y.-R., Ma X., Tang J. Fluorescent metallic nanoclusters: Electron dynamics, structure, and applications // *Part. Part. Syst. Charact.* 2015. V. 32. P. 142–163.

223. Liu J. DNA-stabilized, fluorescent, metal nanoclusters for biosensor development // *Trends Anal. Chem.* 2014. V. 58. P. 99–111.

224. Mathew A., Pradeep T. Noble metal clusters: Applications in energy, environment and biology // *Part. Part. Syst. Charact.* 2014. V. 31. P. 1017–1053.

225. Xie J., Zheng Y., Ying J.Y. Protein-directed synthesis of highly fluorescent gold nanoclusters // *J. Am. Chem. Soc.* 2009. V. 131. P. 888–889.

226. Xu Y., Sherwood J., Qin Y., Crowley D., Bonizzoni M., Bao Y. The role of protein characteristics in the formation and fluorescence of Au nanoclusters // *Nanoscale*. 2014. V. 6. P. 1515–1524.

227. Wang J., Zhang G., Li Q., Jiang H., Liu C., Amatore C., Wang X. *In vivo* self-bio-imaging of tumors through *in situ* biosynthesized fluorescent gold nanoclusters // *Sci. Rep.* 2013. V. 3. 1157.

228. Wang J., Ye J., Jiang H., Gao S., Ge W., Chen Y., Liu C., Amatore C., Wang X. Simultaneous and multisite tumor rapid-target bioimaging through *in vivo* bio-synthesis of fluorescent gold nanoclusters // *RSC Adv.* 2014. V. 4. P. 37790–37795.

229. Chen L.Y., Wang C.W., Yuan Z., Chang H.T. Fluorescent gold nanoclusters: Recent advances in sensing and imaging // *Anal Chem.* 2015. V. 87. P. 216–229.

230. Wu X., He X., Wang K., Xie C., Zhou B., Qing Z. Ultrasmall near-infrared gold nanoclusters for tumor fluorescence imaging *in vivo* // *Nanoscale*. 2010. V. 2. P. 2244–2249.

231. Li G., Jin R. Atomically precise gold nanoclusters as new model catalysts // *Acc. Chem. Res.* 2013. V. 46. P. 1749–1758.

232. Korotcenkov G., Brinzari V., Cho B.K. What restricts gold clusters reactivity in catalysis and gas sensing effects: A focused review // *Mater. Lett.* 2015. V. 147. P. 101–104.

233. Retnakumari A., Setua S., Menon D., Ravindran P., Muhammed H., Pradeep T., Nair S., Koyakutty M. Molecular-receptor-specific, non-toxic, near-infrared-emitting Au cluster-protein nanoconjugates for targeted cancer imaging // *Nanotechnology.* 2010. V. 21. 55103.

234. Wang Y., Chen J., Irudayaraj J. Nuclear targeting dynamics of gold nanoclusters for enhanced therapy of HER2+ breast cancer // *ACS Nano.* 2011. V. 5. P. 9718–9725.

235. Chen H., Li S., Li B., Ren X., Li S., Mahounga D.M., Cui S., Gu Y., Achilefu S. Folate-modified gold nanoclusters as near-infrared fluorescent probes for tumor imaging and therapy // *Nanoscale.* 2012. V. 4. P. 6050–6064.

236. Chen H., Li B., Ren X., Li S., Ma Y., Cui S., Gu Y. Multifunctional near-infrared-emitting nano-conjugates based on gold clusters for tumor imaging and therapy // *Biomaterials.* 2012. V. 33. P. 8461–8476.

237. Ding C., Tian Y. Gold nanocluster-based fluorescence biosensor for targeted imaging in cancer cells and ratiometric determination of intracellular pH // *Biosens. Bioelectron.* 2015. V. 65. P. 183–190.

238. Muthu M.S., Kutty R.V., Luo Z., Xie J., Feng S.S. Theranostic vitamin E TPGS micelles of transferrin conjugation for targeted co-delivery of docetaxel and ultra bright gold nanoclusters // *Biomaterials.* 2015. V. 39. P. 234–248.

239. Ding H., Yang D., Zhao C., Song Z., Liu P., Wang Y., Chen Z., Shen J. Protein-gold hybrid nanocubes for cell imaging and drug delivery // *ACS Appl. Mater. Interfaces.* 2015. 7. 4713–4719.

240. Hembury M., Chiappini C., Bertazzo S., Kalber T.L., Drisko G.L., Ogunlade O., Walker-Samuel S., Krishna K.S., Jumeaux C., Beard P., Kumar C.S., Porter A.E., Lythgoe M.F., Boissière C., Sanchez C., Stevens M.M. Gold-silica quantum rattles for multimodal imaging and therapy // *Proc. Natl. Acad. Sci. USA.* 2015. V. 112. P. 1959–1964.

241. Zhang C., Li C., Liu Y., Zhang J., Bao C., Liang S., Wang Q., Yang Y., Fu H., Wang K., Cui D. Gold nanoclusters-based nanoprobes for simultaneous fluorescence imaging and targeted photodynamic therapy with superior penetration and retention behavior in tumors // *Adv. Func. Mater.* 2015. V. 25. 1314–1325.

242. Jevons M.P., Coe A.W., Parker M.T. Methicillin resistance in staphylococci // *Lancet.* 1963. V. 1. P. 904–907.

243. Deurenberg R.H., Vink C., Kalenic S., Friedrich A.W., Bruggeman C.A. The molecular evolution of methicillin-resistant *Staphylococcus aureus* // *Clin. Microbiol. Infect.* 2007. V. 13. P. 222–235.

244. Millenbaugh N.J., Baskin J.B., DeSilva M.N., Elliott W.R., Glickman R.D. Photothermal killing of *Staphylococcus aureus* using antibody-targeted gold nanoparticles // *Int. J. Nanomedicine.* 2015. V. 10. P. 1953–1960.

245. Khlebtsov B., Tuchina E., Tuchin V., Khlebtsov N. Multifunctional Au nano-clusters for targeted bioimaging and enhanced photodynamic inactivation of *Staphylococcus aureus* // *RSC Adv.* 2015. V. 5. P. 61639–61649.

246. Zhang M., Yilmaz T., Boztas A.O., Karakuzu O., Bang W.Y., Yegin Y., Luo Z., Lenox M., Cisneros-Zevallos L., Akbulut M. A multifunctional nanoparticulate theranostic system with simultaneous chemotherapeutic, photothermal therapeutic, and MRI contrast capabilities // *RSC Adv.* 2016. V. 6. P. 27798–27806.

247. Li N., Li T., Liu Chen., Ye S., Liang J., Han H. Folic acid-targeted and cell penetrating peptide-mediated theranostic nanoplatform for high-efficiency tri-modal imaging-guided synergistic anticancer phototherapy // *J. Biomed. Nanotechnol.* 2016. V. 12. P. 878–893.

248. Tseng Y.-J., Chou S.-W., Shyue J.-J., Lin S.-Y., Hsiao J.-K., Chou P.-T. A versatile theranostic delivery platform integrating magnetic resonance imaging/computed tomography, pH/cis-diol controlled release and targeted therapy // *ACS Nano.* 2016. V. 10. P. 5809–5822.

249. Zhao L., Kim T.-H., Kim H.-W., Ahn J.-C., Kim S.Y. Enhanced cellular uptake and phototoxicity of Verteporfin-conjugated gold nanoparticles as theranostic nano-carriers for targeted photodynamic therapy and imaging of cancers // *Mater. Sci. Eng. C.* 2016. V. 67. P. 611–622.

250. Luo Z., Zheng K., Xie J. Engineering ultrasmall water-soluble gold and silver nano-clusters for biomedical applications // *Chem. Commun.* 2014. V. 50. P. 5143–5155.

Index

Page numbers followed by f and t indicate figures and tables, respectively.

T - #0116 - 111024 - C352 - 234/156/16 - PB - 9780367892210 - Gloss Lamination